Have you been to our website?

For code downloads, print and e-book bundles, extensive samples from all books, special deals, and our blog, please visit us at:

www.rheinwerk-computing.com

Rheinwerk Computing

The Rheinwerk Computing series offers new and established professionals comprehensive guidance to enrich their skillsets and enhance their career prospects. Our publications are written by the leading experts in their fields. Each book is detailed and hands-on to help readers develop essential, practical skills that they can apply to their daily work.

Explore more of the Rheinwerk Computing library!

Johannes Ernesti, Peter Kaiser
Python 3: The Comprehensive Guide
2022, 1036 pages, paperback and e-book
www.rheinwerk-computing.com/5566

Veit Steinkamp
Python for Engineering and Scientific Computing
2024, 511 pages, paperback and e-book
www.rheinwerk-computing.com/5852

Michael Kofler
Scripting: Automation with Bash, PowerShell, and Python
2024, 470 pages, paperback and e-book
www.rheinwerk-computing.com/5851

Thomas Theis
Getting Started with Python
2024, approx. 475 pp, paperback and e-book
www.rheinwerk-computing.com/5876

Philip Ackermann
JavaScript: The Comprehensive Guide
2022, 982 pages, paperback and e-book
www.rheinwerk-computing.com/5554

www.rheinwerk-computing.com

Metin Karatas

Developing AI Applications

An Introduction

Editor Meagan White
Acquisitions Editor Hareem Shafi
German Edition Editor Almut Poll
Translation Winema Language Services, Inc.
Copyeditor Julie McNamee
Cover Design Mai Loan Nguyen Duy
Layout Design Vera Brauner
Production Hannah Lane
Typesetting III-Satz, Germany
Printed and bound in Canada, on paper from sustainable sources

ISBN 978-1-4932-2601-6

© 2024 by Rheinwerk Publishing, Inc., Boston (MA)
1st edition 2024
1st German edition published 2024 by Rheinwerk Verlag, Bonn, Germany

Library of Congress Cataloging-in-Publication Control Number: 2024015323

All rights reserved. Neither this publication nor any part of it may be copied or reproduced in any form or by any means or translated into another language, without the prior consent of Rheinwerk Publishing, 2 Heritage Drive, Suite 305, Quincy, MA 02171.

Rheinwerk Publishing makes no warranties or representations with respect to the content hereof and specifically disclaims any implied warranties of merchantability or fitness for any particular purpose. Rheinwerk Publishing assumes no responsibility for any errors that may appear in this publication.

"Rheinwerk Publishing", "Rheinwerk Computing", and the Rheinwerk Publishing and Rheinwerk Computing logos are registered trademarks of Rheinwerk Verlag GmbH, Bonn, Germany.

All products mentioned in this book are registered or unregistered trademarks of their respective companies.

Contents at a Glance

1	Introduction	15
2	Installation	25
3	Artificial Neural Networks	39
4	Decision Trees	89
5	Convolutional Layers and Images	117
6	Transfer Learning	141
7	Anomaly Detection	151
8	Text Classification	165
9	Cluster Analysis	181
10	AutoKeras	193
11	Visual Programming Using KNIME	203
12	Reinforcement Learning	281
13	Genetic Algorithms	297
14	ChatGPT and GPT-4	311
15	DALL-E and Successor Models	345
16	Outlook	359

Contents

1 Introduction   15

1.1	What Does This Book Offer?	15
1.2	What Is Artificial Intelligence?	17
1.3	The History of AI: A Brief Overview	18
1.4	Development Tools Used in This Book	20
	1.4.1 Python	20
	1.4.2 Jupyter Notebook	22
	1.4.3 KNIME	22
	1.4.4 ChatGPT and GPT-4	23
	1.4.5 DALL-E 2 or DALL-E 3	24

2 Installation   25

2.1	Anaconda Distribution	25
	2.1.1 Windows and macOS	26
	2.1.2 Linux	26
	2.1.3 Configuration and Test	27
2.2	KNIME	30
	2.2.1 Installation	31
	2.2.2 Configuration	34
	2.2.3 Test	37

3 Artificial Neural Networks   39

3.1	Classification	40
3.2	The Recipe	41
	3.2.1 Data Preparation	42
	3.2.2 Building Up the AI	43
	3.2.3 Training the AI	43
	3.2.4 Testing the AI	44
	3.2.5 Using AI	45

3.3	Building ANNs		45
3.4	Structure of an Artificial Neuron		47
3.5	Feed Forward		48
3.6	Back Propagation		51
3.7	Updating the Weights		53
3.8	ANN for Classification		55
3.9	Hyperparameters and Overfitting		63
3.10	Dealing with Nonnumerical Data		65
3.11	Dealing with Data Gaps		67
	3.11.1	Filling Empty Cells with Data	68
	3.11.2	Removing Rows with Empty Cells	68
3.12	Correlation versus Causality		69
3.13	Standardization of the Data		76
3.14	Regression		78
3.15	Deployment		81
	3.15.1	Training, Testing, and Saving	81
	3.15.2	Using the ANN Model	83
3.16	Exercises		85
	3.16.1	Exercise 1: Hyperparameter Optimization for Classification	85
	3.16.2	Exercise 2: Hyperparameter Optimization for Regression	86
	3.16.3	Exercise 3: ANN for Classification	86
	3.16.4	Exercise 4: ANN for Regression	87

4 Decision Trees — 89

4.1	Simple Decision Trees		90
	4.1.1	Decision Tree Classifier	90
	4.1.2	Decision Tree Regressor	96
	4.1.3	Decision Forests	99
	4.1.4	Random Forest Classifier	99
	4.1.5	Random Forest Regressor	100
4.2	Boosting		100
	4.2.1	Gradient Boosting	100
	4.2.2	XGBoost Classifier	103
	4.2.3	Automatic Hyperparameter Setting Using GridSearchCV	107
4.3	XGBoost Regressor		109

4.4	Deployment		110
4.5	Decision Trees Using Orange		111
4.6	Exercises		115
	4.6.1	Exercise 1: XGBoost for Classification	115
	4.6.2	Exercise 2: XGBoost for Regression	116
	4.6.3	Exercise 3: Automatic Hyperparameter Optimization	116

5 Convolutional Layers and Images 117

5.1	Simple Image Classification		118
5.2	Hyperparameter Optimization Using Early Stopping and KerasTuner		123
5.3	Convolutional Neural Network		128
5.4	Image Classification Using CIFAR-10		134
5.5	Using Pretrained Networks		137
5.6	Exercises		140
	5.6.1	Exercise 1: Hyperparameter Optimization for CIFAR-10	140
	5.6.2	Exercise 2: Pretrained VGG19 Model	140

6 Transfer Learning 141

6.1	How It Works		143
6.2	Exercises		150
	6.2.1	Exercise 1: Rock-Paper-Scissors	150
	6.2.2	Exercise 2: Human or Horse	150

7 Anomaly Detection 151

7.1	Unbalanced Data		152
7.2	Resampling		156
7.3	Autoencoders		158
7.4	Exercises		164
	7.4.1	Exercise 1: Anomaly Detection Using XGBoost and Upsampling	164
	7.4.2	Exercise 2: Anomaly Detection Using an Autoencoder	164

8 Text Classification — 165

8.1	Embedding Layer	165
8.2	GlobalAveragePooling1D Layer	168
8.3	Text Vectorization	170
8.4	Analysis of the Relationships	173
8.5	Classifying Large Amounts of Data	177
8.6	Exercises	180
	8.6.1 Exercise 1: Hyperparameter Optimization	180
	8.6.2 Exercise 2: Text Classification	180
	8.6.3 Exercise 3: Text Classification Using Upsampling	180

9 Cluster Analysis — 181

9.1	Graphical Analysis of the Data	182
9.2	The k-Means Clustering Algorithm	186
9.3	The Finished Program	189
9.4	Exercises	192
	9.4.1 Exercise 1: Grouping of Diamonds	192
	9.4.2 Exercise 2: Grouping of Mushrooms	192

10 AutoKeras — 193

10.1	Classification	194
10.2	Regression	195
10.3	Image Classification	196
10.4	Text Classification	199
10.5	Exercises	202
	10.5.1 Exercise 1: Classification	202
	10.5.2 Exercise 2: Regression	202
	10.5.3 Exercise 3: Image Classification	202
	10.5.4 Exercise 4: Text Classification	202

11 Visual Programming Using KNIME — 203

11.1 Simple ANNs — 204
- 11.1.1 Classification — 204
- 11.1.2 Classification Using Python Node — 216
- 11.1.3 Regression — 218
- 11.1.4 Regression Using Python Node — 221

11.2 XGBoost — 223
- 11.2.1 Classification — 223
- 11.2.2 Deployment — 225
- 11.2.3 Regression — 226

11.3 Image Classification Using a Pretrained Model — 227
- 11.3.1 Image Classification Using Keras Node — 227
- 11.3.2 Image Classification Using Python Node — 231

11.4 Transfer Learning — 232
- 11.4.1 Transfer Learning Using Keras Node — 232
- 11.4.2 Transfer Learning Using Python Node — 235

11.5 Autoencoder — 237
- 11.5.1 Autoencoder with Keras Node — 238
- 11.5.2 Autoencoder with Python Node — 242

11.6 Text Classification — 245
- 11.6.1 Text Classification with Keras Node — 245
- 11.6.2 Text Classification with Python Node — 247

11.7 AutoML — 249
- 11.7.1 Installation — 249
- 11.7.2 Classification — 250

11.8 Cluster Analysis — 253
- 11.8.1 Manual Cluster Setting — 253
- 11.8.2 Cluster Setting with a Loop — 254

11.9 Time Series Analysis — 257
- 11.9.1 Recurrent Neural Networks — 257
- 11.9.2 Long Short-Term Memory — 259
- 11.9.3 Prediction of Energy Consumption (Next Hour) Using Keras Node — 260
- 11.9.4 Prediction of Energy Consumption (Next Hour) Using Python Node — 265
- 11.9.5 Prediction of Energy Consumption (Next 500 Hours) Using Keras Node — 267

11.9.6	Prediction of Energy Consumption (Next 500 Hours) Using Python Node	269

11.10 Text Generation 271

11.10.1	Data Preparation	272
11.10.2	Trainings	274
11.10.3	Generation	274

11.11 Further Information on KNIME 277

11.12 Exercises 278

11.12.1	Exercise 1: XGBoost for Classification, Mushrooms	278
11.12.2	Exercise 2: XGBoost for Regression, Diamonds	278
11.12.3	Exercise 3: Image Classification Using InceptionV3	278
11.12.4	Exercise 4: Transfer Learning, Horses or Humans	278
11.12.5	Exercise 5: Anomaly Detection Using an Autoencoder, ECG	278
11.12.6	Exercise 6: Text Classification	278
11.12.7	Exercise 7: AutoML for Regression	279
11.12.8	Exercise 8: Cluster Analysis	279
11.12.9	Exercise 9: Time Series Analysis	279
11.12.10	Exercise 10: Text Generation	279

12 Reinforcement Learning   281

12.1 Q-Learning 282

12.2 Python Knowledge Required for the Game 287

12.2.1	Lists	287
12.2.2	Branches	288
12.2.3	Loops	289
12.2.4	Random Choice	290
12.2.5	Functions	291

12.3 Trainings 292

12.4 Test 294

12.5 Outlook 295

12.6 Exercises 296

12.6.1	Exercise 1: Hyperparameters	296
12.6.2	Exercise 2: Expansion of the Game	296

13 Genetic Algorithms   297

13.1 The Algorithm   298
13.1.1 Start Generation   299
13.1.2 Selection   299
13.1.3 Reproduction   300
13.1.4 Mutation   300
13.1.5 New Generation   301
13.2 Example of a Sorted List   301
13.3 Example of Equation Systems   304
13.4 Real-Life Sample Application   306
13.5 Exercises   309
13.5.1 Exercise 1: Hyperparameter Optimization   309
13.5.2 Exercise 2: System of Equations   309

14 ChatGPT and GPT-4   311

14.1 Prompt Engineering   313
14.1.1 Generating Content   314
14.1.2 Programming   318
14.1.3 Analyzing and Summarizing   324
14.1.4 Final Questions for ChatGPT   326
14.2 The ChatGPT Programming Interface   328
14.2.1 Application Programming Interface Key and First Program 329
14.2.2 Parameters   331
14.2.3 Input Filters   334
14.2.4 Roles   337
14.2.5 Memory   339
14.2.6 User Profiles   340
14.2.7 Playground   341
14.2.8 Speech to Text   341
14.3 Exercise 1: Math Support   344

15 DALL-E and Successor Models — 345

- **15.1 DALL-E 2** — 345
  - 15.1.1 Prompt Engineering — 346
  - 15.1.2 Editing Generated Images — 347
- **15.2 DALL-E 3** — 350
- **15.3 Programming Interface** — 352
  - 15.3.1 Image Creation — 352
  - 15.3.2 Image Variations — 354
  - 15.3.3 Image Processing — 355
- **15.4 Exercise 1: DALL-E API with Moderation** — 357

16 Outlook — 359

Appendices — 361

- **A Exercise Solutions** — 363
- **B References** — 395
- **C The Author** — 397

Index — 399

Chapter 1
Introduction

So, you've decided to familiarize yourself with the topic of "artificial intelligence"? That's a pretty good idea! Let's first get an overview in this chapter.

What This Chapter Is About
- Why knowledge in the field of artificial intelligence (AI) is so important
- Can novice programmers develop AI programs?
- Defining AI
- Subcategories of AI
- A brief overview of the history of AI
- Presentation of the tools used in developing AI

AI has already found its way into our daily lives in medicine, quality assurance, art, or advertising—to name just a few areas—and all signs point to AI becoming even more important in the future. You, dear reader, have obviously recognized this. Whether at school or in training, at university, or at work, you'll encounter AI everywhere.

If you understand how this technology works, you can identify new ways to use it in your environment. Perhaps a lot of data is already available (e.g., in Microsoft Excel), and AI can analyze, group, and classify it for you or provide an indication of future events. It can take work off your hands or support you in your daily work.

Even if you don't develop AI programs professionally, having some knowledge in this area is very helpful. You'll then be able to recognize when and how AI could be used in projects. In addition, you can also better categorize suggestions from AIs because you know the "system" behind them. And don't forget—it's also a lot of fun to develop programs that have the appearance of intelligence.

1.1 What Does This Book Offer?

What is the right strategy for introducing interested parties to this subject area? Should a thorough introduction to Python be given first? With all the required modules? My experience from teaching at the technical college (mechanical engineering technology, mechatronics technology, and business informatics) has made this clear: no, that isn't necessary. Even future mechanical engineers who had little or no programming knowledge were able to download data records from the internet and program AIs during the course. However, the word "programming" isn't quite right; "configuring" is more accurate.

Right at the beginning, we'll look at a finished Python program for a specific problem and discuss it in detail. We'll then modify this program so that we can use it to solve other problems. Later on, we'll continue in this way. We'll also solve some tasks using the KNIME software, including graphical modules. This way, you don't have to rewrite programs from scratch.

The aim isn't to give you a comprehensive introduction to the Python programming language or the modules used. After reading this book, you'll be able to download data records from the internet and develop AIs for them or adapt your existing programs for this purpose. This is a very pragmatic approach. The programming language is only a means to an end. However, if you've worked through the book and want to deepen your knowledge in this area, I recommend that you learn the Python programming language.

You can download, analyze, and run the programs discussed here. Have the courage to change them and take a critical look at the results. The chapters also contain exercises where you can immediately apply what you've learned. Take your time for these exercises. You can copy and adapt an existing program discussed here as a solution. It's important that you don't look at the sample solution straight away. You'll get much more out of the exercise if you try it yourself first. Then questions will arise that you would never have asked yourself otherwise. The analysis of the sample solution will support the learning process even if you were unable to create the program without errors.

Of course, it's perfectly acceptable to research unanswered questions on the internet. But don't make the mistake of simply copying lines of code without understanding their functionality. Adapt the lines of code to your program; for example, the structure and variable names of your program should originate from you. Modify the transferred lines of code so that they fit your program. Don't adopt a line without understanding what it actually does.

You won't learn AI development or programming by reading alone, just like you wouldn't think of learning French just by reading a book. However, reading, analyzing, changing, questioning, and adapting to new tasks will lead to success in this subject area.

Unfortunately, the hard truth is that you won't become an AI professional even if you work through the book conscientiously. However, by the end of the book, you will understand the basics and have already developed your own AIs. The rest of the learning process will be more fun because you won't just have to deal with dry theory. You can also apply what you've learned to new programs.

Feedback from many students has shown me that visual programming using KNIME is particularly fun. The AI program gets put together like a jigsaw puzzle using graphical building blocks. And this isn't just a gimmick; KNIME is widely used in professional development.

You won't experience an AI event without ChatGPT and DALL-E. No wonder, ChatGPT in particular dominates the media coverage when it comes to AIs. We'll not only look at these tools but also use their programming interfaces so that you can easily program your own applications that are capable of impressive performance.

1.2 What Is Artificial Intelligence?

The Encyclopedia Britannica (*www.britannica.com*) defines artificial intelligence (AI) as follows:

> *Artificial intelligence (AI) is the ability of a computer or a robot controlled by a computer to do tasks that are usually done by humans because they require human intelligence and discernment. Although there are no AIs that can perform the wide variety of tasks an ordinary human can do, some AIs can match humans in specific tasks.*

The European Parliament's website (*www.europarl.europa.eu*) also has a definition for this:

> *Artificial intelligence is the ability of a machine to imitate human abilities such as logical thinking, learning, planning and creativity.*

You'll also find similar formulations in the literature. According to these definitions, the navigation device in a car can also be characterized as AI or just a clever algorithm to find the fastest way from A to B. It's just not easy to clearly define AI. However, we can say that the aim of AI is to imitate human intelligence in some way.

Machine learning (ML) is a subset of AI. This includes self-learning programs without a predefined algorithm. Let me illustrate this with an example. Let's suppose you train a program to calculate the sum of two numbers. To do this, you enter many possible combinations one after the other, such as 1 + 3 = 4, 2 + 2 = 4, 2 + 6 = 8, and so on. While you're entering the summands, the program tries to find a link and predict the result you've entered before you type it in. Each time you finally enter the result, the system checks whether the prediction was correct (training phase). If at some point, a correct link is found that also proves to be correct for other summands (test phase), the ML program is ready. Now only the summands have to be entered because the result is calculated automatically. This approach therefore doesn't require an algorithm to be entered to calculate a sum. The AI learns the correct combination itself with the help of the training and test data.

ML can be divided into different subcategories depending on the specialist literature, but we'll agree on the following three subcategories here:

- **Supervised learning**
  Supervised learning needs training data to learn. For example, you have pictures of dogs and cats, and one person has correctly assigned all existing images in advance (this is referred to as *labeling*). The training data therefore consists of images and the corresponding labels. You use this training data to train the program. The program itself is then able to correctly assign new pictures of dogs and cats. The example described earlier with the summands also falls into this category.

- **Unsupervised learning**
  Unsupervised learning is often used to search for and classify patterns in large amounts of data. For example, the "critical" or "noncritical" states can be derived from the sensor data of a machine.

- **Reinforcement learning**
  Reinforcement learning is used, for example, in games or for testing games. A character learns the correct moves independently. Incorrect moves are penalized (e.g., by deducting points), and correct moves are rewarded. Over time, the character learns the correct moves and progresses further and further.

In this book, we'll deal with all three subcategories of ML. The definitions of terms could go on for a long time (e.g., artificial neural network [ANN], decision trees, deep learning), but we'll refrain from doing so here. Rather, we only want to go into this when the corresponding programs have been developed.

1.3 The History of AI: A Brief Overview

People have been exploring the idea of machines with human capabilities for a long time. In his work *Politics*, the Greek philosopher Aristotle (384–322 BC) wrote that one day, automatons could replace slaves:

> *For if it were possible for every tool to accomplish its work at the behest of the gods or beforehand, like the statues of Daedalus or the tripods of Hephaestus, of which the poet says that they entered the assembly of the gods by themselves, and if the weaver's ship wove by itself and the zither played by itself, then neither the artists would need assistants nor the masters slaves.*

The polymath Ismail al-Jazari (1136–1206) constructed a number of mechanical devices for his employers, such as a water-driven peacock and musical automatons with changing facial expressions. There are also sketches by Leonardo da Vinci (1452–1519) of robot knights with articulated joints made from pulleys. It's remarkable what visions these pioneers (and there were many more) had in their time.

Some important milestones in the modern era show the rapid development since the 1940s:

- The "A Logical Calculus of the Ideas Immanent in Nervous Activity" journal article by neuropsychologist Warren McCulloch and logician Walter Pitts, was published in 1943. This groundbreaking publication can be seen as the basis for the principle of ANNs, which you'll get to know in the course of this book.
- In 1956, The Darthmouth Summer Research Project on Artificial Intelligence conference was held in the United States, where researchers were able to exchange ideas on the subject and the term "artificial intelligence" was recorded in writing for the first time.
- Engineer Arthur Samuel gave birth to the term "machine learning" in 1959 with the publication of his "Some Studies in Machine Learning Using the Game Checkers" essay. The program he developed, which was based on ML, could play checkers against humans and itself.
- The BKG 9.8 program developed by Hans Berliner (scientist at Carnegie Mellon University in Pittsburgh) defeated the then reigning world champion in backgammon in 1979. This was the first time that a world champion had been defeated by a computer program.
- In 1997, IBM's Deep Blue computer beat the then world chess champion Garry Kasparov. Deep Blue was able to calculate up to 200 million chess positions per second and more than 6 moves in advance.
- The AlphaGo program from DeepMind beat the then world champion Lee Sedol in the game of Go in 2016. The number of possible moves in this game is greater than the number of atoms in the universe.
- In 2022, the OpenAI company published the chatbot ChatGPT, an AI of a completely new quality. You can have a conversation with this chatbot just like with a human. ChatGPT can, for example, simplify, summarize, compose, program, and philosophize. It can even pass courses at elite universities, as tests have shown.

The development of the past 80 years is remarkable, and AI is providing great services in many areas, such as medicine. However, we must not lose sight of the disadvantages. In social media, this technology is used to ensure that users are presented with news and content that is specifically selected for them. Can we still speak of free will when we've been living in this news bubble since our early youth? Are you objective in your decisions, and can you really block out years of targeted news flow? Do we want to allow AI to classify and evaluate us, for example, in job applications? This often has nothing to do with objectivity, as you'll see when we talk about correlation and causality in Chapter 3. These are just a few points to highlight the dangers.

Nevertheless, we must not close our eyes and think that this is all just a fad. This development is unstoppable and offers great opportunities. AI has now arrived in all sectors of industry. Now, let's concentrate on the positive aspects. Knowledge in this area will also strengthen your judgment about these technologies.

1.4 Development Tools Used in This Book

Many programming languages, libraries, and tools are available for developing AIs. Here is a brief introduction to the selection used in this book. Nothing is installed or programmed yet.

1.4.1 Python

The Python programming language is both very easy to learn and very powerful. Even during development, great importance was attached to readability, platform independence, and simple structures. Even if you program on Windows, the program can then be run on a Linux computer or macOS. These advantages have led to it becoming one of the most popular programming languages of all. In the world of data science and AI, Python has become the de facto standard. As already mentioned, this book isn't intended as an introduction to the Python programming language. For us, it's only a means to an end. The following Python statements are intended to illustrate the simplicity. We won't start programming until the installation has been completed in the next chapter.

The standard "Hello World" example looks like this:

```
print("Hello World!")
```

The `print` function can be used to output data to the console, in this case, "Hello World!". We can also save this character combination (string data type) in a variable first and then output it:

```
output = "Hello World!"
print(output)
```

> **Data Types**
>
> The most elementary data types in Python are as follows:
> - Strings (character strings), which are written in single or double quotation marks (e.g., "Hello World!")
> - Integers (e.g., 72)
> - Floats (floating-point numbers, e.g., 3.14)

> Certain operations are possible depending on the data type. String data such as "Hello" and "World" can be combined to form "Hello World". Integer or float data can be offset against each other, for example, 3 × 8.
>
> We'll only introduce other data types when they are required to solve a specific task.

The same variable can also be reused for an integer (integer data type) or floating-point number (float data type). Python variables can store objects of any type:

```
output = "Hello World!"
print(output)

output = 42
print(output)

output = 42.2
print(output)
```

Listing 1.1 Variables with Python

Here, "Hello World!", 42, and 42.2 are output one after the other. Now let's take a look at a supposedly more complicated program:

```
# Setting the age to 20
age = 20
# Query whether the person is of legal age
if age >= 18:
    print("You are of legal age!")
else:
    print("You are not yet of legal age!")
```

Listing 1.2 If-Else Structure with Python

The # character introduces comments that can help when reading the code, and these lines aren't executed. The number 20 is stored in the age variable. You'll then be asked whether the age is greater than or equal to 18. If that is the case (as in this example), the corresponding output is generated. The else branch is skipped, and the program is ended. Let's clarify the following questions:

1. What effect would it have if the value 16 were stored in the age variable? Answer: The if branch would be skipped, and the else branch would be executed. The output would read "You are not yet of legal age!"

2. In some countries you're only of legal age at 21. What changes are necessary to this program? Answer: In the if query, the comparison is made with the number 21 instead of 18.

3. Can the change to the age of majority query be described as "programming" or as a "configuration"? You can argue about that if you have nothing better to do.

As you can see, it's quite easy to trace the source code in Python.

1.4.2 Jupyter Notebook

We need software into which we can type the Python commands. This software is used for programming, execution, and testing. Software of this type is called an *integrated development environment* (*IDE*). There are numerous very good IDEs available for Python. However, simple Jupyter Notebooks (open source) have established themselves in data science and AI development. The program is programmed in the browser and usually consists of a single file. To install and configure Python and Jupyter Notebook (and many other tools if required), we'll use the free Anaconda distribution, which was developed for this purpose. It's available for Windows, Linux and macOS. Using Anaconda you can set up simple development environments locally on a PC or centrally on a server. The advantage of the server version is that you can easily program in Jupyter Notebook from multiple clients via a browser. No further installation is necessary on these clients. However, if instead of using your own server, you want to use one provided by Anaconda in the cloud, you'll have to pay for it.

> **Development Environment**
>
> An IDE provides important tools that support you in programming. This includes the automatic recognition and completion of commands, auto-correction, and helpful hints in the event of programming errors.

1.4.3 KNIME

The KNIME Analytics Platform software takes a completely different approach in that programming is carried out using graphical blocks, which are dragged and dropped into the workspace, linked together, and configured. This is also known as *visual programming*.

KNIME is based on Eclipse, a widely used development environment for various programming languages. It's open source, free of charge, and platform-independent.

The following example shows a program that reads the contents of an Excel file and plots the numerical values in a graph (see Figure 1.1).

You can see the output of the **Line Chart** module in Figure 1.2. For this purpose, right-click on the block after executing it, and then click on **View: Line Chart**.

If the data source isn't an Excel file, but instead a database or comma-separated values (CSV) file, for example, you only need to replace the **Excel Reader** module.

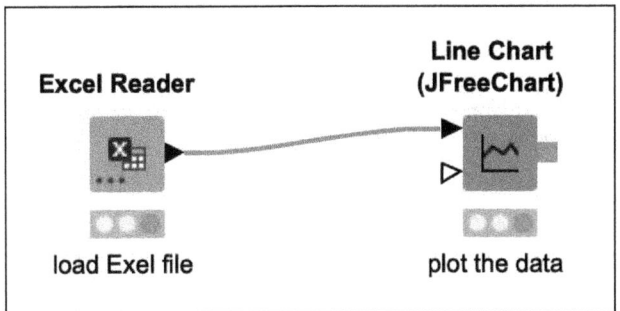

Figure 1.1 Loading an Excel File and Plotting the Contents

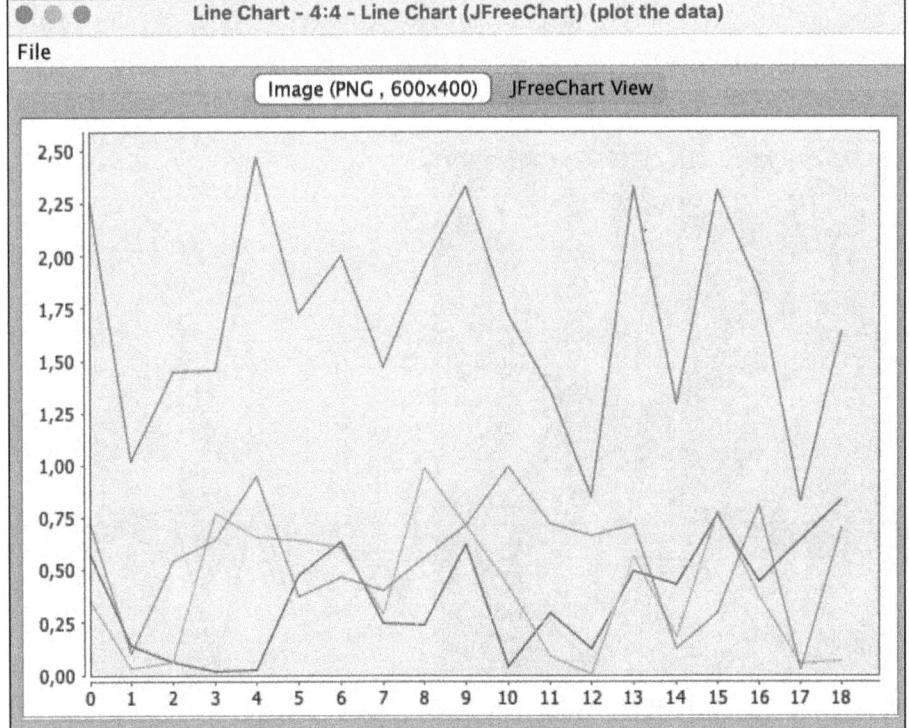

Figure 1.2 Plotting the Contents

1.4.4 ChatGPT and GPT-4

Let's just ask ChatGPT (or its successor GPT-4) itself who or what it is:

User: *Who or what are you? Explain it to me as if I were 12 years old.*

Answer: *Of course! Imagine I'm like an extremely advanced computer program that can answer questions. You can think of me a bit like an interactive encyclopedia or a digital interlocutor who knows a lot of information.*

1 Introduction

I am not really a person and have no thoughts or feelings of my own. I am a tool created by humans to answer questions and provide information based on what was known until 2021.

You can try out this chatbot from OpenAI at *https://chat.openai.com*. We'll take a closer look at this impressive tool later on in this book.

1.4.5 DALL-E 2 or DALL-E 3

DALL-E 2 and its successor DALL-E 3 are based on the ChatGPT model, which has been adapted for text-to-image synthesis. The name is derived from Salvador Dali (important surreal artist) and WALL-E (Pixar film robot).

You can use this tool (*https://labs.openai.com*) to generate images from textual descriptions (see Figure 1.3).

User: *The image of a programmer sitting on an airplane and programming on a laptop.*

Figure 1.3 Image Generated by DALL-E 3

The picture looks as if a real person has been photographed on an airplane. We'll also take a closer look at this tool.

> **DALL-E 2 Is Being Replaced by DALL-E 3**
> OpenAI plans to completely replace DALL-E 2 with DALL-E 3 in the near future.

Chapter 2
Installation

Before we start with the actual development, we'll install and test the main tools used.

> **What This Chapter Is About**
> - Installing the most important tools you need to work with this book
> - Testing those installed tools
> - Installing and configuring KNIME

Setting up a development computer can be very time-consuming, but you should really take this time! We don't want to have to deal with weird error messages later during development.

> **OS Versions**
> The installation statements described in this chapter have been tested on the following computers:
> - Windows 11
> - macOS Ventura 13.2 (Intel)
> - Ubuntu LTS Jammy Jellyfish 22.04

2.1 Anaconda Distribution

Using the Anaconda distribution, we'll install Python and Jupyter Notebook and test the installations. You can download the respective installation file for the Windows, macOS, and Linux operating systems from the following website: *www.anaconda.com/products/distribution*. The operating system will be recognized through the browser, and the correct installation file will be downloaded.

2.1.1 Windows and macOS

Run the installation file, and follow the instructions on the screen. The recommended default settings can be retained. After the installation, you can start the Anaconda Navigator (see Figure 2.1).

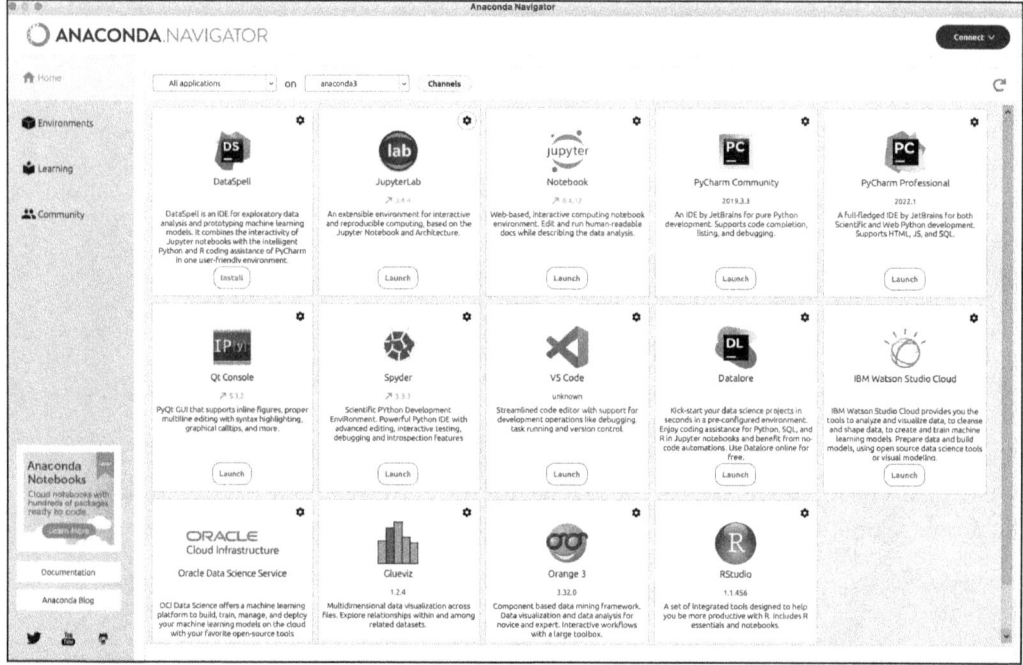

Figure 2.1 Home Screen of Anaconda Navigator

We can now turn our attention to the necessary configuration of the development computer.

2.1.2 Linux

After the file has been downloaded and the installation routine has been performed, you need to open a new terminal window (so that the new environment variables take effect) and enter the following command to start Anaconda Navigator:

```
anaconda-navigator
```

> **Installation Instructions for Anaconda**
> You can find very useful installation instructions for Anaconda at *https://docs.continuum.io/free/anaconda/install/linux*.

This completes the installation. However, you still need to make a few settings.

2.1.3 Configuration and Test

After successful installation, you can start various applications in the home area of Anaconda Navigator using the respective launch button. If you then start Jupyter Notebook from here, you'll be taken to the Jupyter Notebook homepage (see Figure 2.2).

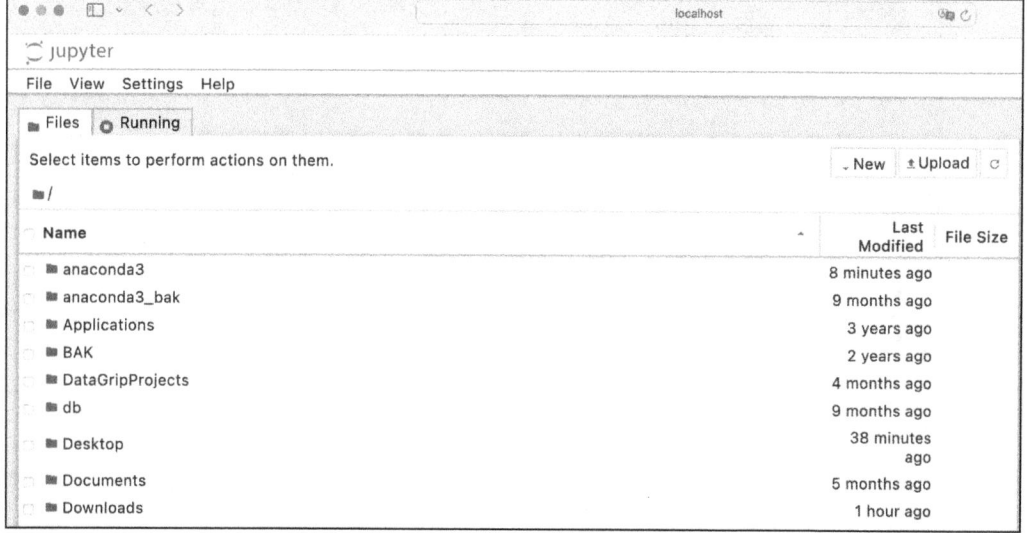

Figure 2.2 Jupyter Notebook Homepage

Then, you go to **New • Python3 (ipykernel)** and start a new notebook (see Figure 2.3).

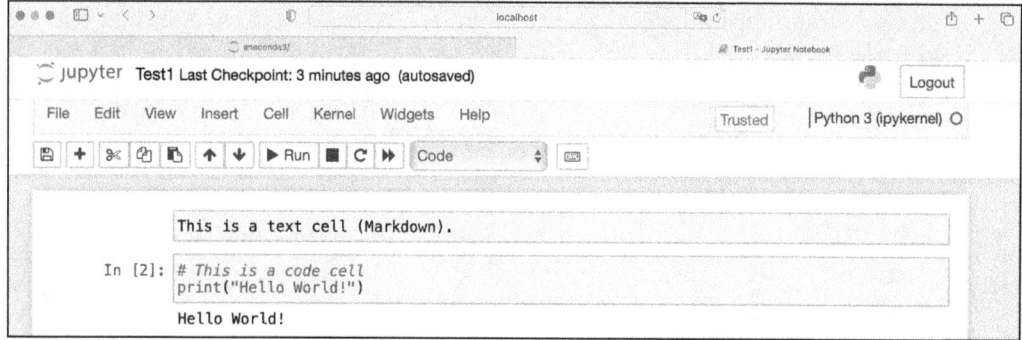

Figure 2.3 First Program in Jupyter Notebook

The program is organized in different cells. You can add a new cell via the **+** button and move the selected cell up or down with the arrow buttons. In the dropdown menu, you can set whether the cell is only used for documentation (**Markdown**) or contains source code (**Code**). The **Run** button executes the selected code cell, while the **Forward** button executes all cells. You can rename the file using **File • Rename** and then specify where the file should be saved via **File • Save as**.

2 Installation

Alternatively, you can start JupyterLab in Anaconda Navigator (see Figure 2.4), which can also run notebooks but provides more functionality.

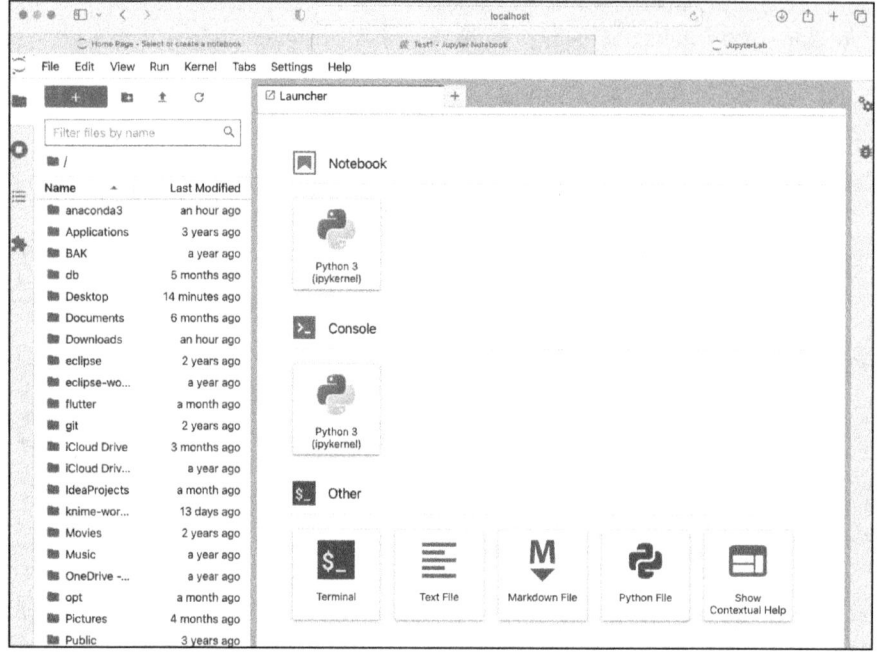

Figure 2.4 JupyterLab Launcher

Here, you can start a new notebook via **Notebook · Python 3 (ipykernel)** (see Figure 2.5) or, for example, open the *Test1.ipynb* file created earlier. The directories and files are listed on the left.

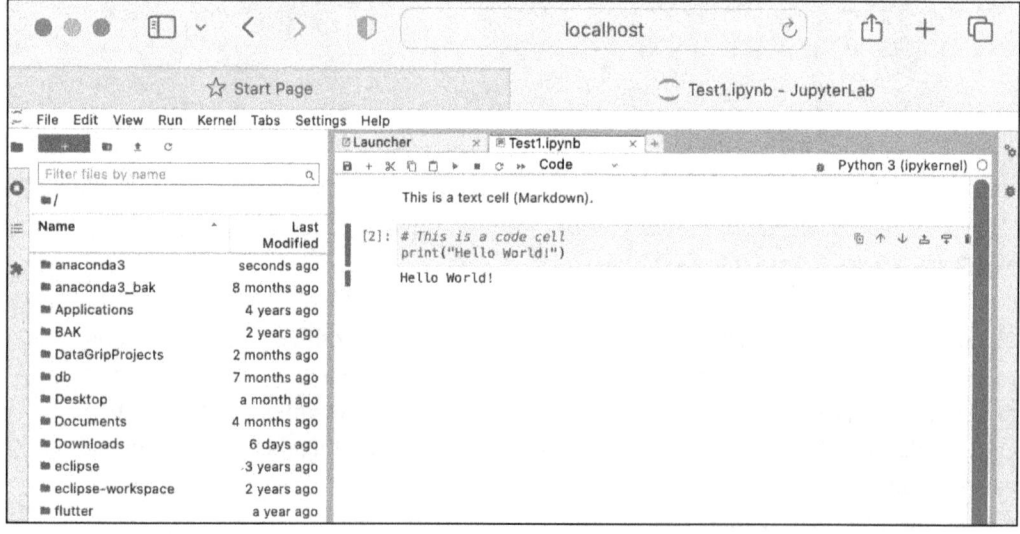

Figure 2.5 First Program Using JupyterLab

2.1 Anaconda Distribution

> **Google Colaboratory**
>
> Google offers an interesting alternative to the installation of Anaconda or Jupyter Notebook: Colaboratory, or Colab (*https://colab.research.google.com*). This solution doesn't require you to install anything locally. Google provides storage and computing capacity free of charge, so all you need is a browser.
>
> As we'll be using KNIME for visual programming later on, you'll also need a local installation. I like using Colab for computationally intensive AI programs because you can easily use a GPU instead of the CPU. This results in an enormous increase in performance. Setting up local GPU support depends on your graphics card and involves some effort.

We'll use this opportunity to install some of the required modules. When you install Python, you don't immediately have access to all the functionalities provided by this programming language. In fact, Python is structured in such a way that you can install additional modules as required. This modular structure ensures that the program doesn't take up an unnecessarily large amount of memory.

Start Anaconda Navigator, and go to the **Environments** menu. Select **All** in the drop-down menu, and use the search field on the right (see Figure 2.6).

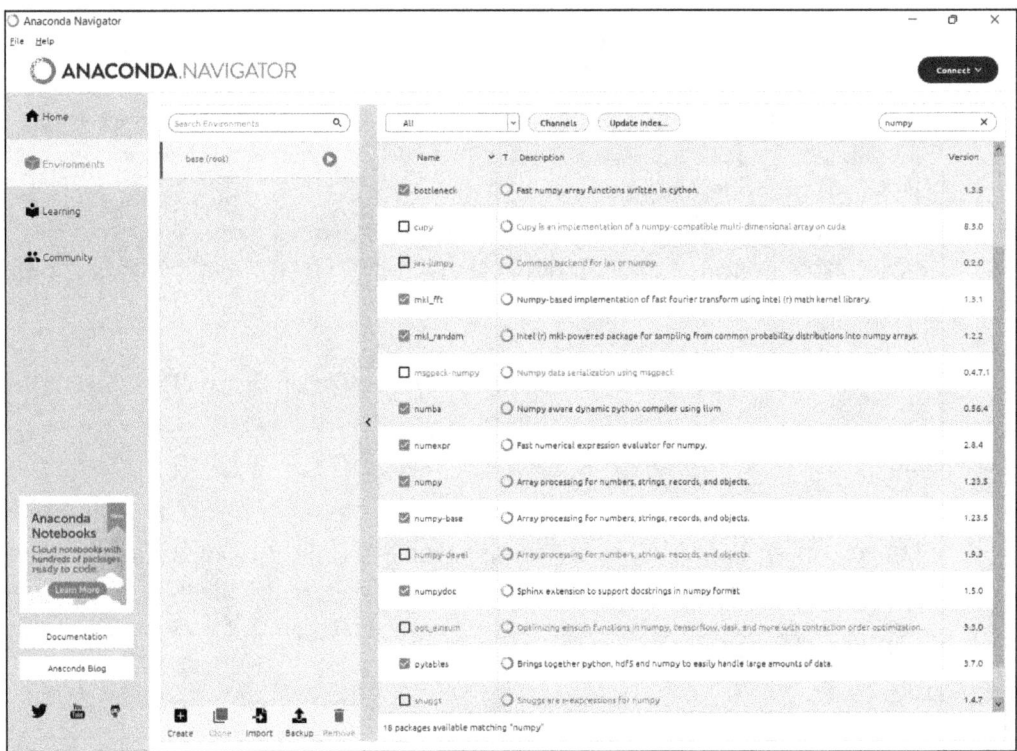

Figure 2.6 Module Search and Installation via Anaconda Navigator

Install the modules listed in Table 2.1 or verify that they are installed.

Module	Explanation
NumPy	NumPy provides data types and functions for an easier handling of complex structures, such as vectors and matrices.
pandas	The pandas module has also been developed for more complex structures and simple handling. One of its strengths is as a great functionality for table structures.
Matplotlib	Matplotlib is used for visual analysis and plotting.
scikit-learn	The scikit-learn module contains many machine learning (ML) algorithms that can be used very easily in your own program.
Keras	Keras can be used to build artificial neural networks (ANNs).
TensorFlow	TensorFlow extends Keras with additional functionalities. It's very performant with large and complex data structures.

Table 2.1 Modules to Be Installed

We'll install additional modules as required.

> **Anaconda and Environments**
>
> In Python, environments are isolated areas in which Python source code can be executed. Each environment can install different packages or versions of packages. The advantage of a separate environment is that the installations in one environment don't affect the other environments.
>
> If you have some experience with Python or Anaconda, you can create your own environment for Jupyter Notebook and install the packages in it. However, you can also use the base environment.

Now that you know how to install the modules, let's move on to install KNIME.

2.2 KNIME

In Chapter 11, we'll solve some tasks using KNIME. You'll get to know an alternative approach, namely visual programming. If you only want to use the Python programming language, you can skip this section. However, I recommend that you familiarize yourself with this type of AI programming.

2.2.1 Installation

You can download the *KNIME Analytics Platform* software for your preferred operating system at *www.knime.com/downloads*. For Windows and macOS, you need to run the installation program and confirm the recommended settings. With Linux, you only need to unpack the archive (in your home directory).

> **Display with Linux Wayland**
>
> When you start KNIME, you'll be notified that you should use Xorg for an optimal display, but you can still try Wayland. If the display isn't satisfactory, you can log out and switch to Xorg in the login window.

After starting the program, a window opens with the instruction to select the workspace in which your programs and configurations are saved (see Figure 2.7). Check the **Use this as the default and do not ask again** box, and click **Launch**.

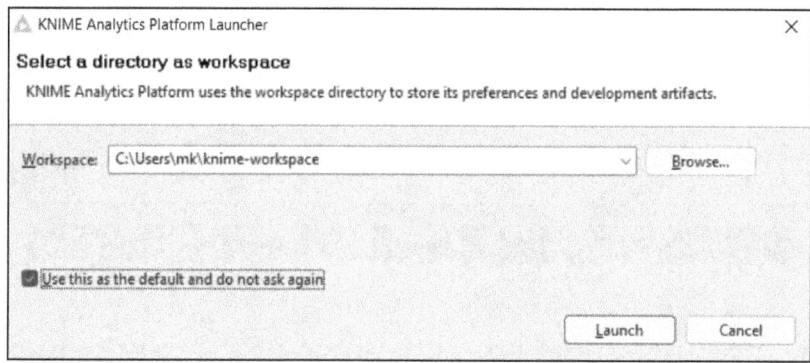

Figure 2.7 Selecting the Workspace

Once you've set or confirmed the workspace, the KNIME Workbench appears.

> **New GUI for Version 5.1**
>
> If you download the current version of KNIME, the user interface (UI) is structured differently by default than shown in the screenshots in this book, but that isn't a problem at all. You only need to make one setting to be able to follow the instructions in this book: on the start page in Figure 2.8, click on the [i] icon at the top right, and select **Switch to KNIME classic user interface** (see Figure 2.9).

2 Installation

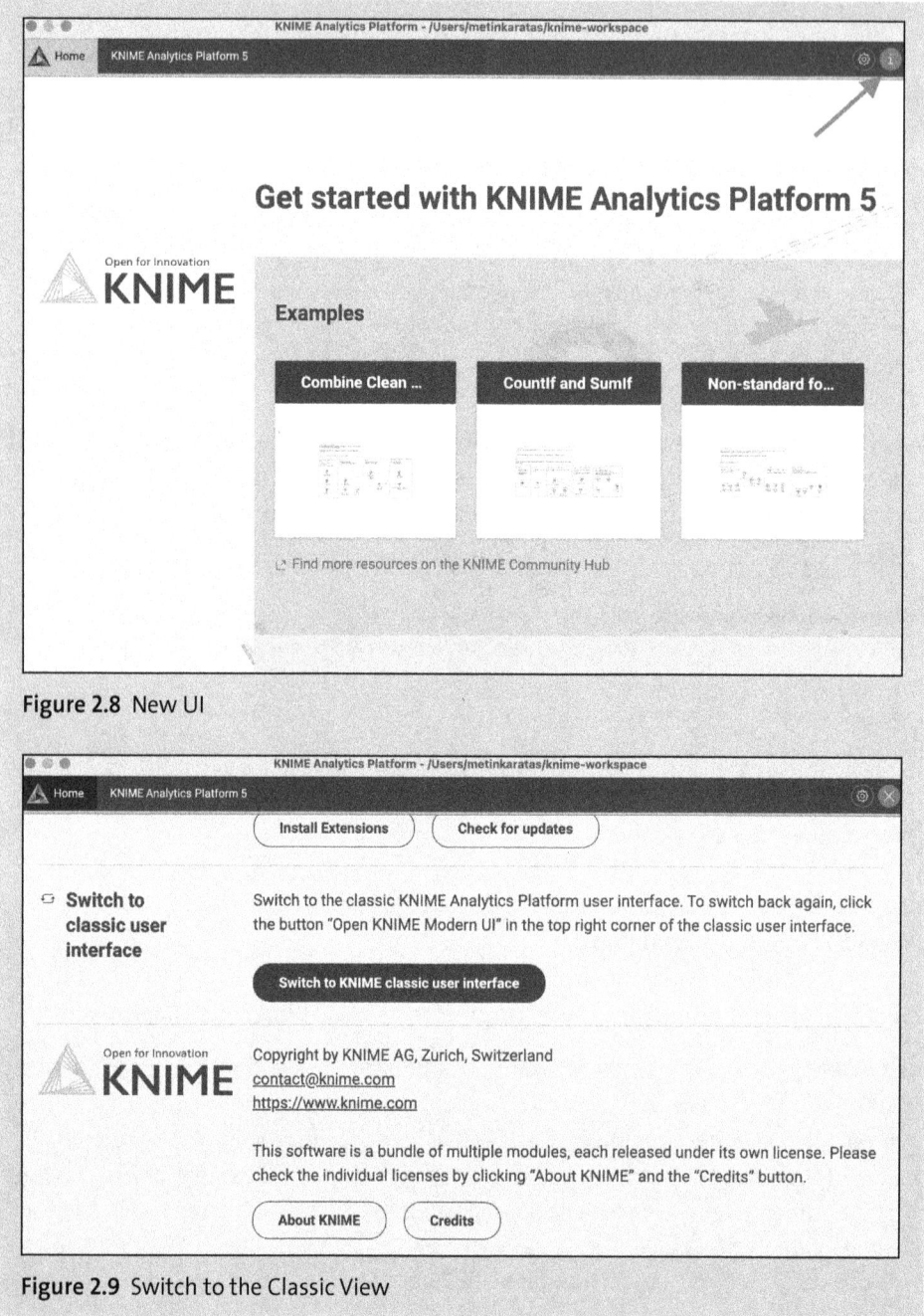

Figure 2.8 New UI

Figure 2.9 Switch to the Classic View

You should get an overview of the structure of the software. If you've never dealt with development environments before, the UI may appear somewhat complex to you (see Figure 2.10). Don't worry, soon you'll be able to routinely create programs and make settings.

2.2 KNIME

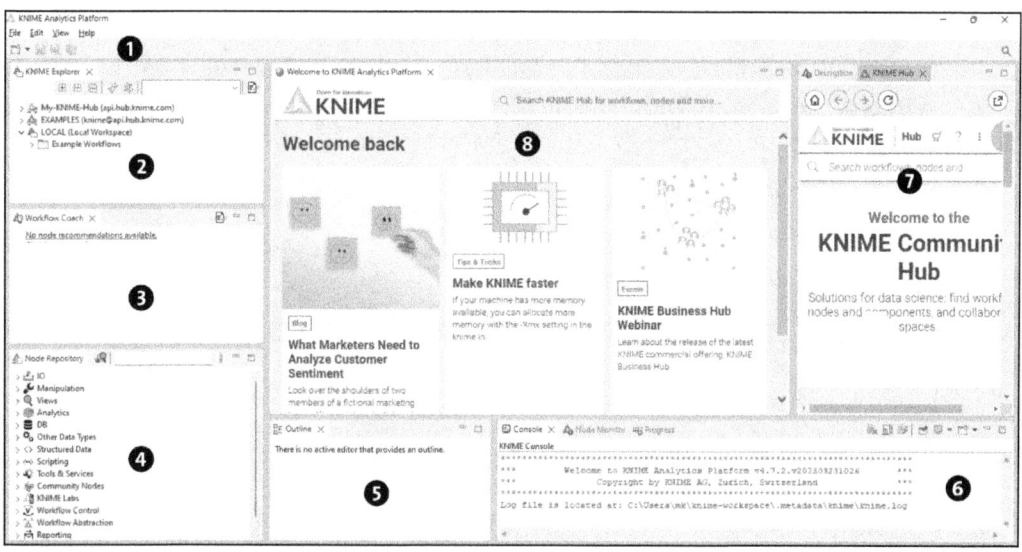

Figure 2.10 KNIME Workbench

Following are areas of the KNIME Workbench:

❶ Menu bar
Make central settings, search for and install updates, and so on.

❷ KNIME Explorer
Access the project structure.

❸ Workflow Coach
Activate this feature by clicking on it, and suitable nodes (e.g., graphic elements or modules) for your program will be recommended. This saves you a lot of time later during development, so activating is recommended.

❹ Node Repository
See an overview of all existing nodes.

❺ Outline
See a complete overview of the workspace. This can be interesting if you have large programs.

❻ Console and **Node Monitor**
Find important issues of interest to you.

❼ Description and **KNIME Hub Search**
Find information on selected nodes in the node description. Access and use ready-made programs (workflows) from the community in the hub search, for example.

❽ Workflow Editor
Access the workspace in which the actual programming work takes place.

2 Installation

2.2.2 Configuration

Next, let's install other required packages by going to **Help • Install New Software** in the menu bar. Under **Work with**, you can set the search to be performed on all pages. Then, search for Python, and select the items shown in Figure 2.11:

- KNIME Conda Integration
- KNIME Python Integration
- KNIME Python Scripting extension

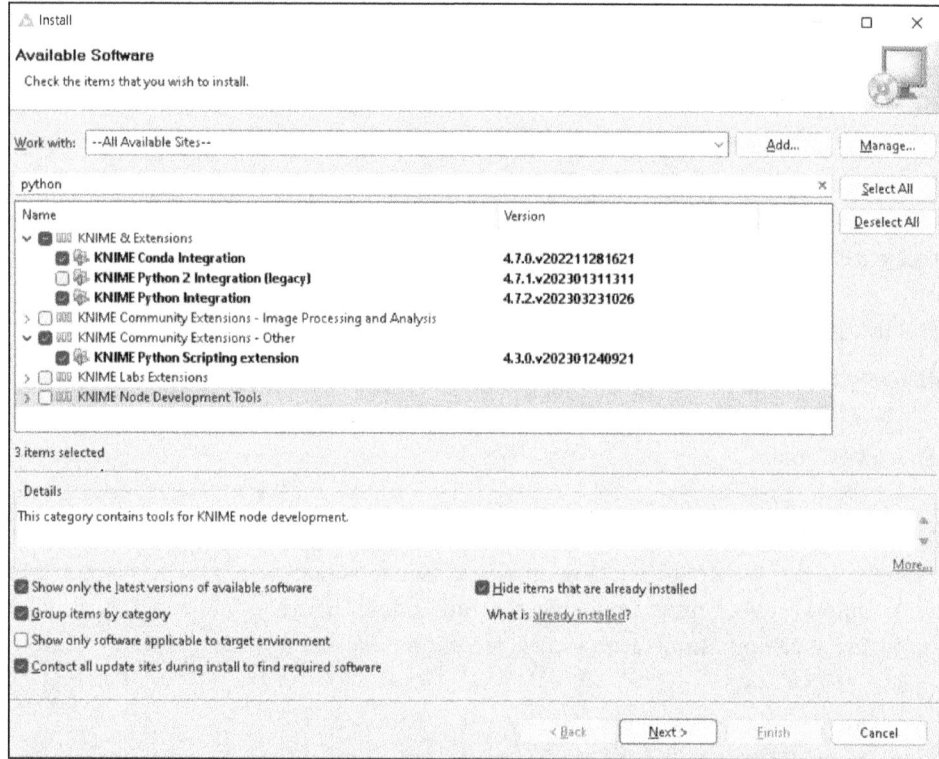

Figure 2.11 Installing Other Packages

Click **Next**, accept the license conditions, and confirm with **Finish**. The selected packages will then be downloaded and installed.

Install the following packages in the same way:

- KNIME Deep Learning—Keras Integration
- KNIME Deep Learning—TensorFlow Integration
- KNIME Deep Learning—TensorFlow2 Integration

- KNIME Text Processing
- KNIME Image Processing

We still need to make some settings for Python, Keras, and TensorFlow. To do this, go to **File • Preferences • KNIME • Python** via the menu bar. Select **Conda** and wait briefly until KNIME has collected all the information about the installation. You'll likely see a warning that some packages are missing. For this reason, we'll now create an environment that contains all the required packages. To do this, click on **New environment**, and keep the suggested name for the environment (here **py3_knime**), as shown in Figure 2.12.

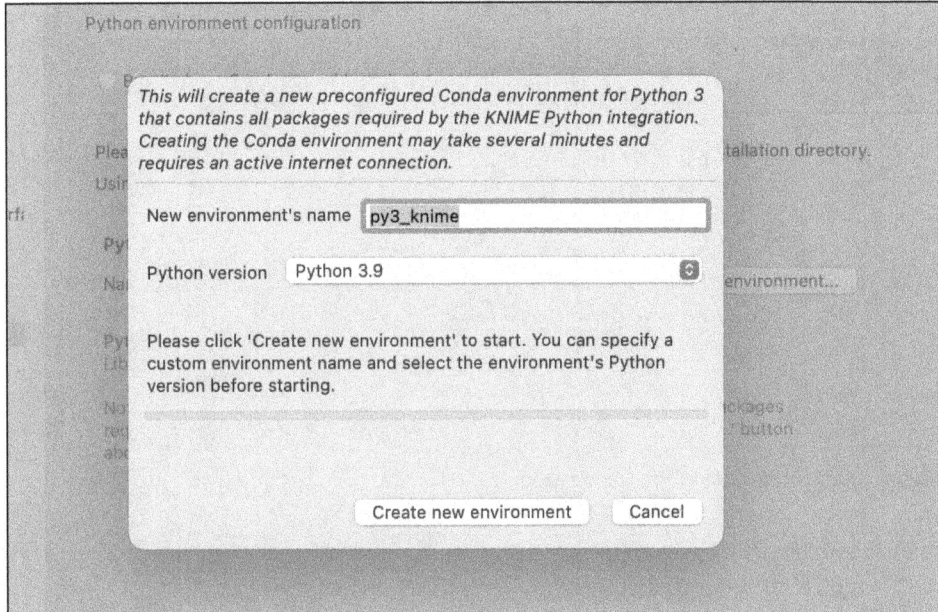

Figure 2.12 Setting for Python

To ensure that programs with ANNs can be created later, you must create another environment for this by going back to **File • Preferences • KNIME • Python Deep Learning** via the menu bar. Then, select **Use special Deep Learning as defined below**, and choose **Keras** and **Conda**. You'll then have to wait again until KNIME has collected information about the Python installation. Now click on **New environment** under **Keras**, accept the suggested name, and click on **Create new CPU environment**. Close the menu (Figure 2.13) by clicking on **Apply and Close**.

Create an environment for TensorFlow 2 in the same way. This prepares the development environment for visual programming.

2 Installation

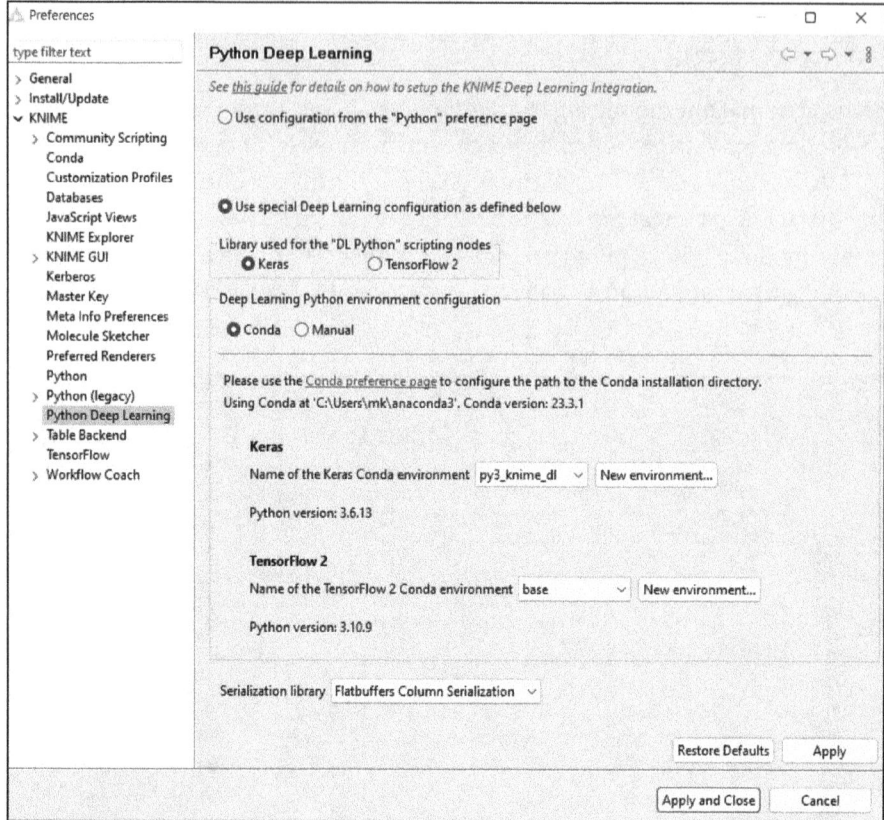

Figure 2.13 Setting for Python Deep Learning

> **Troubleshooting**
>
> Unfortunately, creating environments via KNIME doesn't always work smoothly and depends on the versions of the operating system, Anaconda Navigator, Python, and KNIME. If you receive error messages, try updating Anaconda first. If the problem occurs when you create the environment for Python, you can select an older Python version (e.g., 3.9) from the dropdown menu before you select **Create New Environment**.
>
> If you encounter problems when creating an environment for Python Deep Learning, and you can't solve them yourself, it's not the end of the world. First, take a look at the Python examples. By the time you've worked your way through to Chapter 11, there will certainly be more up-to-date versions of the required tools. And if it still doesn't work, I'll show you how you can use Python modules in KNIME. Graphical modules are therefore used for data processing, while the ANN is set up in a Python module with a few lines of code. To do this, it's necessary (if you were unable to create an environment for deep learning) to change the environment for Python to Base, as the packages for Keras and TensorFlow have been installed in this environment.

2.2 KNIME

> It's advisable to consult the KNIME forum (*https://forum.knime.com*) if you have any problems with KNIME.

2.2.3 Test

We'll test the installation with an Excel file, which you can create using Microsoft Excel or LibreOffice Calc, for example (see Figure 2.14).

	A	B	C	D	E
1	Column 1	Column 2	Column 3	Column 4	
2	10	20	30	40	
3	11	21	31	41	
4	12	22	32	42	
5	13	23	33	43	
6					

Figure 2.14 Table with Numerical Values

In **KNIME Explorer**, right-click on **LOCAL**, select **New Workflow Group** from the context menu, name it "Test", and click **Finish**. Then, create a workflow named "Test-1" in this workflow group. After that, you need drag and drop the XLSX file into the workflow group (this is a folder) to copy it there. The folder structure should look like Figure 2.15.

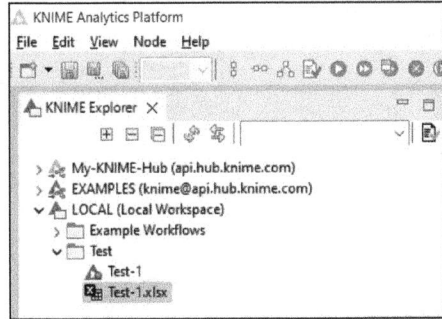

Figure 2.15 File Structure in KNIME Explorer

Now, you can drag and drop the XLSX file into the workflow editor again. A creation wizard appears, as shown in Figure 2.16, which usually recognizes all settings correctly. All you have to do now is confirm by clicking **OK**.

The **Excel Reader** module or node appears with a yellow dot. You can execute the module by right-clicking and then clicking **Execute**. The dot changes to green, and the content of the file gets displayed in the output (see Figure 2.17). To reset the module, you need to right-click and then select **Reset**.

2   Installation

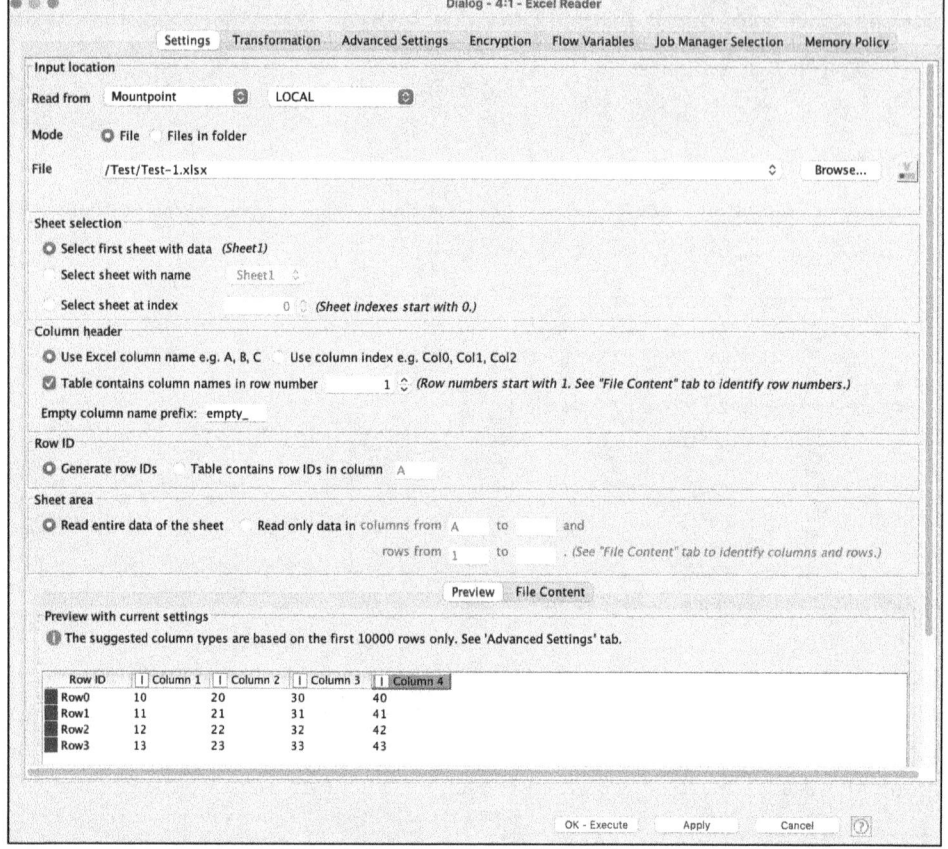

Figure 2.16 KNIME Creation Wizard

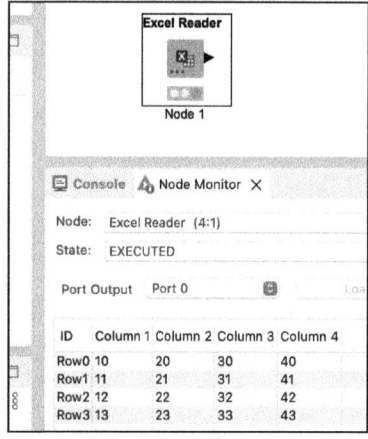

Figure 2.17 Output of the "Excel Reader" Node

Your development machine is now configured, tested, and ready for use. Now you can start developing AI models.

Chapter 3
Artificial Neural Networks

An artificial neural network (ANN) is an ingenious machine learning (ML) concept that is often highly romanticized as a mystery. In this chapter, we'll reveal the secret and immerse you into the world of ANNs.

What This Chapter Is About
- Introduction to classification
- Introduction to regression
- Structure and function of artificial neural networks (ANNs)
- Mathematical consideration of the algorithm
- Presentation of a "recipe" for development
- First programs with ANNs
- Parameter setting options
- Difference between correlation and causality
- Dealing with data gaps
- Importance of data standardization

The greatest progress and most impressive results in ML have recently been achieved in the field of artificial neural networks (ANNs). The structure and mode of operation aren't particularly complicated. However, if you want to familiarize yourself with this subject area, there is a risk of getting lost in the details. We'll therefore approach this topic step-by-step and layer-by-layer. Let's look at the whole thing in the context of a specific task.

Classification means that we assign an object to a specific category. The result you're looking for can have one of several states. Examples of this type include the following:

- Recognition of dogs and cats in pictures (categories "dog" and "cat")
- Evaluation in quality control (categories "good" and "bad")

In the field of artificial intelligence (AI), *regression* is a method that can be used to predict specific numerical values. The result is a constant value. Examples include the following:

- Predicting the purchase price for a specific product
- Determining the monthly earnings of potential customers

For some tasks, both variants are possible. Do you really want to know what customers earn (regression), or is it enough to know whether it's more or less than $x (classification with the categories ">=x" and "<x")?

3.1 Classification

A widely used dataset for getting started with AI development is the *Iris dataset*. This table contains only 5 columns and 150 rows and is therefore well suited as an introduction to this topic. The data was collected by British biologist Ronald Fisher in 1936 to evaluate mathematical methods for classification.

The iris has three subspecies: *Iris setosa*, *Iris versicolor*, and *Iris virginica*. Each iris can be assigned to a subspecies based on the length and width of the *sepal* and *petal*. The dataset contains four columns with information on the size of the leaves, length and width of the sepal, and length and width of the petal. The fifth column contains information on the specific subspecies.

The table is complete and easy to understand. The question might now arise as to why AI should be developed. Imagine you go out into the meadow and measure the length and width of the leaves of an iris. There is no entry or line in the table that corresponds exactly to the measured values. So how are you supposed to know which subspecies this iris belongs to? We'll therefore develop an AI based on an ANN, which will be trained with the dataset. It should then be able to determine the specific subspecies for the four measured (and therefore still unknown) input variables (length and width of the leaves), even if the measured values don't correspond exactly to an entry in the table.

We won't specify any rules or algorithms in the program structure. The program won't contain a query or statement such as "If the sepal is longer than x cm, then ...". The AI itself is supposed to derive correlations from the dataset, which is the main difference compared to conventional programming.

Without knowledge of AI, you would have to find an algorithm to classify irises yourself. This requires a more detailed analysis of the available data. After determining the relationships, you could say, for example, that the *Iris setosa* subspecies is present if the dimensions of the leaves lie within certain ranges. This would be feasible with such a clearly structured dataset. But imagine you have a table of user behavior on a homepage with 50 columns and 100,000 rows. Who is supposed to analyze all that? The great advantage of AI development over conventional development is precisely this:

you set up the AI and train it with data, after which the AI can handle new, unknown data itself. Let's take a look at exactly how this works with the following task: an AI is to be developed that determines the specific subspecies of iris based on the length and width of its leaves (see Figure 3.1).

Figure 3.1 Iris: Petalum and Sepalum

There are four input variables for this task: sepal length, sepal width, petal length, and petal width. We can refer to the result (output or target column) as class or species.

3.2 The Recipe

The procedure for tasks of this type (supervised ML) is always similar. In the dataset, we know which columns are to be used as input and which output they should lead to. We'll therefore create a "recipe" that you can use to transfer what you've learned to other, similar datasets. So, step-by-step, let's look at the AI as a black box and see what happens around it. This is followed later by a look inside the black box.

3.2.1 Data Preparation

Data must be collected and, if necessary, processed. A biologist could go out into the fields, measure and categorize different irises and then record the results in a table. In our example, there is a table with 5 columns and 150 rows, as shown in Table 3.1.

Sepal Length	Sepal Width	Petal Length	Petal Width	Class
5.1	3.5	1.4	0.2	Iris setosa
7.0	3.2	3.5	1.0	Iris versicolor
6.3	3.3	6.0	2.5	Iris virginica
...	...	...	...	...

Table 3.1 Iris Dataset

The class contains nonnumeric values. For calculations to be made, these must be converted into numerical values. Mathematically, you can't do much with strings such as "Iris setosa". One approach would be to substitute the class names with integers, such as iris setosa → 0, iris versicolor → 1, and iris virginica → 2.

The data must also be divided into training and test data. Usually, 80% of the data is used for training and the rest for testing.

> **Division of Data**
>
> For teaching purposes, the data is often divided up as follows:
> - 80% training data
> - 20% test data
>
> If, on the other hand, you're developing for a production system, you also need data for the evaluation. This is described in detail later in this section.

The result of the division isn't two tables but four tables. You must separate the input variables from the output variables for both the training and test data, so you have the following:

- 4 columns and 120 rows of training data
- 1 column and 120 rows of training classes
- 4 columns and 30 rows of test data
- 1 column and 30 rows of test classes

It's very important that you know how to split a given table (see Figure 3.2). You should be able to calculate the number of rows and columns for all result tables.

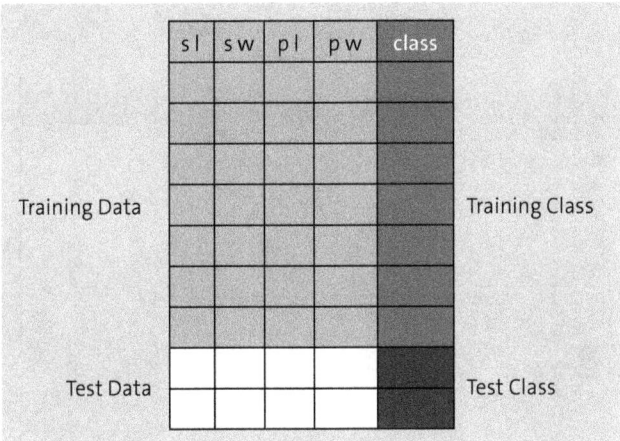

Figure 3.2 Division of the Table

3.2.2 Building Up the AI

Once the data is prepared, you can build up the AI. However, we said earlier we would take a look inside at a later date. Let's start by imagining the AI as a black box with one input and one output (see Figure 3.3). Further details at this point would be a hindrance and rather confusing, so we'll leave those out.

Figure 3.3 AI Viewed as a Black Box

3.2.3 Training the AI

The AI can now be trained using the training data. In the training process, the training data is fed into the AI after having run in an iteration process through the training table row by row. This procedure generates a proposal as output for each input dataset (four variables with the length and width of the leaves). The proposed class is compared with the actual entry in the table, and the parameters within the AI are updated with the deviation or difference. Which parameters these are will follow later (don't forget, we'll look inside the black box later). The iteration process runs multiple times from top to bottom (*n* runs or epochs), and the process gets repeated. The goal is to adjust the parameters in the AI so that the prediction improves over time and the deviation becomes smaller and smaller (see Figure 3.4).

3 Artificial Neural Networks

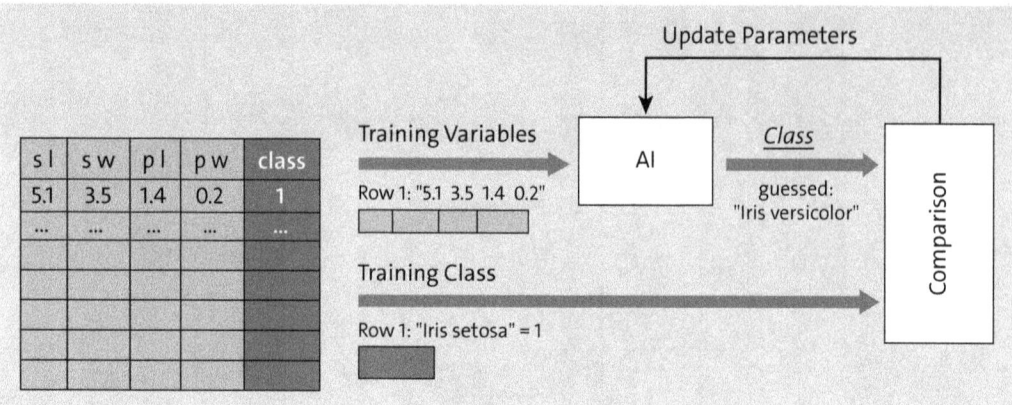

Figure 3.4 AI Training

3.2.4 Testing the AI

The AI is now trained but still needs to be tested with new, unknown data. We must verify that the AI returns reliable results with new data. This is where the test data comes into play. Here, too, the table is iterated through, the variables are fed to the AI as input, and the output is compared with the value in the table. However, there are two important differences compared to the training procedure:

- There is only one iteration run from top to bottom through the table.
- The deviation no longer causes a parameter update within the AI. Instead, the AI developer must look at the deviation and assess whether the result is acceptable or not.

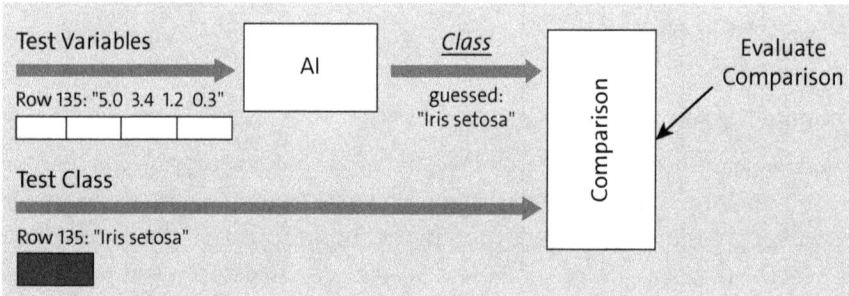

Figure 3.5 AI Test

All steps in the recipe are repeated from the beginning until you're satisfied with the result of the test. For example, you can add data, omit other data, and so on.

3.2.5 Using AI

If the test results are satisfactory, the development of the AI is complete. Now you no longer need a biologist to classify irises (but the data collection did require specialist knowledge). With this AI, even nonexperts can measure the leaves, feed the values into the AI, and receive the corresponding subspecies as a result. The model has therefore been trained and tested. The result is an AI based on ANNs that can classify the three subspecies of iris.

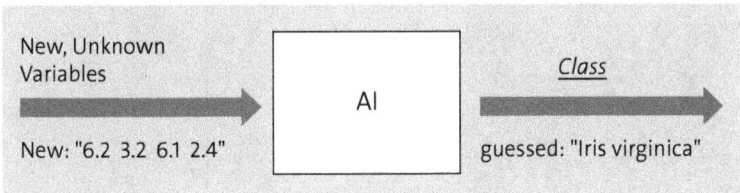

Figure 3.6 AI Usage

But if the AI makes a false statement, someone without any knowledge of irises won't even notice. The AI must therefore be well trained and tested. However, you'll have to come to terms with the fact that AI programs are very rarely 100% reliable. This doesn't mean that these programs are unusable. It always depends on the specific area of application. For example, if a person classifies 95% of all irises correctly, this correct classification rate can serve as a benchmark for the use of AI.

3.3 Building ANNs

Now it's time to take a look inside the AI black box. A simple algorithm is used for the explanation. Most software libraries have developed the process further, but the principle has remained the same.

This simple example has three inputs and three outputs. It also consists of three *layers*, each with three nodes. The procedure is as follows: three values are entered simultaneously on the left and processed internally, and the result (in this example, three values) gets output on the right. Each node is connected to all subsequent nodes in the next layer. The connections between the nodes are weighted and marked with $w_{j,k}$. For the sake of clarity, only a few weights are shown as examples. These weights are a measure of how strong a connection between the nodes is compared to other connections. The weight $w_{2,3}$ indicates, for example, the connection between node 2 of one layer and node 3 of the following layer. One input and one output layer are always required. In between, there are *hidden layers*). If there are several hidden layers, this is also referred to as *deep learning*.

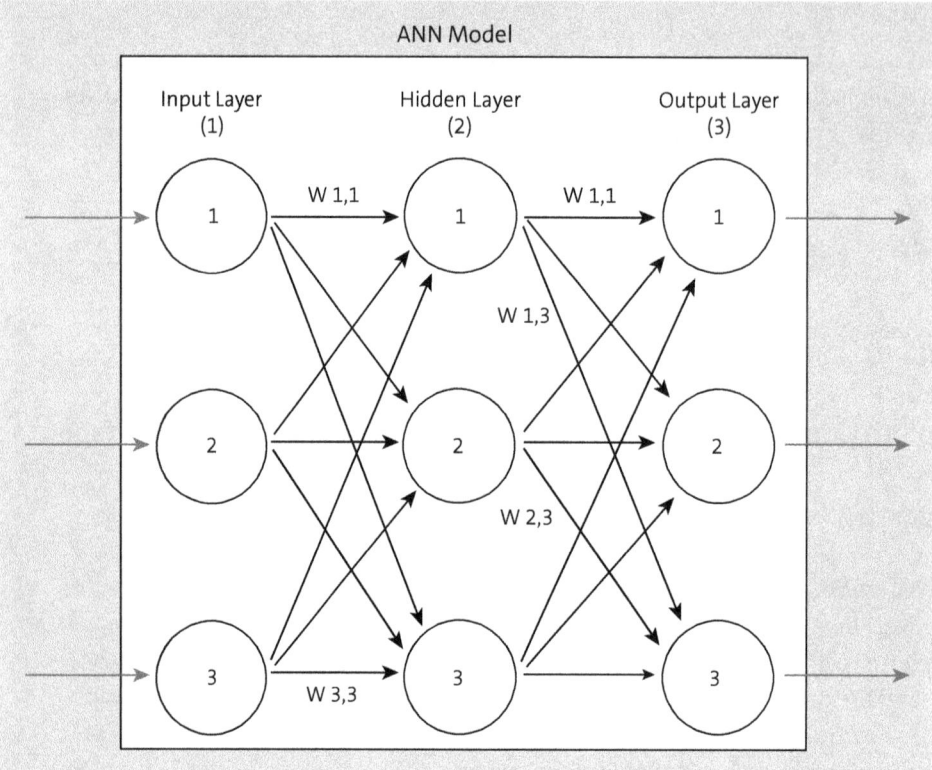

Figure 3.7 ANN Structure

The following questions now arise for our task:

1. How many nodes must input and output layers have for the classification of irises?
2. How many hidden layers are required, and how many nodes should each of these layers have?

We can answer the first question in a comprehensible way. The input layer must have four nodes, as the ANN is "fed" with four variables (sl, sw, pl, pw).

An obvious answer for the number of nodes in the output layer is 1. The ANN could then be used for the four variable contents (e.g., 5.1, 3.5, 1.4, 0.2) provide an output (1 for Iris versicolor). But we still can't be satisfied with that because we want to know the probability of the answer being 1. This gives us more information and allows us to update the parameters in a more targeted manner. It makes a difference whether the prediction is close or way off the mark. There is a very elegant solution for this: We build the output layer from three nodes, one for each class. The ANN outputs three probabilities for each input (5.1, 3.5, 1.4, 0.2), one for each class (class 0 with 0.2, class 1 with 0.7, class 2 with 0.1). This input data therefore classifies the *Iris versicolor* subspecies with a probability of 70%. We were therefore able to answer question 1.

There is no clearly correct answer to question 2. The number of hidden layers and the number of nodes per hidden layer depend on the task and are left to the decision and experience of the AI developer. If the number is too small, the ANN might not be trained well and won't assign classes correctly (*underfitting*). If, on the other hand, the number is too large, the ANN takes a very long time to train and memorizes the training data, but can't correctly assign unknown test data. This phenomenon is referred to as *overfitting*. In this case, the ANN did not understand the "system" to classify new data, but merely memorized the test data. The weights are set in such a way that the corresponding subtype is output for each set of input data from the table (in our example, the four variables for the length and width of the leaves). The training data can also be used to achieve a correct classification rate of 100%. The weights are too optimized for this training data. This state can be recognized by the fact that the correct classification rate of the test data deteriorates, which is another example of overfitting.

We want to design several different ANNs in the future, so that you'll have some experience in determining the number of layers and nodes (hidden layers) depending on the problem. For our task, we could first define that we use two hidden layers (32 and 64 nodes).

3.4 Structure of an Artificial Neuron

Now that we've gained an overview of the ANN, let's go one layer deeper and take a look at the structure of a single artificial neuron.

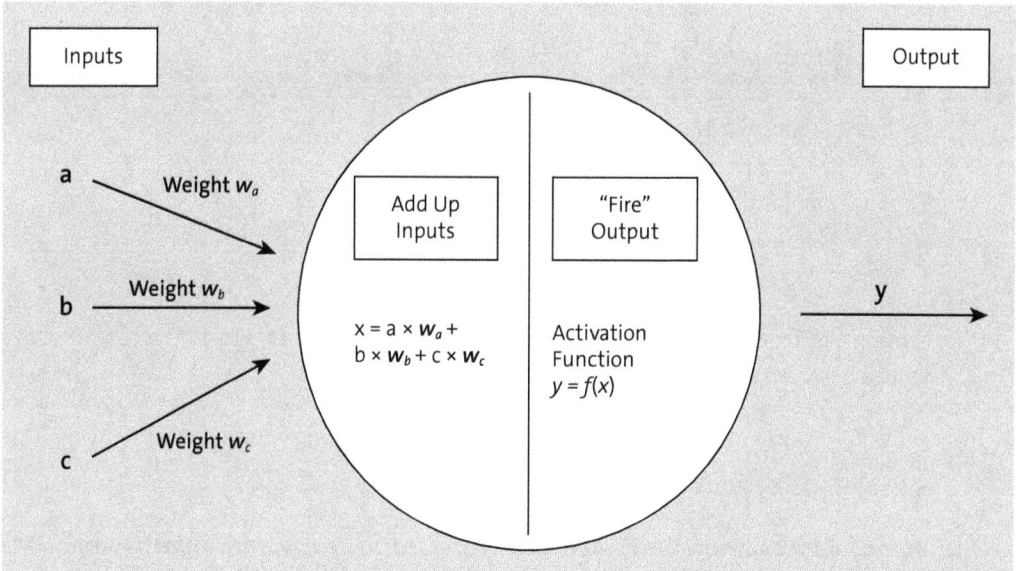

Figure 3.8 Structure of an Artificial Neuron

3 Artificial Neural Networks

All inputs are multiplied by the respective weights of the connections and added up in the left half of the node. However, this sum isn't simply output as an output variable. Inspired by neurons, this example checks whether a certain threshold value has been exceeded and outputs a value accordingly. The strict implementation of this procedure would mean that a 0 is output if the value falls below the threshold value, and the cumulated variable *x* is output if the threshold value is reached or exceeded. We want to follow the natural model here and make the initial function (*activation function*) smoother. One way to achieve this is to use the sigmoid function as an activation function. The output variable is passed on to all neurons in the subsequent layer.

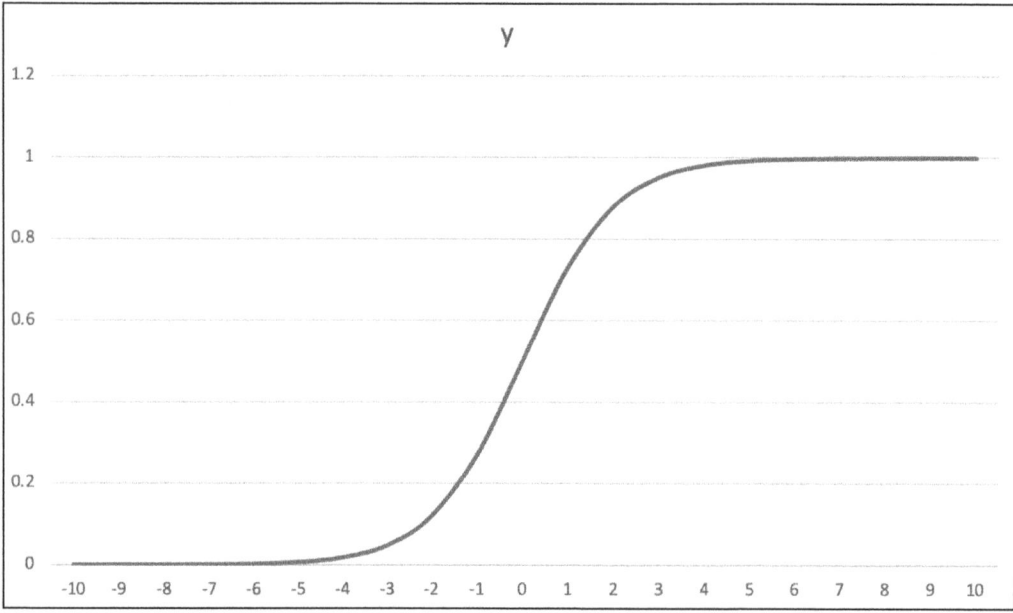

Figure 3.9 Sigmoid Activation Function

The formula for the sigmoid function is as follows:

$$y = \frac{1}{1 + e^{-x}}$$

You can see from the function flow that the output value of this function can only take values between 0 and 1.

3.5 Feed Forward

Let's take a look at how the input data is processed on its way from input to output. The ANN in Figure 3.10 is given again.

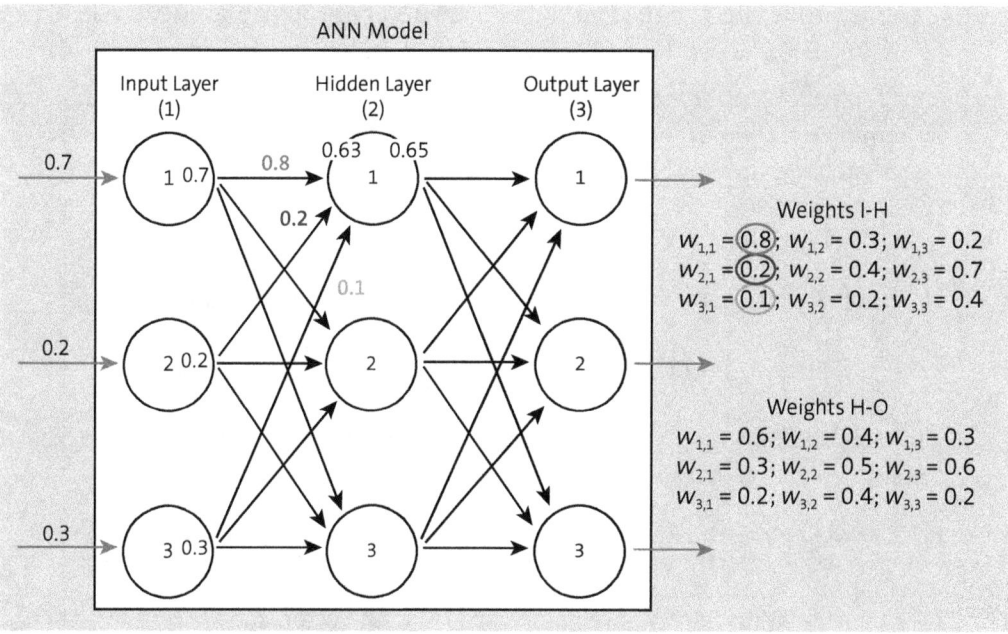

Figure 3.10 Calculation during the Data Flow from Left to Right

The numbers (from top to bottom, nodes 1 to 3) 0.7, 0.2, and 0.3 are entered at the inputs of the input layer. These values aren't processed by the input layer but are passed on directly to the nodes of the subsequent layer. The input layer is the exception; the activation function is used for all other layers. Let's enter these input values at the outputs of the input nodes in the graphic. The weights between input layer and hidden layer are as follows:

$w_{1,1} = 0.8; w_{1,2} = 0.3; w_{1,3} = 0.2$

$w_{2,1} = 0.2; w_{2,2} = 0.4; w_{2,3} = 0.7$

$w_{3,1} = 0.1; w_{3,2} = 0.2; w_{3,3} = 0.4$

Let's enter three weights (numerical values) in the graph as examples. The values of the input layer can now be multiplied by the respective weights and added up, and then x can be calculated. Let's enter the result at the input of the first node in the hidden layer (0.63). We can calculate the node output using the sigmoid function $y = sigmoid(0.63)$. We enter the result (0.65) at the output of the first node in the hidden layer. You can continue the calculations for the other nodes of the hidden layer.

Now this procedure gets repeated for the next layer. The intermediate results (outputs of the hidden layer) are multiplied by the weights between the hidden layer and output

3 Artificial Neural Networks

layer and added up, and then result in the inputs of the output layer. After the activation function, you get the results for the output. As an exercise, you can calculate all values up to the outputs of the nodes in the output layer yourself.

Alternatively, you can solve the calculation using matrix multiplication. A matrix is a contiguous group of numbers:

$$\begin{pmatrix} 1 & 2 \\ 3 & 4 \end{pmatrix}$$

Two matrices can easily be multiplied with each other as follows (dot product):

$$\begin{pmatrix} 1 & 2 \\ 3 & 4 \end{pmatrix} \cdot \begin{pmatrix} 5 & 6 \\ 7 & 8 \end{pmatrix} = \begin{pmatrix} (1 \cdot 5 + 2 \cdot 7) & + & (1 \cdot 6 + 2 \cdot 8) \\ (3 \cdot 5 + 4 \cdot 7) & + & (3 \cdot 6 + 4 \cdot 8) \end{pmatrix} = \begin{pmatrix} 19 & 22 \\ 43 & 50 \end{pmatrix}$$

To calculate the output values (also intermediate output values) for ANN, we have to multiply a matrix by a column vector, which is even simpler:

$$\begin{pmatrix} 1 & 2 \\ 3 & 4 \end{pmatrix} \cdot \begin{pmatrix} 5 & 6 \\ 7 & 8 \end{pmatrix} = \begin{pmatrix} (1 \cdot 5 + 2 \cdot 7) & + & (1 \cdot 6 + 2 \cdot 8) \\ (3 \cdot 5 + 4 \cdot 7) & + & (3 \cdot 6 + 4 \cdot 8) \end{pmatrix} = \begin{pmatrix} 19 & 22 \\ 43 & 50 \end{pmatrix}$$

A simple notation with W = weight matrix, i = input vector, and x = result vector is as follows:

$$x = W \cdot i$$

The sigmoid function is then applied to each element of the result matrix:

$$o = sigmoid(x)$$

To calculate the preceding task, you need to follow these steps:

1. Calculate $x = W \cdot i$ with W as the weight matrix given previously between the input and hidden layer (with 3 · 3 elements) and i as input data (column vector with three elements) at the input. First write the equation with variables, and then with numbers. Then carry out the calculation according to the preceding schema. The result is a column vector with three elements.

2. Calculate o = sigmoid(x) by applying the sigmoid function to each element. The result is again a column vector with three elements.

3. This last column vector is the input i, while W is the weight matrix between the hidden and output layers. Calculate $x = W \cdot i$ again, and then finally o = sigmoid(x). o is the output as a column vector. The result must match that of the previous task (calculation without a matrix).

You won't have to carry out any calculations yourself for the programming. These calculations are only intended to aid your understanding.

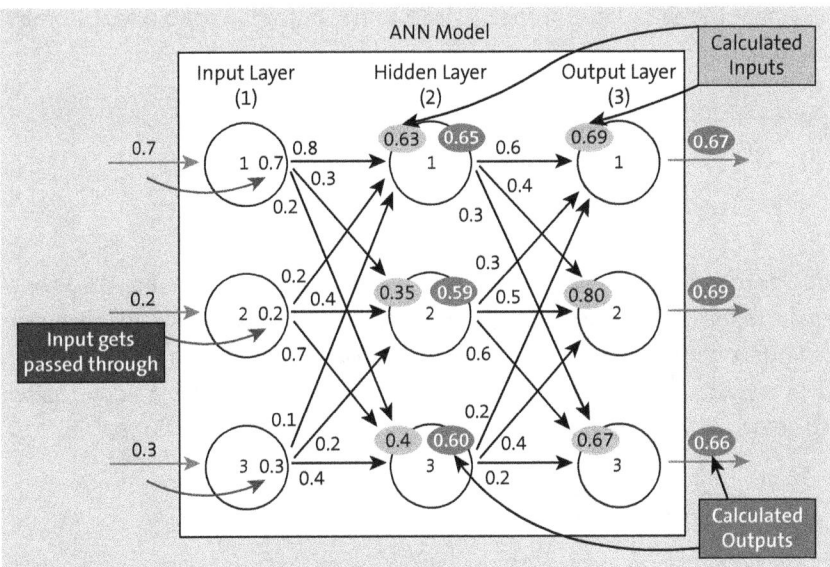

Figure 3.11 ANN with Feed Forward Results

3.6 Back Propagation

Let's clarify the question of what happens if the calculated output values don't match the target values. Remember that during the training phase, these values are compared with each other, and the parameters are adjusted if there is a difference. The parameters that are adjusted are the weights. Before we can update weights, we have to divide the difference proportionally to the outputs of all previous layers because the outputs of the inner nodes can't be compared with target values at the output of the AI box. Let's take a look at a smaller example so that the calculations remain clear.

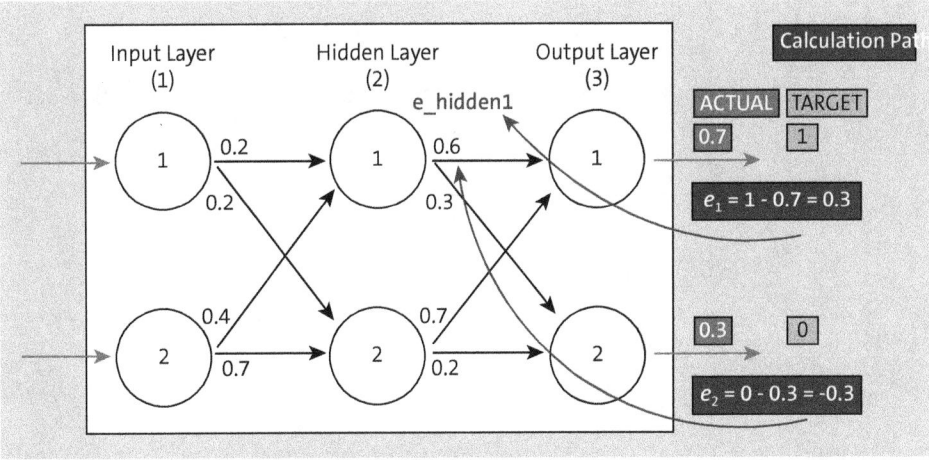

Figure 3.12 Calculation with Back Propagation: From Right to Left

3 Artificial Neural Networks

You can distribute the difference proportionally to the outputs of the nodes step-by-step from right to left:

$$e_{hidden_1} = \frac{w_{1,1}}{w_{1,1} + w_{2,1}} \cdot e_1 + \frac{w_{1,2}}{w_{1,2} + w_{2,2}} \cdot e_2$$

$$= \frac{0.6}{0.6 + 0.7} \cdot 0.3 + \frac{0.3}{0.3 + 0.2} \cdot (-0.3) = -0.042$$

Pay attention to the fraction in the calculation. The numerator contains the weight between the deviation (difference or error) at the output and the respective node, whereas the denominator contains the sum of the weights or paths across which the deviation is distributed. In this way, you can distribute the error proportions to the previous nodes. You can also use matrices here. Put into matrix form, the formula is as follows:

$$e_{hidden_1} = \begin{pmatrix} \frac{w_{1,1}}{w_{1,1} + w_{2,1}} & \frac{w_{1,2}}{w_{1,2} + w_{2,2}} \\ \frac{w_{2,1}}{w_{2,1} + w_{1,1}} & \frac{w_{2,2}}{w_{2,2} + w_{1,2}} \end{pmatrix} \cdot \begin{pmatrix} e_1 \\ e_2 \end{pmatrix}$$

However, our weight matrix from before is given in the following form (pay attention to the indexes):

$$\begin{pmatrix} w_{1,1} & w_{2,1} \\ w_{1,2} & w_{2,2} \end{pmatrix}$$

How can we convert them into the form we need for the formula just mentioned:

$$\begin{pmatrix} w_{1,1} & w_{1,2} \\ w_{2,1} & w_{2,2} \end{pmatrix}$$

The answer is simply by transposing them, which means that we form columns from rows of a matrix:

$$\begin{pmatrix} 1 & 2 & 3 \\ 4 & 5 & 6 \\ 7 & 8 & 9 \end{pmatrix}^T = \begin{pmatrix} 1 & 4 & 7 \\ 2 & 5 & 8 \\ 3 & 6 & 9 \end{pmatrix}$$

The weights must be scaled according to the paths for the proportional reduction of the deviation:

$$\begin{pmatrix} w_{1,1} & w_{1,2} \\ w_{2,1} & w_{2,2} \end{pmatrix} \rightarrow \begin{pmatrix} \frac{w_{1,1}}{w_{1,1} + w_{2,1}} & \frac{w_{1,2}}{w_{1,2} + w_{2,2}} \\ \frac{w_{2,1}}{w_{2,1} + w_{1,1}} & \frac{w_{2,2}}{w_{2,2} + w_{1,2}} \end{pmatrix}$$

Therefore, the following applies:

$$e_{hidden} = W^T_{hidden_{output}} \cdot e_{output}$$

You can calculate e_{input}, and so on in the same way in this example.

3.7 Updating the Weights

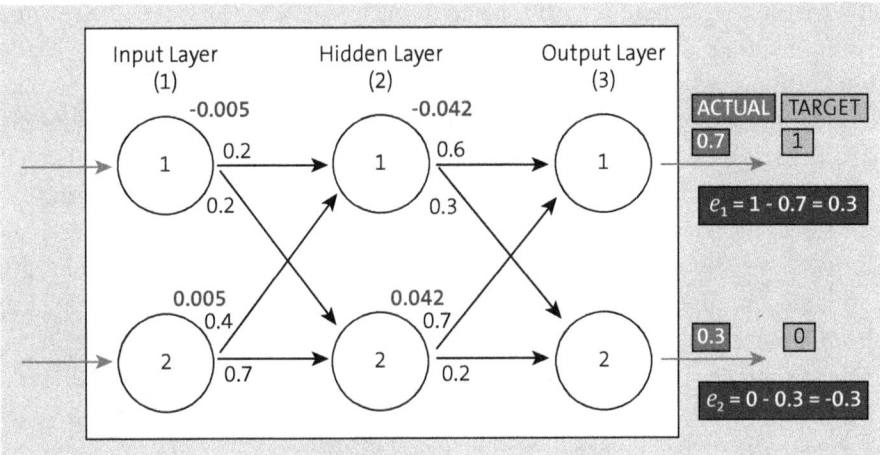

Figure 3.13 Calculation for a Back Propagation

The following principle applies here: to program AIs, you don't have to calculate anything yourself, just use the corresponding modules.

3.7 Updating the Weights

The deviations at the outputs of all nodes of the output layer were passed back proportionally to the outputs of all previous nodes. But how should these values be offset against the weights so that the deviation at the output is ideally smaller for each run? After all, at the end of our training phase, we want the outputs to deliver exactly what we expect, that is, the value (target value) with which we compare the output value (actual value). The closer the actual value is to the target value, the better. In the case of our iris AI, it means that the output should deliver exactly the right class. How can we achieve this? The *gradient method* will help us.

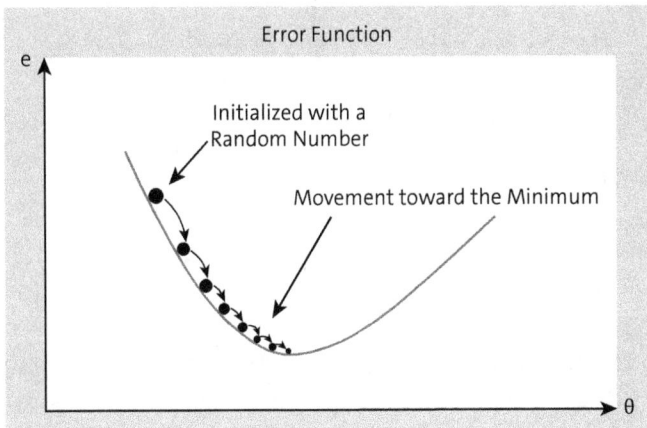

Figure 3.14 Gradient Method

53

We initialize the weight randomly at the beginning and look at the error. The red line (the error function) is invisible to us. If we knew it, we could immediately jump to the minimum (valley). We consider the flying point in the air, which has the coordinates θ (theta, here for weight) and e (deviation). To reduce the error, we move one step to the right (i.e., increase theta) and compare whether the error becomes larger or smaller. If it gets smaller, we go a little to the right again, as we seem to be on the right path. The entire process gets repeated until we take a step that increases the error. Then we realize that in the previous step we were in the valley and have determined the minimum. Thus, our aim is to find the minimum of this function because we want to create the smallest difference. But we have to be careful here: If the jump is too big, we'll be on the other side of the valley and have missed the minimum. If the jump is too small, it will take forever to reach the valley. For this reason, the differential weight is multiplied by a small, constantly set factor: *learning rate* α (e.g., 0.001).

The formula for updating the weights is (negative, as we're moving against the gradient) as follows:

$$w_{j,k} = \text{alt}_{w_{j,k}} - \alpha \cdot \Delta w_{j,k}$$

The gradient method attempts to determine a $\Delta w_{j,k}$ so that the deviation from the target value is smaller after the calculation using the preceding formula.

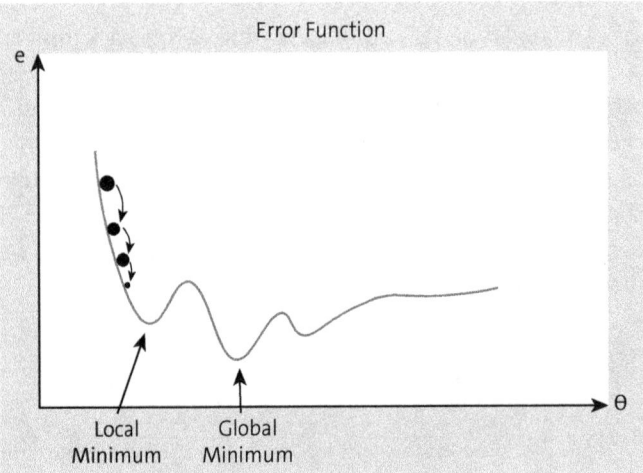

Figure 3.15 Search for a Global Minimum

What happens if the error function doesn't work out so well (of course, it doesn't in reality), and there is a risk of getting stuck in the local minimum or progressing too slowly? One possibility is to reset the parameters of the weights at certain intervals to obtain different starting points. You can also include momentum in the calculation to overcome local lowest points. There are now many optimization methods, and everything revolves around $\Delta w_{j,k}$. The calculation of $\Delta w_{j,k}$ isn't trivial, but it's sufficient if you understand the principle. You don't need to understand the higher mathematics behind it because even then, you can't tell in advance from a formula for an AI task whether this exact mathematical method is the most suitable. Again, it's the experience you gain over time that counts. An optimizer we'll often use in the practical examples is called Adam, which is a powerful algorithm for optimization.

> **Adam (Adaptive Moment Estimation)**
> Adam is an algorithm that is often used in the world of ML. Unlike other optimizers, Adam adjusts the learning rate for each parameter individually during the training phase.

The function graphs we've shown so far are only dependent on one variable. Two variables result in a three-dimensional graph with even more valleys.

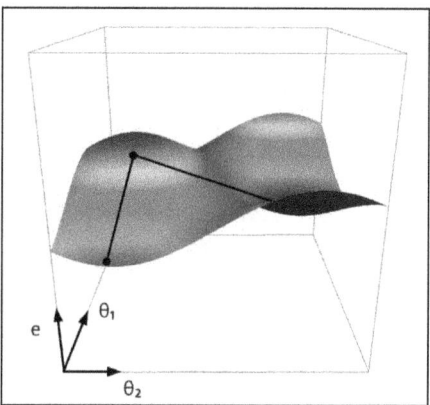

Figure 3.16 Search for the Minimum with Two Variables

Our simple example of the iris already has four variables, has many nodes or weights, and is no longer easy to represent graphically. The procedure for determining $\Delta w_{j,k}$ is generally very complex as far as computations are concerned.

> **Bias**
> Many implementations for ANNs add a value (*bias*) to the individual nodes in the layers to obtain better results. This bias value is also optimized using the gradient method.

3.8 ANN for Classification

The iris data is stored in a comma-separated values (CSV) file (*iris.csv*). Open the file in a spreadsheet or simple text editor to get an overview of the structure before we start programming.

```
Import required modules and load the file. The first lines are displayed.

import tensorflow as tf
import pandas as pd
from sklearn.model_selection import train_test_split

path = "../Data/iris.csv"
data = pd.read_csv(path, delimiter=',')
print(data.head())

Prepare data.

# What is to be predicted? Store column name in a variable.
col_name = 'species'

# Zeichenkette in Zahlen umwandeln
data[col_name] = data[col_name].astype('category')
data[col_name] = data[col_name].cat.codes

# Here, the division into two tables occurs (input=data and output=col).
col = data[col_name]
data = data.drop([col_name], axis = 1)

build ANN

# Four tables are created from the two tables.
train_data, test_data, train_col, test_col = train_test_split(data,col, test_size=0.2)

# build ANN
model = tf.keras.Sequential([
    tf.keras.layers.Dense(32, activation=tf.nn.sigmoid, input_dim=4),
    tf.keras.layers.Dense(64, activation=tf.nn.sigmoid),
    tf.keras.layers.Dense(3, activation=tf.nn.softmax)
])
# configure ANN
model.compile(optimizer='adam', loss='sparse_categorical_crossentropy', metrics=['accuracy'])

train

# 30 iterations
model.fit(train_data, train_col, epochs=30)

test

test_loss, test_acc = model.evaluate(test_data, test_col)
print('Test accuracy:', test_acc)
```

Figure 3.17 Complete Source Code in Jupyter Notebook

> **CSV File**
>
> In a CSV file, data is saved in a table structure. It's a simple text file in which the columns are separated by defined separators.

Let's take a detailed look at the first programming example, *K3_iris-1.ipynb*, the ANN for the iris dataset. Open the program via Jupyter Notebook or JupyterLab and run it from top to bottom, cell by cell.

```
import tensorflow as tf
import pandas as pd
from sklearn.model_selection import train_test_split
```

Listing 3.1 Import of Required Modules

First, the required modules such as TensorFlow are loaded. The first two lines load the modules and provide access to them via shortcuts. If you want to use the layers in the module, for example, you can access them using tf.keras.layers.Dense. The module is structured hierarchically; the dot notation can be used to address the hierarchy level. Only train_test_split is imported from the sklearn module. This function can later be called directly without a shortcut.

```
path = "../Data/iris.csv"
data = pd.read_csv(path, delimiter=',')
print(data.head())
```

Listing 3.2 Loading the CSV File

The path to the CSV file (located in the neighboring *Data* folder) is then saved in a variable, and the file is opened using the read_csv function, which is located in the pandas module. This function is called with two parameters, the path to the file and the separator within the file. If the file were saved in the same subfolder, the path would have to be ./iris.csv. The first lines of the dataset are displayed in the subsequent line.

```
  sepal.length  sepal.width  petal.length  petal.width species
0          5.1          3.5           1.4          0.2  Setosa
1          4.9          3.0           1.4          0.2  Setosa
2          4.7          3.2           1.3          0.2  Setosa
3          4.6          3.1           1.5          0.2  Setosa
4          5.0          3.6           1.4          0.2  Setosa
```

Figure 3.18 Output of the First Five Rows

You can use print(data) to output all data. Feel free to experiment with the source code.

```
# What should be predicted? Save column name in variable.
col_name = 'species'

# Transform string to numbers
data[col_name] = data[col_name].astype('category')
data[col_name] = data[col_name].cat.codes
```

Listing 3.3 Defining the Target Column and Converting the Data Type

Once trained and tested, the AI should be able to predict a class. In this example, we save the name of this column in a variable. In the next step, the data type of this column is converted to 'category'. The actual transformation follows in the next line. We could also assign the column to a new variable, but instead we overwrite the existing variable. The variable named data[col_name] now contains the coded categories and no longer strings. You can convince yourself of the transformation by calling print(data[col_name]).

> **List, Pandas Series, and Pandas DataFrame**
>
> Related data can be saved in lists in the Python programming language:
>
> list = [1,2,'anything',3.4]
>
> The complete list can be output using print(list), but you can also address a specific element in the list. The corresponding index is specified in square brackets, and the numbering starts at 0 (for the first element). The print(list[1]) call outputs the number 2, the second element of the list. Lists can contain elements of different data types and can be subsequently changed.
>
> In Pandas Series, related data can also be stored as columns; data['species'] is an example of this. For columns in tables without column headings, you can address the column with an index. In this case, the leftmost column (the first column) has the index 0.
>
> Pandas DataFrame (e.g., data) has a table structure and consists of columns (Pandas Series).

```
# Here the division into two tables takes place (Input=data and Output=col).
col = data[col_name]
data = data.drop([col_name], axis=1)
```

Listing 3.4 Division into Two Tables

The data[col_name] column is saved in the new variable col, and the remaining columns are saved again in the data variable. The call means to remove the specified column from the table and assign the result back to data again. With axis=1, you specify that a column (and not a row) is meant. Check the results for yourself by using print(col) and print(data).

```
# Create four tables from the two tables
train_data, test_data, train_col, test_col =
 train_test_split(data,col, test_size=0.2)
```

Listing 3.5 Division into Four Tables

The train_test_split function splits the table and column given as parameters and returns four tables (or two tables and two columns) as a result, with 20% test data and 80% training data. You can see the results for yourself again using print.

> **Division of the Table**
>
> You should sketch on a sheet of paper how the original table is divided into four tables or divided into two tables and two columns. This knowledge is very important.

3.8 ANN for Classification

```
# Build ANN
model = tf.keras.Sequential([
    tf.keras.layers.Dense(32, activation=tf.nn.sigmoid, input_dim=4),
    tf.keras.layers.Dense(64, activation=tf.nn.sigmoid),
    tf.keras.layers.Dense(3, activation=tf.nn.softmax)
])
```
Listing 3.6 Building the Artificial Neural Network

This is where the actual ANN is built. The input layer has 4 nodes, the first hidden layer has 32 nodes, the second hidden layer has 64 nodes, and the output layer has 3 nodes. The first two hidden layers have the sigmoid activation function, while the output layer has softmax. This function ensures that the total of all probabilities is exactly 1. The formula for the softmax function is as follows:

$$S_{y_i} = \frac{e^{y_i}}{\sum_j^n e^{y_j}}$$

The formula looks more complicated than it is, which becomes clear with an example:

$$y = \begin{bmatrix} 2.0 \\ 1.0 \\ 0.1 \end{bmatrix}, S_{y_i} = \frac{e^{y_i}}{\sum_j^n e^{y_j}} = \begin{bmatrix} 0.7 \\ 0.2 \\ 0.1 \end{bmatrix}$$

Calculate the denominator by inserting the values of y one after the other as the exponent of Euler's number and adding them up. Then, enter "2.0" for the numerator, and you get 0.7. You can insert the other two values of y one after the other and calculate the result. Let's take another look at the source code.

```
model.compile(optimizer='adam', loss='sparse_categorical_crossentropy',
    metrics=['accuracy'])
```
Listing 3.7 Configuration of the Learning Process

Adam is used as the optimization function for the weights. A function is also required to determine the deviation of the output from the target values: the *loss function*. Here, `sparse_categorical_crossentropy` is used for the numerically categorized data (you'll get to know other loss functions). The loss function for categorized data is generally calculated as follows:

$$loss = -\sum_{i=1}^{n} y_i \cdot ln(\hat{y}_i)$$

Here, n is the number of categories (or number of nodes in the output layer), y_i is the actual value, and $\hat{y}_i$ is the predicted value. In our iris example, the actual value is 1 for one subspecies and 0 for the remaining two subspecies. This means that *loss* for all other nodes is also 0. For example, the prediction for this first subspecies is 0.7, so the calculation looks as follows:

$$loss = -1 \cdot ln(0.7) = 0.357$$

3 Artificial Neural Networks

The *accuracy* is used to specify that the correct classification rate will be calculated. The calculation reads as follows:

$$\text{accuracy} = \frac{\text{number of correct predictions}}{\text{number of all predictions}}$$

```
# 30 runs
model.fit(train_data, train_col, epochs=30)
```

Listing 3.8 Training Process

This is where the actual training process takes place. It's iterated 30 times through the training data. I've plotted the values for *loss* and *accuracy* (see Figure 3.19) so that you have an overview of the progression. Note that as the number of epochs increases, the deviation decreases, and the correct classification rate increases (albeit with dips). Figure 3.20 illustrates how the values change during training.

Once a trained model is available, it's interesting to determine how well new and unknown data is classified. This is where the test data comes into play.

```
test_loss, test_acc = model.evaluate(test_data, test_col)
print('Test accuracy:', test_acc)
```

Listing 3.9 Test Process

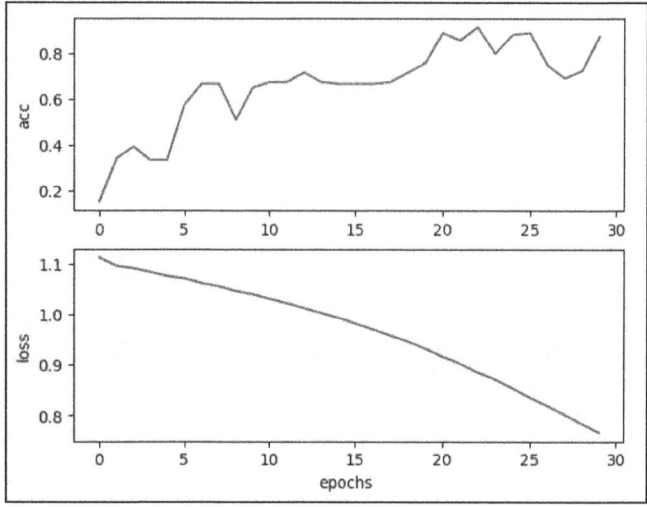

Figure 3.19 Value Progression of "Loss" and "Accuracy" during the Training

The trained ANN is evaluated with the test data. This method returns two values: the result of the loss function and the correct classification rate. We only output the latter. In this first example, the prediction is relatively good at just under 80%, but there is room for improvement.

3.8 ANN for Classification

Try to achieve a better rate by adjusting the number of epochs. During training, the value of the loss function should decrease, and the correct classification rate should increase. Pay particular attention to the result of the test data because that's what matters in the end.

Next, let's take a look at an AI that predicts the quality of red wine (0 to 10, i.e., 11 classes) based on its chemical properties. This *K3_redwine-1.ipynb* program differs only minimally from the *K3_iris-1.ipynb* program.

> **Practice: Create the Next Program Yourself**
>
> You can create this program yourself by making the following changes cell by cell:
>
> 1. Make a copy of *K3_ris-1.ipynb*, and rename it to *K3_redwine-1b.ipynb*.
> 2. Open the *winequality-red.csv* file with a text editor, and find out which separator is set and how many columns there are.
> 3. Adjust the path and the separator in your program.
> 4. Which column values should be predicted? Replace the column name in the program.
> 5. Update the ANN. How many nodes must have input and output layers?

The *K3_redwine-1.ipynb* program is pretty similar to the *K3_iris-1.ipynb* program, so let's just look at the differences.

```
path = "../Data/winequality-red.csv"
data = pd.read_csv(path, delimiter=';')
```

Listing 3.10 Loading the CSV File

The filename and the separator must be adapted to the new file.

```
   fixed acidity  volatile acidity  citric acid  residual sugar  chlorides  \
0            7.4              0.70         0.00             1.9      0.076
1            7.8              0.88         0.00             2.6      0.098
2            7.8              0.76         0.04             2.3      0.092
3           11.2              0.28         0.56             1.9      0.075
4            7.4              0.70         0.00             1.9      0.076

   free sulfur dioxide  total sulfur dioxide  density    pH  sulphates  \
0                 11.0                  34.0   0.9978  3.51       0.56
1                 25.0                  67.0   0.9968  3.20       0.68
2                 15.0                  54.0   0.9970  3.26       0.65
3                 17.0                  60.0   0.9980  3.16       0.58
4                 11.0                  34.0   0.9978  3.51       0.56

   alcohol  quality
0      9.4        5
1      9.8        5
2      9.8        5
3      9.8        6
4      9.4        5
```

Figure 3.20 Output of the First Five Rows

Have a look at the columns. The name of the target column must also be updated in the source code:

```
col_name = 'quality'
```

This AI is designed to predict the quality. The corresponding column name is therefore saved in the col_name variable.

```
model = tf.keras.Sequential([
    tf.keras.layers.Dense(32, activation=tf.nn.sigmoid, input_dim=11),
    tf.keras.layers.Dense(64, activation=tf.nn.sigmoid),
    tf.keras.layers.Dense(11, activation=tf.nn.softmax)
])
```

Listing 3.11 Building the ANN

The ANN now has an input layer with 11 nodes. The table has a total of 12 columns, and the values of 1 column will be predicted. The output layer has a total of 11 nodes because the result has 11 categories (0 to 10). You can increase the number of epochs to 70. Can you understand how the number of input and output nodes comes about? If you're still unsure, sketch the table on a sheet of paper, and think about which columns should be used as input and which column should be output.

With these changes to the source code, you've created a new AI that now classifies red wine. The correct classification rate isn't particularly good (around 55%). But how good can the forecast really be? People have tasted and rated the wines. This isn't an exact science. For this reason, we won't achieve any great improvement with further optimization options, which you'll learn about later.

Do we need 11 categories? Or is the classification into "bad", "good", and "very good" sufficient? You can open the file in a spreadsheet and replace the 11 categories with 3 ("poor" = 0, "good" = 1, and "very good" = 2). Then, export the file back to a CSV file. In the source code (*K3_redwine-2.ipynb*), you must pay attention to the number of nodes in the output layer. Let's have a look at the changes in the source code.

```
path = "../Data/winequality-red-2.csv"
```

Remember to update the filename in the code.

```
# Build ANN
model = tf.keras.Sequential([
    tf.keras.layers.Dense(32, activation=tf.nn.sigmoid, input_dim=11),
    tf.keras.layers.Dense(60, activation=tf.nn.sigmoid),
    tf.keras.layers.Dense(3, activation=tf.nn.softmax)
])
```

Listing 3.12 Building the ANN

There are now three nodes in the output layer. With these changes, you can increase the correct classification rate by around 10%.

3.9 Hyperparameters and Overfitting

We can tweak a number of adjustment options in the program. For example, we can add another layer, adjust the number of accounts, and change the activation function to relu (negative values are set to zero; positive values remain unchanged). Further activation functions can be found at *www.keras.io/api/layers/activations*.

In our iris example, we've selected an epoch number of 30. But why did we do that? Why not 100, 200, or 1,000? All these parameters (number of layers and nodes, activation function, number of epochs, etc.) are called hyperparameters. You can adapt an existing program to new tasks relatively easily. Finding the ideal hyperparameters, on the other hand, can be time-consuming. You change something, and then train, test, evaluate, and repeat the whole thing with other hyperparameters.

You could decide to always select a very large number of layers and nodes and set the number of epochs, for example, to 100,000. The sky is the limit. But be careful! An ANN that is too large and a number of epochs that is too high lead to the phenomenon of *overfitting*, as mentioned earlier. The ANN memorizes the training data. In this case, you'll therefore achieve very good training results. However, when you use new, unknown data for testing, the prediction will be very poor. An example should clarify the entire situation: You want to teach elementary school children arithmetic and practice three calculations, 1 + 1 = 2, 2 + 2 = 4, and 3 + 4 = 7, together over a long period of time. In exercises, you'll notice that, at some point, all children can solve these tasks. Then, follows a test with new tasks, such as 7 + 9 or 6 + 2. This approach wouldn't be successful because the children simply memorized the two previous calculations without understanding the system. The knowledge cannot be transferred to other, new calculations.

But how should you go about finding good settings for hyperparameters? First, you should use similar programs as a basis. Adjust the layers, nodes, epochs, and so on until the program is in overfitting. You'll notice that the training results are much better than the test results. Then, go back slightly with the epoch number, for example. Over time, you'll gain a wealth of experience in optimization options. In the end, the test results are always important and need to be optimized.

Let's make a copy of *K3_redwine-1.ipynb*, rename it to *K3_redwine-3.ipynb*, and look at the relevant parts of the program.

```
# Create four tables from the two tables
train_values, test_values, train_col, test_col =
 train_test_split(data, col, test_size=0.2, random_state=42)
```

Listing 3.13 Division into Four Tables

The `train_test_split(data,col,test_size=0.2)` function call not only splits the data but also merges it line by line. If you set parameters and run the program repeatedly, the random order of the data can lead to an improvement. However, you could assume that the improvement is due to your current settings. To avoid this, there is a way to mix the data reproducibly. If you add another parameter to the function call (namely, `random_state=42`), the data gets shuffled in the same way. The call is then `train_test_split(data, col, test_size=0.2, random_state=42)`. The numerical value for `random_state` is irrelevant. It's only important that it isn't changed between calls. But why is the order of the data relevant?

Here, too, we want to use the example of the elementary school. Let's suppose you practice addition with the children for a week and then subtraction for a week. A test follows in the third week. Many children would no longer remember the addition well. If the data isn't mixed or is poorly mixed, the ANN would first update the weights for one data group, then for the other, and so on. The weights would then no longer be optimized for the previous group, and the ANN would "forget" them over time.

> **Mixing the Data Is Important**
> To verify the significance of data blending, you can temporarily deactivate blending in the *iris-1.ipynb* example. To do this, add a further (here the last) parameter to the `train_test_split(data, col, test_size=0.2, shuffle=False)` function. The test results will be very poor as the data in this CSV file is sorted by class.

Let's analyze the source code further and see which parameters we can still set:

- The ANN is built up:

    ```
    model = tf.keras.Sequential([
        tf.keras.layers.Dense(32, activation=tf.nn.relu, input_dim=11),
        tf.keras.layers.Dense(120, activation=tf.nn.relu),
        tf.keras.layers.Dense(60, activation=tf.nn.relu),
        tf.keras.layers.Dense(10, activation=tf.nn.softmax)
    ])
    ```

    Increase the number of hidden layers to three, and try to achieve better test results.

- The number of epochs is increased:

    ```
    # 50 runs
    model.fit(train_data, train_col, epochs=50)
    ```

The number of epochs has been set to 50. With these few changes, the correct classification rate gets increased to approximately 57%.

The weights between the nodes are optimized for the training data. There is a risk that the hyperparameters set are only optimized for the test data, and the ANN doesn't provide a good prediction with other new data. Production systems are therefore eventually tested with additional evaluation data. The breakdown of the data could be as follows: 70% training data, 20% test data, and 10% evaluation data.

3.10 Dealing with Nonnumerical Data

Remember that character strings have to be converted into numerical values so that calculations can take place. Many datasets contain character strings such as "large", "blue", "defective", and so on.

You learned about one way of converting nonnumerical data in *K3_iris-1.ipynb*.

```
# What should be predicted? Save column name in variable.
col_name = 'species'

# Transform string to numbers
data[col_name] = data[col_name].astype('category')
data[col_name] = data[col_name].cat.codes
```

Listing 3.14 Definition of the Target Column and Conversion of the Data Type

You can view the result by using `print(data[col_name])`. Here, the iris species has been replaced by an integer (0 to 2). The problem is that this results in an unwanted weighting. One type weighs more or has a higher numerical value than the other. An alternative transformation option is *OHE* (*one hot encoding*). This is a binary coding that has as many digits as there are classes. All digits except one have the value 0, and the position of the first digit varies depending on the class. For our example with three classes, the code could look like that shown in Table 1.2.

Iris	Coding
Iris setosa	001
Iris versicolor	010
Iris virginica	100

Table 3.2 OHE of Species

Let's take a look at the relevant parts of the *K3_iris-2.ipynb* file.

```
# What should be predicted? Save column name in variable.
col_name = 'species'

# Convert string to OHE, result in new table
col = pd.get_dummies(data[col_name], dtype=float)

# The species are removed from the original table
data = data.drop([col_name], axis = 1)
#Thus two tables are available

# Create four tables from the two tables
train_data, test_data, train_col, test_col = train_test_split(data,col,
 test_size=0.2, random_state=42))
```
Listing 3.15 Preparation and Division of Data

The iris species are OHE-encoded (data type float) using a function from the pandas module (get_dummies), and the original column is removed from the table. Again, we have the two tables we need and can pass them to the train_test_split function.

```
# Build ANN
model = tf.keras.Sequential([
    tf.keras.layers.Dense(32, activation=tf.nn. sigmoid, input_dim=4),
    tf.keras.layers.Dense(64, activation=tf.nn.sigmoid),
    tf.keras.layers.Dense(3, activation=tf.nn.softmax)
])
# Configuration of the learning process
model.compile(optimizer='adam', loss='categorical_crossentropy',
 metrics=['accuracy'])
```
Listing 3.16 Building the ANN and Configuring the Learning Process

If you OHE-encode input variables, you must adjust the number of nodes in the input layer, as the ANN requires correspondingly more inputs. If there are too many categories at the input, you should consider whether or not the substitution should take place using numbers. Otherwise, you would have a lot of input nodes.

An ANN with OHE-encoded output data (the column or columns to be predicted) requires categorical_crossentropy as a loss function.

```
# 50 runs
model.fit(train_data, train_col, epochs=50)
```
Listing 3.17 Training Process

The number of eras is increased to 50.

```
test_loss, test_acc = model.evaluate(test_data, test_col)
print('Test accuracy:', test_acc)
```

Listing 3.18 Test Process

In this way, you should achieve a correct classification rate of up to 100%, whereby the increase in the number of epochs was the decisive factor in this example. You can test the program with different epoch numbers.

3.11 Dealing with Data Gaps

In reality, you often have to deal with datasets that are incomplete. Not all chemical properties are available for every red wine sample. When it comes to user data, some users have information about their education, whereas others don't. Next, I'll show you two simple ways of dealing with this.

The data in the *iris_empty.csv* file is incomplete. Using the *K3_iris-3.ipynb* program, we'll look at how you can proceed in such cases.

```
path = "../Data/iris_empty.csv"
data = pd.read_csv(path, delimiter=',')

print(data.head())
print("Empty columns: ", data.columns[data.isnull().any()])
```

Listing 3.19 Loading the CSV File and Outputting the Column Names with Data Gaps

This opens the modified iris file with a few empty cells. The `data.columns[data.isnull().any()]` expression uses pandas functions to output all column names that contain empty cells. The output is shown here:

```
Empty columns:  Index(['sepal.length', 'petal.length', 'species'], dtype='object')
```

You can include this output in all of your programs in the future.

> **Pandas Module**
>
> This isn't a textbook on pandas. The pandas module is very powerful and provides numerous functions for processing large amounts of data with high performance. If you're familiar with the Python programming language, it's worth taking a closer look at this module. We'll use pandas functions that you can transfer to other tables with minor changes or no changes. However, it's also possible to import the CSV file using a spreadsheet. You can then fill empty cells with the average values of the columns, for example, or delete the corresponding rows. Then, you can export the data to a CSV file. If you want to deal with this subject area in the long term, you should learn Python, including the pandas module.

3.11.1 Filling Empty Cells with Data

In our case, you can see in the output that sepal.length, petal.length, and species contain empty cells.

```
# Fill empty cells with mean values of the columns
data['sepal.length'] = data['sepal.length'].fillna(data['sepal.length'].mean())
data['petal.length'] = data['petal.length'].fillna(data['petal.length'].mean())
data['species'] = data['species'].fillna("empty")
```

Listing 3.20 Filling Empty Cells with Mean Values of the Columns

Here, you can see a way to fill empty cells with average values of the column using pandas. Data gaps in the category are filled with the empty string. Remember, however, that you have four categories in the output variable: the three types of iris and the empty string. You must therefore adjust the number of nodes in the output layer. If the AI can't make a prediction, empty will be classified:

```
tf.keras.layers.Dense(4, activation=tf.nn.softmax)
```

The following principle applies here: the purpose of these lines of code is to present these functions to you as clearly as possible so that you can adapt them to new CSV files with minor changes and little knowledge of pandas.

3.11.2 Removing Rows with Empty Cells

Sometimes, you get better results if incomplete data is removed from the dataset. In the *K3_iris-4.ipynb* file, we'll remove rows containing empty cells as an alternative approach. You can also use a spreadsheet to remove the incomplete rows. The following program demonstrates the procedure using pandas functions.

```
# Remove empty cells
data.dropna(inplace=True)
print("Empty columns: ", data.columns[data.isnull().any()])
```

Listing 3.21 Deleting Rows with Empty Cells

All relevant rows are removed here. The changes are made to the existing data. This is a short notation for the following:

```
data = data.dropna()
```

The system checks whether the data still contains empty cells. With this approach, the number of nodes in the output layer remains at three. The rest of the source code remains unchanged.

For data with empty cells, you can test both variants (fill with mean values or remove empty cells) and choose the one that leads to better results.

3.12 Correlation versus Causality

AI (e.g., based on supervised learning) is able to recognize correlations. However, you must not take the results as absolute truth without reflection because correlation isn't the same as causality. For example, if you collect data on body height and income, you can see that tall people earn more money than short people. Now one could assume that the height is the decisive factor for the level of income. You can also create a nice graphic using a spreadsheet. Unfortunately, however, women still earn less than men (on average), and women are often smaller than men. But with the data collected, you can't recognize this problem. For this purpose, you would also have to include the gender in the data collection. A correlation is therefore not clear proof of causality, but causality is always a correlation.

Correlation

Correlation is a statistical term that describes the relationship between two variables. It can take values between -1 and 1. If something is more expensive the heavier it is (e.g., precious metals), the correlation tends toward 1. The correlation between grade and score tends toward -1, as the grade improves (i.e., decreases) as the score increases. The correlation can generally be calculated using the following formula:

$$r_{x,y} = \frac{\sum_{i=1}^{n}(x_i - \bar{x})(y_i - \bar{y})}{\sqrt{\sum_{i=1}^{n}(x_i - \bar{x})^2 + \sum_{i=1}^{n}(y_i - \bar{y})^2}}$$

Again, this looks more complicated than it is. Let's take a look at this using an example created in a spreadsheet.

	Grade	h	x	y	x*y	x^2	y^2	
	1	20	-1.8	5.6	-10.08	3.24	31.36	
	2	17	-0.8	2.6	-2.08	0.64	6.76	
	1	22	-1.8	7.6	-13.68	3.24	57.76	
	2	20	-0.8	5.6	-4.48	0.64	31.36	
	3	15	0.2	0.6	0.12	0.04	0.36	
	2	18	-0.8	3.6	-2.88	0.64	12.96	
	3	12	0.2	-2.4	-0.48	0.04	5.76	
	4	10	1.2	-4.4	-5.28	1.44	19.36	
	5	5	2.2	-9.4	-20.68	4.84	88.36	
	5	5	2.2	-9.4	-20.68	4.84	88.36	
	3	13	0.2	-1.4	-0.28	0.04	1.96	
	4	11	1.2	-3.4	-4.08	1.44	11.56	
	3	14	0.2	-0.4	-0.08	0.04	0.16	
	3	16	0.2	1.6	0.32	0.04	2.56	
	1	18	-1.8	3.6	-6.48	3.24	12.96	
Mean	2.8	14.4			-90.8	24.4	371.6	Sum
			Correlation		-0.954			

Figure 3.21 Correlation between the Grade and the Learning Effort

> In the first two columns, the grades and hours for the learning effort are entered. The mean values are calculated for both columns. In the third column (x), the difference between the grade and the mean value is calculated; in the fourth column (y), the difference between the learning effort and the mean value is calculated. The other columns are self-explanatory. The totals are calculated for the last three columns. Finally, the following applies to the correlation:
>
> $$\frac{-90.8}{\sqrt{24.4 + 371.6}} = -0.954$$
>
> Thus, there is a strong negative correlation between learning effort and grade.
>
> A correlation close to 0 indicates that there is no connection between the variables.

Let's take another scenario. If you train an AI with application and recruitment data and expect it to make recommendations for applicants, the result should also be viewed critically. If, for example, women or minorities have been disadvantaged in recruitment in the past, the AI recognizes the correlation in the data and rates applications from women and minorities lower than those from men. Or, if there is much more data on male applicants, the prediction is therefore worse for the other groups. AI is only as good as the data it has been trained with. The negative distortion is also referred to as bias. You've come across this term before (but with a different meaning), namely when calculating weights.

Attributes (in CSV files, these are columns) that don't correlate at all can be removed from the training and test data. When predicting the rating of champagne, one could neglect the label color. But perhaps the test subjects were unconsciously influenced by this when collecting the data? You can find this out by calculating the correlation. If there is no correlation, this attribute can be ignored. If the correlation exists, it must be weighed up: Should you keep the attribute to improve the prediction? Or should you, when collecting data, carry out the test with standardized bottles so that the test subjects aren't influenced?

In the following *K3_adult-1.ipynb* example, we want to try to predict the income (more or less than $50,000 per year) of people based on certain characteristics. The focus here is only on the development of AIs. The *adult.csv* dataset is often used for teaching purposes (see https://archive.ics.uci.edu/ml/datasets/adult) but usually without reference to ethical issues. We want to make up for that here.

If you're developing programs for productive use, you need to be careful. Let's assume you want to use AI to find out whether a person earns more than $50,000 a year and is a potential customer for your product. As some people are disadvantaged because of their gender, ethnic origin, or sexual orientation, for example, they may earn less. There are many reasons for this (e.g., fewer educational opportunities, discrimination in application procedures). Your program, if it uses data without reflection, would reinforce the discrimination. Even members of minorities who earn more would be disadvantaged by your program if a large part of this group earns less. The AI would make discriminatory predictions, as it only recognizes correlations and not causalities. This still requires common sense. Bear this in mind when handling data. Discriminatory decisions aren't the fault of the AI; rather, developers haven't given enough thought to the issue or have made mistakes.

In the following example, we'll open or import the CSV file (*adult.csv*) in a spreadsheet and replace empty cells that contain a question mark. We'll also insert column headers and save the result as *adult-2.csv*.

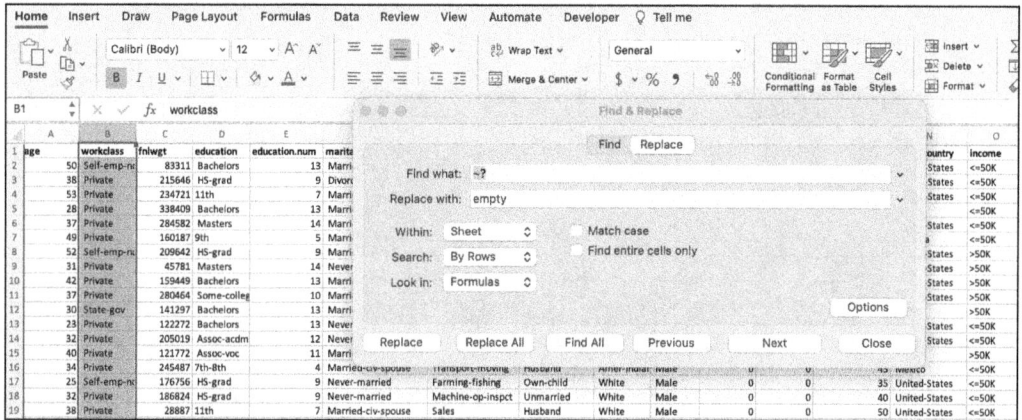

Figure 3.22 Column-by-Column Search and Replace of the "?" Character

The question mark is a placeholder in Excel, so you have to "defuse" it with the tilde character to actually interpret it as a question mark. You can now proceed column by column and check whether there are any cells with question marks. In our example, only columns with string values contain this character, so we replace it with the `empty` string. For columns with numerical values, you could find the mean value for each of these columns and fill the empty cells with these values. An alternative would be to delete the lines entirely, but there would be a lot of lines here, and we don't want to do without this data. The result is exported to a CSV file.

```
import tensorflow as tf
import pandas as pd
from sklearn.model_selection import train_test_split

path = "../Data/adult-2.csv"
data = pd.read_csv(path, delimiter=';')
print(data.head())
print("Empty columns: ", data.columns[data.isnull().any()])
```

Listing 3.22 Loading Modules and the File and Outputting Empty Cells

3 Artificial Neural Networks

The required modules are imported here again, and the CSV file is loaded. Let's take a look at the output and get an overview in Figure 3.23.

At this point, you should think about which columns should be converted and how.

```
# Output of the correlations
correlations = data[data.columns].corr(numeric_only=True)
#print(correlations)
print('All correlations')
print('-' * 30)
correlations_abs_sum = correlations[correlations.columns].abs().sum()
print(correlations_abs_sum)
print('Weakest correlations')
print('-' * 30)
print(correlations_abs_sum.nsmallest(5))
```

Listing 3.23 Determining Correlations

```
      age           workclass  fnlwgt   education  education.num  \
0     50     Self-emp-not-inc   83311   Bachelors             13
1     38              Private  215646    HS-grad              9
2     53              Private  234721       11th              7
3     28              Private  338409   Bachelors             13
4     37              Private  284582     Masters             14

        marital.status         occuaption   relationship   race     sex  \
0   Married-civ-spouse    Exec-managerial        Husband  White    Male
1             Divorced  Handlers-cleaners  Not-in-family  White    Male
2   Married-civ-spouse  Handlers-cleaners        Husband  Black    Male
3   Married-civ-spouse      Prof-specialty           Wife  Black  Female
4   Married-civ-spouse    Exec-managerial           Wife  White  Female

   capital.gain  capital.loss  hours.per.week  native.country  income
0             0             0              13   United-States   <=50K
1             0             0              40   United-States   <=50K
2             0             0              40   United-States   <=50K
3             0             0              40            Cuba   <=50K
4             0             0              40   United-States   <=50K
```

Figure 3.23 Output of the First Five Rows

In the first row, the correlation is calculated for all columns with numerical values. This could also be done using a spreadsheet, but the result would be very confusing. For this reason, an implementation with Python is suitable for this example. If you want to output all correlations, you can do this via `print(correlations)`.

3.12 Correlation versus Causality

```
                 age     fnlwgt  education.num  capital.gain  capital.loss  \
age         1.000000  -0.076646       0.036527      0.077674      0.057775
fnlwgt     -0.076646   1.000000      -0.043159      0.000437     -0.010259
education.num  0.036527  -0.043159   1.000000      0.122627      0.079932
capital.gain   0.077674   0.000437   0.122627      1.000000     -0.031614
capital.loss   0.057775  -0.010259   0.079932     -0.031614      1.000000
hours.per.week 0.068756  -0.018770   0.148127      0.078409      0.054256

                hours.per.week
age                   0.068756
fnlwgt               -0.018770
education.num         0.148127
capital.gain          0.078409
capital.loss          0.054256
hours.per.week        1.000000
```

Figure 3.24 Correlations between All Columns

Did you notice anything in the places that have the value 1? The correlation with itself is always 1.

The output is confusing because we now have to go through entry by entry and recognize rows or columns that correlate with other rows or columns. Therefore, the absolute values (i.e., amounts) are calculated for all columns of this correlation table and then the total. We calculate the absolute values because negative and positive column values would otherwise cancel each other out when added. Positive and negative correlations are important (because there is obviously a connection) and should be retained in the dataset.

```
All correlations
---------------------------
age              1.317378
fnlwgt           1.149271
education.num    1.430372
capital.gain     1.310761
capital.loss     1.233836
hours.per.week   1.368318
dtype: float64
Weakest correlations
---------------------------
fnlwgt           1.149271
capital.loss     1.233836
capital.gain     1.310761
age              1.317378
hours.per.week   1.368318
dtype: float64
```

Figure 3.25 Summed Correlation Values

All results are displayed, followed by the five columns with the lowest correlations. You can simply copy the complete code cell for future programs. If you want to calculate the correlations for all columns, you must also convert all columns into numerical values.

```
data.drop(['fnlwgt','capital.loss','capital.gain'], axis = 1, inplace=True)
```

Listing 3.24 Deleting Columns with Small Correlation Values

The columns with the lowest summed correlation values are removed from the dataset. You can see that these are new hyperparameters. Whether and which columns should be removed must be determined experimentally. Removing the columns can lead to an improvement or deterioration in the result or prediction.

The following source code again emphasizes clarity and reusability.

```
# income is to be predicted
col = data['income']
col = pd.get_dummies(col, dtype=float)
data = data.drop(['income'], axis = 1)
```

Listing 3.25 Definition and Data Type Conversion of the Target Column

The `income` column is OHE-encoded, assigned to a new table, and removed from the original table.

```
# convert these columns into numerical values 0...n
conv_num =
['workclass', 'education', 'marital.status', 'occuaption','relationship',
 'hours.per.week', 'native.country']

data[conv_num] = data[conv_num].astype('category')
data[conv_num] = data[conv_num].apply(lambda x: x.cat.codes)
```

Listing 3.26 Conversion of Data Types (Numerical Values) of Multiple Columns

The data type of the listed columns is converted to `category` and all entries are then converted to numerical values (integers). In future, you can copy these rows and simply adapt the listed columns to new data sources.

```
# perform OHE for this data
conv_ohe = ['race', 'sex']
data = pd.get_dummies(data, columns = conv_ohe, dtype=float)
```

Listing 3.27 Conversion of Data Types (OHE) of Multiple Columns

All listed columns are OHE-encoded here. You can also copy and adapt these lines for future programs. In the next example, the `sex`, `race`, and `native.country` columns are completely removed from the data, as the use of these attributes is very questionable. However, this is only about technical details.

Using print(data.shape), the number of rows and columns of the training data (data) is output in a tuple, here (32560, 16). You can access the first entry via data.shape[0] and the second one by using data.shape[1]. You should test it using a print function.

> **Tuple Data Type**
>
> The tuple data type can contain related data of different types, similar to the list:
>
> my_tuple = (1,2,'something',3.4)
>
> The differences are that a tuple can't be subsequently changed and the values are enclosed in parentheses.

Now an ANN is created, configured, trained, and tested.

```
# Create four tables from the two tables
train_data, test_data, train_col, test_col = train_test_split(data,col,
 test_size=0.2, random_state=42)

# Build ANN
model = tf.keras.Sequential([
    tf.keras.layers.Dense(128, activation=tf.nn.relu, input_dim=data.shape[1]),
    tf.keras.layers.Dense(256, activation=tf.nn.relu),
    tf.keras.layers.Dense(64, activation=tf.nn.relu),
    tf.keras.layers.Dense(2, activation=tf.nn.softmax)
])
```

Listing 3.28 Division of the Tables and Building the ANN

You can now use data.shape[1] for the number of nodes in the input layer, as the number of columns is stored here. The advantage of this variant is that you no longer have to update this row if the number of columns changes.

```
# Configuration of the learning process
model.compile(optimizer='adam', loss='categorical_crossentropy',
 metrics=['accuracy'])

# 10 runs
model.fit(train_data, train_col, epochs=10)

test_loss, test_acc = model.evaluate(test_data, test_col)
print('Test accuracy:', test_acc)
```

Listing 3.29 Test and Training Process

The correct classification rate is relatively good at approximately 83%. Try to increase this value by changing the hyperparameters.

3.13 Standardization of the Data

The standardization of the data is a simple way to (possibly) improve the prediction of the ANN. We'll also look at this using an example: An AI is supposed to use various attributes (living space, floor area, number of rooms, number of floors, etc.) to determine whether a house is expensive or not. If the living space or floor area attributes vary by small amounts, this won't have a major impact on the purchase price. However, if the number of rooms or floors changes by small amounts, this has a significant impact on the purchase price. The numerical values of the various inputs of the ANN have different effects on the weights during the training process. During standardization, an attempt is made to change the numerical values of the attributes in such a way that the weighting isn't distorted too much.

One option is to scale the data between minimum and maximum values. The formula is as follows:

$$x_{norm_i} = \frac{x_i - x_{min}}{x_{max} - x_{min}}$$

This type of scaling is used if there is a fixed value range. Pixel values, for example, can only have values between 0 and 255. This data is suitable for *min-max scaling*.

If data has a Gaussian normal distribution, it should be standardized using the *standardization* or *z-transformation*. Examples of usage for this are age, size, and income. The outliers aren't so significant in this calculation. The formula is as follows:

$$z_i = \frac{x_i - \mu}{\sigma}$$

Here, μ is the mean value, and σ is the standard deviation. Let me explain the standard deviation by using an example. Take a look at the grades as the result of a performance record: 1, 2, 2, 3, 4, 6. The arithmetic mean is calculated as follows:

$$\bar{x} = \frac{\text{sum of grades}}{\text{number of grades}} = \frac{1+2+2+3+4+6}{6} = 3$$

The variance can be obtained by calculating the difference between each grade and the mean value, squaring this difference, and adding up all the individual results:

$$\sigma^2 = \frac{(3-1)^2 + (3-2)^2 + (3-2)^2 + (3-3)^2 + (4-3)^2 + (6-3)^2}{6} = 2.67$$

This result can be used to caculate the standard deviation:

$$\sigma = \sqrt{\sigma^2} = \sqrt{2.67} = 1.63$$

3.13 Standardization of the Data

The average deviation of the grades from the mean is ±1.63. Of all grades, 68% are in the range 3 ± 1.63, that is, between 1.37 and 4.63. The better outliers are below this value range, and the worse ones are above.

Let's apply what we've just learned to the *K3_adult-2.ipynb* example. It won't change the result much here (as there are no large numerical values), but you'll see how you can standardize data with these options.

```
import tensorflow as tf
import pandas as pd
from sklearn.model_selection import train_test_split
from sklearn.preprocessing import StandardScaler
from sklearn.preprocessing import MinMaxScaler
```

Listing 3.30 Import of Required Modules

You also want to import `StandardScaler` and `MinMaxScaler`.

```
data.drop([
'fnlwgt','capital.loss','capital.gain', 'race', 'sex', 'native.country'],
 axis = 1, inplace=True)
```

Listing 3.31 Removal of Unnecessary Data from the Dataset

We're now also removing gender and ethnic origin from the data. Thanks to the optimizations, the result isn't worse, and we've made sure that the AI isn't obviously racist or sexist. But is it really ethically sound? If the other attributes in the data imply that a minority earns less (e.g., because they have fewer educational opportunities), the AI can continue to make racist statements. We also classify people in general, for example, according to their educational qualifications. People who have managed to earn a lot of money despite having lower educational qualifications are discriminated against by the AI. On the other hand, it's also critical when lower-income families are sent unsolicited catalogs for luxury watches. For each use case, which attributes should be used and which ones should not must be carefully considered. As you can see, it remains complicated as long as people are involved.

```
# Create objects
s_scaler = StandardScaler()
m_scaler = MinMaxScaler()
```

Listing 3.32 Instantiation of the Scalers

Here, you create an object each (`s_scaler` and `m_scaler`). You can use these "tools" to adjust and standardize the data.

```
# Columns for StandardScaler
cols_to_s_scale = [
 'age', 'workclass', 'education', 'education.num', 'marital.status',
 'occuaption', 'relationship']
data[cols_to_s_scale] = s_scaler.fit_transform(data[cols_to_s_scale])
```

Listing 3.33 Scaling Selected Columns Using StandardScaler

Here, you can list the columns to which the scaling should be applied. If there are multiple columns, only the variable in the first row needs to be adjusted (e.g., `cols_to_s_scale = ['age', weight']`). The `s_scaler` object can be used to call the `fit_transform` method.

```
# Columns for MinMaxScaler
cols_to_m_scale = ['hours.per.week']
date[cols_to_m_scale] = m_scaler.fit_transform(date[cols_to_m_scale])
```

Listing 3.34 Scaling Selected Columns Using MinMaxScaler

We proceed in the same way for standardization. You can also copy and reuse these lines for your future programs.

In this example, the standardization of the data increased the correct classification rate by 1% or 2%. That's not much. It's very often the case that you can create a functioning AI for new datasets relatively quickly by adapting existing programs. However, optimizing these programs is time-consuming.

When developing for production systems, you only set up the scaler with the training data. The scaler is initialized using `fit_transform` and applied to the training data. You can then use the same scaler for test data using `transform`. We're not yet ready to implement professional development for production systems. First of all, it's important to understand the basic principles. In Chapter 11, we'll take a closer look at this by using graphical blocks, which makes it particularly easy to understand the data flow.

3.14 Regression

Up to this point, our ANN has always done of the work of classification. We can also program an ANN with small changes, which predicts continuous numerical values. To do this, we'll experiment with the iris data in *K3_iris-5.ipynb*. The length of the sepal should be predicted on the basis of the other attributes.

```
import tensorflow as tf
import pandas as pd
from sklearn.model_selection import train_test_split
from sklearn.preprocessing import StandardScaler
```

```
path = "../Data/iris.csv"
data = pd.read_csv(path, delimiter=',')
print(data.head())
print("Empty columns: ", data.columns[data.isnull().any()])

# Output of the correlations
correlations = data[data.columns].corr(numeric_only=True)
print('All correlations')
print('-' * 30)
correlations_abs_sum = correlations[correlations.columns].abs().sum()
print(correlations_abs_sum)
print('Weakest correlations')
print('-' * 30)
print(correlations_abs_sum.nsmallest(5))
```

Listing 3.35 Loading the Modules and the CSV File and Analyzing the Data

The output is as follows:

```
All correlations
------------------------------
sepal.length    2.807265
sepal.width     1.912136
petal.length    3.263059
petal.width     3.146932
dtype: float64
Weakest correlations
------------------------------
sepal.width     1.912136
sepal.length    2.807265
petal.width     3.146932
petal.length    3.263059
```

The width of the sepal actually has a low correlation. Who would have thought that? You always have to experiment to see whether omitting attributes actually leads to a better result.

```
# possibly drop data['sepal.width'], has low correlation -> test
data = data.drop(['sepal.width'], axis = 1)

# This column is to be predicted
col = data['sepal.length']
data = data.drop(['sepal.length'], axis = 1)

# perform OHE for this data
conv_ohe = ['species']
```

3 Artificial Neural Networks

```
data = pd.get_dummies(data, columns = conv_ohe, dtype=float)

# Create object
s_scaler = StandardScaler()
# Columns for StandardScaler
cols_to_s_scale = ['petal.length', 'petal.width']
data[cols_to_s_scale] = s_scaler.fit_transform(data[cols_to_s_scale])

# Create four tables from the two tables
train_data, test_data, train_col, test_col = train_test_split(
 data,col, test_size=0.2, random_state=42)
```
Listing 3.36 Preparation of the Data and Division into Four Tables

The four tables (or two tables and two columns) are created. You should now always be able to specify the number of columns and rows in the four tables if you know the number of columns and rows in the data source. As you can see, it's always about preparing the data and creating these tables. The rest of the source code always remains similar.

```
# Build ANN
model = tf.keras.Sequential([
    tf.keras.layers.Dense(32, activation=tf.nn.sigmoid,
 input_dim=data.shape[1]),
    tf.keras.layers.Dense(64, activation=tf.nn.sigmoid),
    tf.keras.layers.Dense(1)
])
```
Listing 3.37 Building the ANN

The output layer has only one node and no activation function. The continuous number is output unchanged via the node.

```
model.compile(optimizer='adam', loss='mae',metrics=['mae'])
```
Listing 3.38 Configuration of the Learning Process

We use mae as loss and metrics. The output is therefore not the length of the sepal, but the mean absolute deviation or the error of the prediction for each dataset. The formula is very simple:

$$\text{mae} = \frac{1}{n}\sum_{i=1}^{n}|y_i - \hat{y}_i|$$

In each case, y_i stands for the correct value, and $\hat{y}_i$ stands for the predicted value. If there is only one node, then $n = 1$.

```
test_loss, test_mae = model.evaluate(test_data, test_col)
print('Test mae:', test_mae)
```
Listing 3.39 The Test Process

This ANN can predict the length of the sepal with an accuracy of approximately 3 mm (0.12 inches).

3.15 Deployment

We've only trained and tested the ANN so far. But how can we actually apply it in the field to a new dataset? The procedure will be demonstrated using a small example via *iris.csv* again.

3.15.1 Training, Testing, and Saving

The *K3_iris-6a.ipynb* file is quite similar to the previous ones.

```
import tensorflow as tf
import pandas as pd
from sklearn.model_selection import train_test_split
from sklearn.preprocessing import LabelEncoder, StandardScaler
import joblib

path = "../Data/iris.csv"
data = pd.read_csv(path, delimiter=',')
```
Listing 3.40 Loading the Modules and the CSV File

We now also import the `LabelEncoder` from the sklearn module to encode the subtype numerically (integer).

```
# Output of the correlations
correlations = data[data.columns].corr(numeric_only=True)
print('All correlations')
print('-' * 30)
correlations_abs_sum = correlations[correlations.columns].abs().sum()
print(correlations_abs_sum)
print('Weakest correlations')
print('-' * 30)
print(correlations_abs_sum.nsmallest(5))
```
Listing 3.41 Determination of Correlations

From now on, we'll always output the correlations.

3 Artificial Neural Networks

```python
# species must be predicted
le = LabelEncoder()
col = le.fit_transform(data['species'])
data = data.drop(['species'], axis = 1)
```
Listing 3.42 Preparation of the Target Column

We create an object named le and use it to transform the species column. This column will be removed from data.

```python
# Create object
s_scaler = StandardScaler()
# Columns for StandardScaler
data = s_scaler.fit_transform(data)

# Create four tables from the two tables
train_data, test_data, train_col, test_col = train_test_split(
  data,col, test_size=0.2, random_state=42)
```
Listing 3.43 Scaling and Division of Data

You should now be familiar with the division of the data. There are no changes to the sections for setting up, configuring, training, and testing the ANN.

```python
# Save ANN, scaler, and LabelEncoder
model.save('model1.h5')
joblib.dump(s_scaler,'scaler1.joblib')
joblib.dump(le,'le1.joblib')
```
Listing 3.44 Saving Required Data

If you're satisfied with the result, you can save the scaler, LabelEncoder, and the model. The *.h5* extension indicates a Hierarchical Data Format (HDF) file, which is used to store large amounts of data in a hierarchically structured form. The *scaler1.joblib* and *le1.joblib* files are simple binary files. The development of the ANN is now complete.

You can also use the OneHotEncoder of the sklearn module instead of the LabelEncoder. The *K3_iris-7a.ipynb* file uses OHE encoding. Let's take a look at the relevant places.

```python
# species must be predicted
oe = OneHotEncoder()
col = oe.fit_transform(data[['species']])
col = col.toarray()
data = data.drop(['species'], axis = 1)
```
Listing 3.45 Preparing the Target Column

The column is transformed and converted into the data type array (similar to the list) using the `toarray()` function. This is necessary when you use `OneHotEncoder`.

```
model.compile(optimizer='adam', loss='categorical_crossentropy',
 metrics=['accuracy'])
```

Listing 3.46 Configuring the Learning Process

The `loss` function must be adapted again, as the output data is OHE-encoded.

```
# Save ANN, scaler, and LabelEncoder
model.save('model2.h5')
joblib.dump(s_scaler,'scaler2.joblib')
joblib.dump(oe,'oe2.joblib')
```

Listing 3.47 Saving Required Data

The data is saved in the corresponding files.

3.15.2 Using the ANN Model

The model has been trained and tested. Predictions can now be made for new and unknown data. The AI is therefore in use.

The *K3_iris-6b.ipynb* file is intended for the prediction and opens a test file with only four entries. When you measure the leaves of the iris, you don't know which subspecies it belongs to.

```
import tensorflow as tf
import pandas as pd
from sklearn.model_selection import train_test_split
from sklearn.preprocessing import LabelEncoder, StandardScaler
import numpy as np
import joblib
```

Listing 3.48 Loading Required Modules

We also import the numpy module, which provides numerous functions for the simple calculation of arrays (similar to lists).

```
path = "../Data/iris_test.csv"
data = pd.read_csv(path, delimiter=',')
print(data.head())
```

Listing 3.49 Loading the CSV File

This file contains the four sizes, while the specific subspecies isn't included. The aim is for the AI to provide this information. The output is as follows:

```
   sepal.length  sepal.width  petal.length  petal.width
0           5.1          3.4           1.5          0.1
```

Let's find out which subspecies the iris belongs to.

```
# Create object
s_scaler = joblib.load('scaler1.joblib')
# scale data
data = s_scaler.transform(data)
# LabelEncoder
le = joblib.load('le1.joblib')
```

Listing 3.50 Instantiating the Encoder and Scaler

You open the saved scaler and apply it to the data. This scaler stores the mean value and the standard deviation, which were determined using the larger amount of data during development. The LabelEncoder is also loaded from the file. It contains the assignment of the number to the subspecies.

```
model = tf.keras.models.load_model('model1.h5')
pred = model.predict(data)
print("prediction:",pred)
```

Listing 3.51 Loading and Applying the Pretrained Model

The saved model gets opened. You don't have to create a layer, and so on, by yourself. Then, the prediction is made. The output provides three elements (output of the softmax function) with the respective probabilities. In this case, the output is as follows:

```
prediction: [[0.92637813 0.0724499 0.00117189]]
```

The first subspecies applies with a probability of approximately 93%. Let's analyze this prediction.

```
num = np.argmax(pred)
print("number:",num)
```

Listing 3.52 Determining the Result as an Index

The NumPy function is used here to determine the index of the largest number. The output is as follows:

```
number: 0
```

The first element with the index 0 has the highest value. But which subtype has the 0 encoding?

```
spec = le.inverse_transform([num])
print("species:",spec)
```

Listing 3.53 Determining the Result as a Character String

This is where the prediction for the specific subspecies is made. The output is as follows:

```
Species: ['Setosa']
```

The `LabelEncoder` has saved the "Number → subtype" assignment. We can use the `inverse_transform` function to determine the subtype. This data refers to *Iris setosa*.

A similar procedure is necessary if the data is OHE-encoded (see *K3_iris-7b.ipynb*).

```
pred = model.predict(data)
print("prediction:",pred)

num = np.argmax(pred)
print("number:",num)

spec = oe.inverse_transform(pred)
print("species:",spec)
```

Listing 3.54 Determining the Result

The transformer doesn't receive the number as a parameter, but instead receives the three probabilities, and then outputs the corresponding subtype.

You've now become familiar with further options for transforming nonnumerical data using the sklearn module. In addition, you now also know how to save your ANN and reuse it as an application. As already mentioned, when developing for production systems, you should test your model with new data. There is a risk that the hyperparameters have only been optimized for the test data. You can do this by generating specific predictions in the application (as in the preceding example) for unknown evaluation datasets and comparing these with the entries in the evaluation data. With a little knowledge of Python, this is easily possible. The only difference compared to "field use" is that you can evaluate the prediction because you know the result.

3.16 Exercises

3.16.1 Exercise 1: Hyperparameter Optimization for Classification

Make a copy of the *K3-iris-2.ipynb* file, rename it to *K3_solution-1.ipynb*, change some hyperparameters, and try to increase the correct classification rate. Follow these steps:

1. Change the number of hidden layers and nodes. It has become established that powers of 2 are used for the number of nodes in the hidden layer, that is, 4, 8, 16, 32, and so on.
2. Test various activation functions.
3. Adjust the number of epochs. Make sure that you don't enter the overfitting area. You can tell by the fact that the test results will become much worse than the training results.
4. The result of the test data is always very important.

3.16.2 Exercise 2: Hyperparameter Optimization for Regression

Create a copy of the *K3-iris-5.ipynb* file, rename this copy to *K3_solution-2.ipynb*, change some hyperparameters, and try to minimize the deviation. Follow the instructions in the previous task.

3.16.3 Exercise 3: ANN for Classification

Program an AI with an ANN to classify mushrooms into "edible" and "poisonous". Use the *mushrooms.csv* file. Information on the attributes can be found at *www.kaggle.com/datasets/uciml/mushroom-classification*. You can also search for information on the data on other web pages. Proceed step by step, and don't continue until you've completed all the previous steps. Follow these steps:

1. Output the first rows. Pay attention to the correct separator.
2. Determine the number of empty cells.
3. Divide the data into two tables. The target column should be OHE-encoded.
4. Simply transform all remaining columns at the same time using `LabelEncoder` to save typing:

   ```
   le = LabelEncoder()
   data = data.apply(le.fit_transform)
   ```

5. Determine the correlations. The values are calculated for all columns as they have been converted into numerical values.
6. Scale all the data at the same time to save typing:

   ```
   s_scaler = StandardScaler()
   data = s_scaler.fit_transform(data)
   ```

   Pay attention to the sequence of the work steps. After scaling, you'll lose the column headings. You can't then calculate the correlations (as presented here) because you'll need the column headings again.

7. Set up, train, and test the ANN model.

8. Test whether removing one or more columns with low correlation values leads to an improvement in the result.
9. Try to improve the correct classification rate by adjusting the hyperparameters.

3.16.4 Exercise 4: ANN for Regression

Program an AI with ANN to determine the price of diamonds. Use the *diamonds.csv* file. Information on the attributes can be found at *www.kaggle.com/datasets/shivam2503/diamonds*, for example. Follow these steps:

1. Output the first rows. Pay attention to the correct separator.
2. Determine the number of empty cells.
3. Divide the data into two tables. In addition, remove the *Unnamed: 0* column from the data.
4. Convert nonnumeric values to numeric values.
5. Determine the correlations.
6. Scale numerical columns.
7. Set up, train, and test the ANN model.
8. Test whether removing one or more columns with low correlation values leads to an improvement in the result.
9. Try to minimize the deviation by adjusting the hyperparameters.

Chapter 4
Decision Trees

When we talk about artificial intelligence, many people immediately think of artificial neural networks. However, technologies based on decision trees often deliver better results in some areas.

> **What This Chapter Is About**
> - Introduction to decision trees and decision forests
> - Introduction to gradient boosting and XGBoost
> - Classification and regression using XGBoost as an alternative to artificial neural networks
> - Automatic setting of hyperparameters
> - Visual programming using Orange

What effort did we put into making predictions based on tabular data? We've built and tested an artificial neural network (ANN) with various hidden layers, nodes, and activation functions. We've also standardized the data. Nevertheless, we could not even say with certainty whether there was still potential to improve the result. So, we tested, evaluated, reconfigured, and tested again different configurations. We weren't sure about the number of epochs either. It wasn't until we were in the overfitting area that we knew we had exceeded the limit and had to reduce the number of epochs. Is there no easier way, with questions that can be answered with a simple "yes" or "no" and thus lead step-by-step or question-by-question to the goal? As a matter of fact, that's very well possible, but not without new hyperparameters. However, you'll also learn a method for automatically testing and evaluating various hyperparameter settings.

In the following sections, you'll learn new techniques (as an alternative to ANNs) to develop AI programs. Remember the model of the AI black box from Chapter 3, Section 3.2, where the recipe for AI development was presented? This model can be reused here. You can simply replace the inside of the black box (ANN) with the decision trees or forests presented here and the techniques based on them.

4 Decision Trees

4.1 Simple Decision Trees

The previous chapter presented AI models based on ANNs. We'll use the simple, clear Iris dataset again that we used before, but we'll develop the AIs using a completely different approach: decision trees.

4.1.1 Decision Tree Classifier

Let's start with simple trees to get to know the process better. You can see what the individual questions are in a generated tree and thus trace the data flow. The questions are formulated in such a way that you can answer them with yes or no and continue in the left or right branch accordingly. The algorithm is simple and easy to understand. With ANNs, we know how the weights are updated, but a trained ANN is a black box. In retrospect, we can't say why the weights have their specific values. These are set after an iterative process of optimization using the gradient method. Even if you view the weights of a trained ANN, you can't easily recognize which decision or prediction a data input will lead to. With decision trees, on the other hand, this is possible; the AI's decisions and predictions are transparent and can be predicted and understood by humans.

So let's use the iris data again that we already know. What should the questions be to determine the correct subspecies? Which question should we start with? And what questions should we continue with? An example is shown in Figure 4.1.

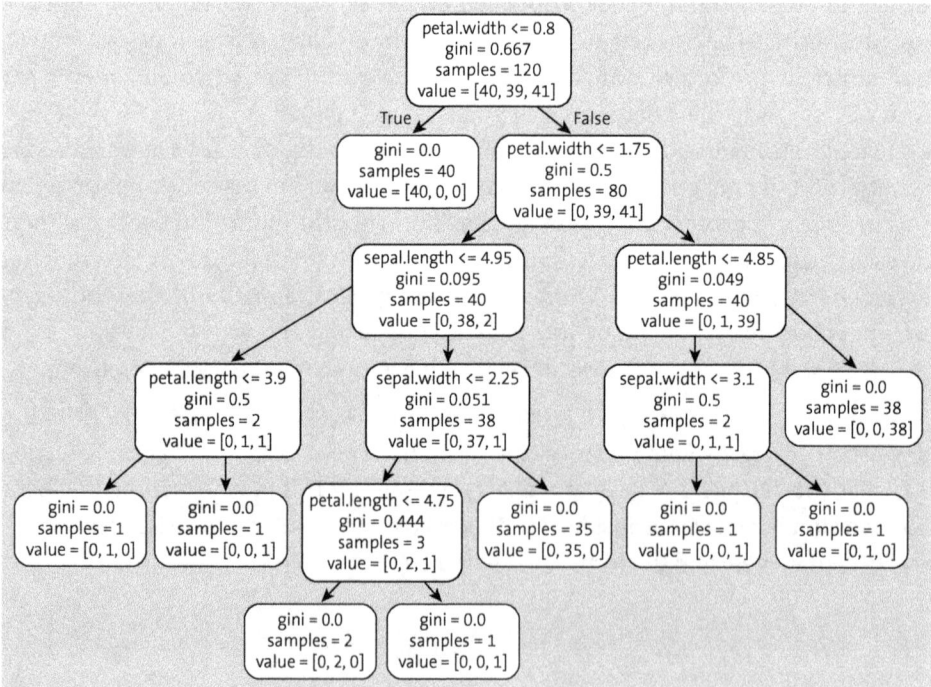

Figure 4.1 Decision Tree Classifier for Irises

At the top, you can see the first question, which is also referred to as the *root node*. Here, you're asked whether the petal.width variable is less than or equal to 0.8. If the answer is yes, the left branch continues to the last node or *leaf node*. If the answer is no, it continues in the right branch, and a few more questions follow. Let's look at the information in the nodes to better understand the structure.

The first entry is the question, which can be answered with yes or no. The third entry (*Samples*) specifies the number of datasets to which this node applies. In the root node, there are 120 (rows of our comma-separated values [CSV] file), which represents our 80% training data. The last entry indicates the distribution. In the root node, there are 40 setosa, 39 versicolor, and 41 virginica. The distribution of the leaves is interesting. In the first level after the root node, for example, you'll see the distribution 40, 0, 0. This means that all 40 datasets belong to setosa. If you've arrived at this leaf when answering the questions, it's therefore the setosa subspecies. Now take a look at the other nodes, especially the leaves. There are several leaves for one subspecies. It's also interesting that the same attribute is used in different levels to formulate the question, each with different numerical values.

But how do you know what the question should be or which attribute you should start with? Why does the root node get compared with 0.8 and no other number? This is where the second entry in the node comes into play, which we've deliberately skipped so far, the *Gini impurity*, which measures the inequality of a distribution. The formula is as follows:

$$\text{Gini} = 1 - \sum_{i=1}^{n} p_i^2$$

Here, p is the probability of i for n classes. Let's carry out the calculations for the exercise with the given values in the root node (see Figure 4.1). There are three classes (three subspecies of iris). You'll find the total number in the denominator and the number of the respective subspecies in the numerator:

$$\text{Gini} = 1 - \left(\frac{40}{120}\right)^2 - \left(\frac{39}{120}\right)^2 - \left(\frac{41}{120}\right)^2 = 0.67$$

You can verify the Gini impurity of other nodes by way of calculation as an exercise. The smaller the value, the more uneven the distribution. In the case of a node, we want to achieve an uneven distribution to reach the destination quickly. In our program, we ideally want to correctly assign a complete subspecies with one answer. The target for each node is therefore a small value.

Let's look at another example, namely the leaf node at the top left in Figure 4.1:

$$\text{Gini} = 1 - \left(\frac{40}{40}\right)^2 - \left(\frac{0}{40}\right)^2 - \left(\frac{0}{40}\right)^2 = 0$$

4 Decision Trees

This Gini impurity with a value of 0 is an extreme example of unequal distribution. The process is best illustrated with categorized data, as listed in Table 4.1.

Students	Subject	Absence > 10%	Exercises > 50%	Passed
1	Math	Yes	Yes	No
2	Ethics	Yes	Yes	Yes
3	IT	No	No	Yes
4	Math	No	No	No
5	Math	No	Yes	Yes
6	IT	Yes	No	No
7	Ethics	No	No	Yes
8	Ethics	No	No	Yes
9	IT	Yes	Yes	Yes

Table 4.1 Student Data on Degrees

In this example, information was collected on nine students. A decision tree will be modeled, which predicts whether the passing of a subject is at risk or not. The information collected includes the following points:

- Subject name.
- Was the absence greater than 10% of the total teaching time?
- Have more than 50% of the exercises in the subject been completed?
- Has the subject been successfully completed?

There are two possible result categories: "passed" with six data records, and "failed" with three data records.

$$\text{Gini}_{\text{total}} = 1 - \left(\frac{6}{9}\right)^2 - \left(\frac{3}{9}\right)^2 = 0.44$$

Take a look at the first subject attribute. There are three categories here: Math (three entries), Ethics (three entries), and IT (three entries). For all these categories, we have to calculate the Gini impurity related to the information of passed or failed. In math, there is one "pass" and two "fail," which are the values in the numerator. The denominator is the total number of passed and failed entries:

$$\text{Gini}_{\text{math}} = 1 - \left(\frac{1}{3}\right)^2 - \left(\frac{2}{3}\right)^2 = 0.44$$

We're continuing this for the other subjects accordingly:

$$\text{Gini}_{\text{ethics}} = 1 - \left(\frac{3}{3}\right)^2 - \left(\frac{0}{3}\right)^2 = 0$$

$$\text{Gini}_{\text{IT}} = 1 - \left(\frac{2}{3}\right)^2 - \left(\frac{1}{3}\right)^2 = 0.44$$

We use these values to calculate the weighted average for the subject attribute:

$$\text{Gini}_{\text{subject}} = \text{Gini}_{\text{math}} \cdot \frac{n_{\text{math}}}{n_{\text{total}}} + \text{Gini}_{\text{ethics}} \cdot \frac{n_{\text{ethics}}}{n_{\text{total}}} + \text{Gini}_{\text{IT}} \cdot \frac{n_{\text{IT}}}{n_{\text{total}}}$$

$$\text{Gini}_{\text{subject}} = 0.44 \cdot \frac{3}{9} + 0 \cdot \frac{3}{9} + 0.44 \cdot \frac{3}{9} = 0.29$$

Let's repeat the whole thing for the absence and exercise attributes:

$$\text{Gini}_{\text{absence\_yes}} = 1 - \left(\frac{2}{4}\right)^2 - \left(\frac{2}{4}\right)^2 = 0.5$$

$$\text{Gini}_{\text{absence\_no}} = 1 - \left(\frac{4}{5}\right)^2 - \left(\frac{1}{5}\right)^2 = 0.32$$

$$\text{Gini}_{\text{absence}} = 0.5 \cdot \frac{4}{9} + 0.32 \cdot \frac{5}{9} = 0.4$$

$$\text{Gini}_{\text{exercises\_yes}} = 1 - \left(\frac{3}{4}\right)^2 - \left(\frac{1}{4}\right)^2 = 0.38$$

$$\text{Gini}_{\text{exercises\_no}} = 1 - \left(\frac{4}{5}\right)^2 - \left(\frac{1}{5}\right)^2 = 0.32$$

$$\text{Gini}_{\text{exercises}} = 0.38 \cdot \frac{4}{9} + 0.32 \cdot \frac{5}{9} = 0.35$$

Let's summarize the results of the attributes:

$\text{Gini}_{\text{subject}} = 0.29$

$\text{Gini}_{\text{absence}} = 0.4$

$\text{Gini}_{\text{exercises}} = 0.35$

According to these results, the subject should be asked about in the first node. Ethics has the lowest value among the subjects, so the question in the root node should be whether the subject is ethics. Then it can already be said that the subject has been passed, and the branching would end there. If the answer to the question is no, the entire procedure must be repeated with the new, reduced table for the new node.

4 Decision Trees

Students	Subject	Absence > 10%	Exercises > 50%	Passed
1	Math	Yes	Yes	No
3	IT	No	No	Yes
4	Math	No	No	No
5	Math	No	Yes	Yes
6	IT	Yes	No	No
9	IT	Yes	Yes	Yes

Table 4.2 Reduced Student Data for the Next Iteration

These calculations are intended to aid your understanding. Here, too, you don't need to calculate anything yourself when programming. During training, the tree is constructed according to the Gini impurity and used with the test data during testing. The source code is even clearer than with ANN. Let's start with *K4_iris-1.ipynb*:

```
import pandas as pd
from sklearn.model_selection import train_test_split
from sklearn import metrics
from sklearn import tree


path = "../Data/iris.csv"
data = pd.read_csv(path, delimiter=',')
```

Listing 4.1 Loading Required Modules and the CSV File

The required modules are imported, the file is loaded, and the first rows are output.

```
# What should be predicted? Save column name in variable.
col_name = 'species'

# Here the division into two tables takes place (Input=data and Output=col).
col = data[col_name]
data = data.drop([col_name], axis = 1)

# Create four tables from the two tables
train_data, test_data, train_col, test_col = train_test_split(
  data,col, test_size=0.2, random_state=42)
```

Listing 4.2 Definition of the Target Column and Division of the Table

This is followed by the division into four tables, as training and testing is required. So far, no line of code is really new. Do you remember how the division into four tables should be calculated?

We'll go through the remaining source code line by line:

```
# Create tree
tr = tree.DecisionTreeClassifier()
```

A tr object of the DecisionTreeClassifier type is created here. We can use tr to call various methods of the tree:

```
#train
tr.fit(train_data, train_col)
```

The two tables (or one table and one column) are used to train or build the tree:

```
# prediction
predicted_col = tr.predict(test_data)
```

A column with predictions is generated here using the test data (third table):

```
# evaluate
score = metrics.accuracy_score(test_col, predicted_col)
print(score)
```

This generated predicted_col column is compared with the test_col test column (fourth table) and the accuracy gets calculated (the formula is known from the ANN). With these few lines of source code, you have an AI that achieves a correct classification rate of 93%. You can even output the generated tree (this is how Figure 4.1 came about):

```
tree.plot_tree(tr)
```

What adjustments can we make here to possibly improve the prediction? The documentation page of scikit-learn (*http://r-wrk.de/9763-scikit-l*) can help you. Let's take a look at two possibilities as examples:

- **max_depth**
  You can use max_depth to set the maximum number of levels in the tree. Too many levels can lead to overfitting. The value 3 is recommended as a starting value.

- **min_samples_leaf**
  You can use min_samples_leaf to set a minimum number of datasets per leaf. For a few classes you should try using the value 1; otherwise, use 5.

You can set the parameters when you create an object. To do this, you need to write them in parentheses in the right-hand expression (for the *constructor*) when creating the object:

4 Decision Trees

```
tr = tree.DecisionTreeClassifier(max_depth=3)
```

If you want to set multiple parameters, you must separate them with a comma:

```
tr = tree.DecisionTreeClassifier(max_depth=3, min_samples_leaf=1)
```

Experiment with the two parameters, and take a look at the graphical output of the tree. The correct classification rate is already very good without any special parameter settings.

4.1.2 Decision Tree Regressor

Decision trees can also be used to predict continuous values. To determine the node point, the mean squared error (MSE) is used instead of the Gini impurity. The formula is as follows:

$$MSE = \frac{1}{n}\sum_{i=1}^{n}(y_i - \hat{y}_i)^2$$

y_i is the actual value; $\hat{y}_i$ is the prediction. Let's take another look at the procedure in Figure 4.1 using sample data, in this case the first five rows of the *iris.csv* file listed in Table 4.3.

sepal.width	petal.length	petal.width	species	sepal.length
2.90	1.4	0.2	Setosa	4.4
3.00	1.4	0.2	Setosa	4.9
3.10	1.5	0.2	Setosa	4.6
3.20	1.3	0.2	Setosa	4.7
3.40	1.4	0.3	Setosa	4.6

Table 4.3 The First Five Rows of the iris.csv File

The aim is again to determine the length of sepals based on the four attributes (length and width of the petal, width of the sepal and the subspecies); we had already implemented this using an ANN. First, the character strings of the subtype (species) are converted into numerical values. We then start with the first column and sort the values in ascending order. Then, we calculate the mean value for each two consecutive values, starting with *(2.9 + 3) ÷ 2 = 2.95*. The data is divided accordingly: into sepal.width < 2.95 (true for one row) and sepal.width >= 2.95 (true for four rows). The mean values are calculated for both resulting sections, which are to be regarded as predicted values, as shown in Table 4.4.

sepal.width	petal.length	petal.width	species	sepal.length	Prediction
2.90	1.4	0.2	0	4.4	4.4
3.00	1.4	0.2	0	4.9	4.7
3.10	1.5	0.2	0	4.6	4.7
3.20	1.3	0.2	0	4.7	4.7
3.40	1.4	0.3	0	4.6	4.7

Table 4.4 With the Additional Column for Mean Values for Prediction

The mean values now form the prediction for this iteration step. The MSE formula can now be applied:

$$\text{MSE} = \frac{1}{5}[(4.4-4.4)^2 + (4.9-4.7)^2 + (4.6-4.7)^2 + (4.7-4.7)^2 + (4.6-4.7)^2]$$

$$\text{MSE} = 0.012$$

For the sepal.width < 2.95 condition, the MSE is 0.012. The procedure is repeated for the other values in this column (3.0 and 3.1; 3.1 and 3.2; etc.) and other columns, and the MSE is calculated in each case. The division with the smallest MSE value is used for node formation. Note that categorized data must first be transformed into integer values. OHE (one hot encoding) isn't absolutely necessary as no weights need to be updated. This means that no column or attribute can have a heavier effect than others, as no $\Delta w_{j,k}$ is calculated. The question within the nodes only needs to be answered with yes or no. Therefore, a transformation of character strings into numerical values is completely sufficient.

Now, let's take a look at the relevant places in the *K4_iris-2.ipynb* source code.

```
# What should be predicted? Save column name in variable.
Col_name = 'sepal.length'

# Here the division into two tables takes place (Input=data and Output=col).
Col = data[col_name]
data = data.drop([col_name], axis = 1)

le = LabelEncoder()
data['species'] = le.fit_transform(data['species'])
```

Listing 4.3 Determining the Target Column and Converting the Data Type

Nothing should surprise you with these lines. If anything is unclear to you, take another look at the preparation of the data using an ANN.

4 Decision Trees

```
# Create four tables from the two tables
train_data, test_data, train_col, test_col = train_test_split(data,
 col, test_size=0.2)

tr = tree.DecisionTreeRegressor()
tr.fit(train_data, train_col)

predicted_col = tr.predict(test_data)

score = metrics.mean_absolute_error(test_col, predicted_col)
print(score)
```

Listing 4.4 Division of the Table, Structure (Training), and Application (Test) of the Tree

The regressor is used here instead of the classifier because the model is supposed to predict a number. The MSE is used to evaluate the prediction. With this AI in the form of a decision tree, the prediction is off by around ±3 mm (0.12"). You can plot the tree again. You'll notice that the tree is much bigger this time.

You can also set the familiar hyperparameters for the regressor. In the following sections, however, we'll focus on other technologies based on decision trees that deliver better results (especially XGBoost). These simple decision trees, which you've learned about in this chapter, are primarily intended to familiarize you with the procedure. If you need an algorithm with an easy-to-understand decision-making process, use decision trees.

Using decision trees, it's particularly easy to understand why you should reconsider the use of some attributes, such as gender, during development. Imagine an AI is developed to select applicants for certain positions, and the gender of the candidates is used in the dataset. In the decision tree, the gender is asked for in a node, and it continues to the left or right accordingly. That would be fatal. But what is unthinkable for an AI in job application processes can be useful for medical purposes. If the AI is supposed to generate information on certain diseases based on the hemogram, it makes sense to use the gender attribute.

In this context, however, I'll point this out once again: just because gender doesn't appear explicitly in the dataset doesn't mean that the AI can't make sexist judgments. If the dataset comes from sexist recruitment processes or career opportunities, the AI can also make sexist predictions based on the correlations. For example, people who have been on parental leave for a longer period of time have poorer chances of promotion. It should be clear to everyone that this is unfair. However, the AI recognizes this correlation in the dataset and rates people who have had a longer parental leave negatively (without moral judgement). However, as it's predominantly women who take longer parental leave, they are discriminated against by the AI, and the injustice is reinforced.

4.1.3 Decision Forests

Why settle for one tree when you can have an entire forest? This method generates several trees with randomly selected attributes (*ensemble learning*). The majority of these trees determine which class our iris belongs to (*bagging*).

AIs with decision forests often lead to better results than AIs with decision trees, but the results are no longer as easy to understand.

4.1.4 Random Forest Classifier

The sklearn module documentation shows that 100 trees are generated for this forest by default. Each tree has a randomly selected number and combination of attributes (columns). Each tree is also trained with a randomly selected number of datasets (rows). Let's look at this again using an example (classification of irises), but here only the relevant parts (*K4_iris-3.ipynb*).

```
tr = ensemble.RandomForestClassifier()
tr.fit(train_data, train_col)

predicted_col = tr.predict(test_data)

score = metrics.accuracy_score(test_col, predicted_col)
print(score)
```

Listing 4.5 Development (Training) and Application (Testing) of the Decision Forest

A forest is created here instead of a tree, and that's all there is to it. You can display the trees individually and see which attributes (column index) and number of data records were used.

```
# 100 trees, 0 to 99
tree.plot_tree(tr.estimators_[0])
```

Listing 4.6 Output of a Tree of the Forest

As 100 trees are used by default, you can select indexes 0 to 99 for output. The documentation contains information on possible hyperparameters, such as the maximum number of trees, maximum depth of branching, and so on.

In the next section, we'll take a look at regression with decision forests. However, we don't want to spend too much time on decision forests; instead, I would like to move you in the direction of XGBoost. This module delivers particularly good results with little effort. However, it's worth looking at the evolution of the development from a simple tree to XGBoost, which will make you more confident in using this technology.

4.1.5 Random Forest Regressor

The procedure is the same as for classification, except that at the end, the mean value of the values on the leaves of all trees is calculated to determine the continuous value. Here, too, only the relevant lines of code for irises will be considered (*K4_iris-4.ipynb*).

```
tr = ensemble.RandomForestRegressor()
tr.fit(train_data, train_col)

predicted_col = tr.predict(test_data)

score = metrics.mean_absolute_error(test_col, predicted_col)
print(score)
```

Listing 4.7 Development (Training) and Application (Testing) of the Decision Forest

As you can see, once again, the tree is replaced by the forest. The decision forests aren't as suitable as decision trees if the structure is supposed to be comprehensible for humans. However, it's possible to view the individual trees. The data also doesn't need to be standardized, as the amount isn't relevant for the questions. There are only yes or no answers.

The forests provide better results, but you have to take multiple trees into account if you want to visualize the result. Technologies based on decision trees are well advanced. If you attach importance to good results and not to visualization and comprehensibility, then you should definitely familiarize yourself with boosting methods, which you'll learn about in the next section.

4.2 Boosting

Boosting methods also belong to the group of ensemble learning methods, as do decision forests. However, not many trees are set up and trained at the same time. Instead, the trees are "planted" one after the other—each time with the aim of compensating for the error or deviation of the predecessor. The idea behind this is that many *weak learners* add up to one *strong learner*. This usually provides better results than decision trees or decision forests.

4.2.1 Gradient Boosting

Gradient boosting is a special boosting method that we want to look at in more detail. The first tree is set up and applied as usual, and the deviation from the target is determined. The aim of the second tree is to determine the deviation of the first tree. This tree therefore doesn't even know the original target value. The result of this second tree is the deviation from the deviation of the first tree. The procedure is continued with

4.2 Boosting

this deviation. This may sound complicated, but if you look at the source code, it should be understandable.

Let's program gradient boosting (*K4_iris-5.ipynb*) ourselves to understand the idea behind it. The aim is to determine the length of the sepal.

```
# Create four tables from the two tables
train_data, test_data, train_col, test_col = train_test_split(data,
 col, test_size=0.2)

# first tree with training data, as usual
tr_1 = tree.DecisionTreeRegressor(max_depth=3, random_state=42)
tr_1.fit(train_data, train_col)
predicted_col_1 = tr_1.predict(train_data)
```
Listing 4.8 Division of the Data and Structure of the First Decision Tree

The four tables are created, and the first decision tree is set up with training data.

```
# form difference between target value and prediction
train_col_2 = train_col - predicted_col_1

# second tree
tr_2 = tree.DecisionTreeRegressor(max_depth=3, random_state=42)
tr_2.fit(train_data, train_col_2)
predicted_col_2 = tr_2.predict(train_data)
```
Listing 4.9 Determination of the Deviation and Structure of the Second Decision Tree

The second tree is built using the training data, but with the difference that the target column is now the difference between the last prediction and the actual target values.

```
# form difference to last prediction
train_col_3 = train_col_2 - predicted_col_2

# third tree
tr_3 = tree.DecisionTreeRegressor(max_depth=3, random_state=42)
tr_3.fit(train_data, train_col_3)
```
Listing 4.10 Determination of the Deviation and Structure of the Third Decision Tree

The third tree is constructed in the same way.

```
# 3 predictions with the test data
predicted_final_1 = tr_1.predict(test_data)
predicted_final_2 = tr_2.predict(test_data)
predicted_final_3 = tr_3.predict(test_data)
```

```
# add up the individual predictions
predicted_final = predicted_final_1 + predicted_final_2 + predicted_final_3

score = metrics.mean_absolute_error(test_col, predicted_final)
print(score)
```

Listing 4.11 Use of the Trees and Determination of the Overall Result

The three trees provide predictions for the test data or deviations, and the results are totaled. Positive and negative individual predictions add up in the final result to an overall prediction close to the target.

GradientBoostingRegressor from the sklearn module applies this procedure (*K4_iris-6a.ipynb*), so you don't need to determine the deviations individually and build the trees accordingly.

```
# Create four tables from the two tables
train_data, test_data, train_col, test_col = train_test_split(data,
  col, test_size=0.2)

gbr = ensemble.GradientBoostingRegressor(max_depth=3, n_estimators=3,
  random_state=42, learning_rate=1.0)

gbr.fit(train_data, train_col)
predicted_col = gbr.predict(test_data)

score = metrics.mean_absolute_error(test_col, predicted_col)
print(score)
```

Listing 4.12 Regression with "GradientBoostingRegressor"

These few lines of source code replace our previous program with three consecutive trees. However, three trees aren't enough for practical use. Let's increase the number to 500 (*K4_iris-6b.ipynb*). You'll notice that the result gets worse (the deviation is greater). Now also set the learning_rate in the constructor to 0.01, which should lead to a noticeable improvement in the result (lower deviation). Using the learning rate, you can scale down the predictions based on deviations. If we were to increase the number of trees in our self-programmed gradient booster to 500, for example, the calculation for determining the total prediction would be as follows:

```
predicted_final = predicted_final_1 + predicted_final_2 * 0.01 +
  predicted_final_3 * 0.01 + predicted_final_4 * 0.01 etc.
```

The influence of each individual weak learner is therefore limited. In real life, you have several trees (default for GradientBoostingRegressor is 100) and a small learning rate (default for `GradientBoostingRegressor` is `0.1`). These parameters may need to be coordinated with each other for each task.

As far as the selection of decision trees and procedures based on them is concerned, we haven't yet reached the end. In fact, the XGBoost library is the crowning glory.

4.2.2 XGBoost Classifier

The XGBoost library is based on gradient boosting methods that you learned about in the previous section and provides a number of strengths:

- Very high-performance and can use several CPU cores
- Algorithms for categorical and missing data
- Used to win many ML competitions

Tianqi Chen and Carlos Guestrin published the results of their research project at the University of Washington in their 2016 paper, "A Scalable Tree Boosting System" (*https://arxiv.org/abs/1603.02754*). The procedure is similar to gradient boosting in that each node candidate of an attribute is assigned a score according to a mathematical procedure. The node candidate with the highest score finally makes it to the node. After a tree has been built up completely from top to bottom, the number of nodes is compared with an adjustable parameter γ from the leaves upwards, and weak branches are removed. This reduces the likelihood of overfitting and results in fewer branches.

> **Installing the XGBoost Module**
> You must install the XGBoost module via **Anaconda Navigator · Environments**.

Using XGBoost, you can achieve impressive results with little effort. Let's take a look at this with the Iris dataset (*K4_iris-7.ipynb*).

```
import pandas as pd
from sklearn.model_selection import train_test_split
from sklearn import metrics
from sklearn.preprocessing import OneHotEncoder
from xgboost import XGBClassifier

path = "../Data/iris.csv"
data = pd.read_csv(path, delimiter=',')
print(data.head())
```

Listing 4.13 Required Modules and the CSV File Loaded

The preceding lines should be familiar to you by now. If anything is unclear, take another look at the previous sections.

4 Decision Trees

```
# species must be predicted
oe = OneHotEncoder()
col = oe.fit_transform(data[['species']])
col = col.toarray()

data = data.drop(['species'], axis = 1)
```
Listing 4.14 Preparing the Data

The target column is OHE-encoded and the two tables known to us are created. A transformation into numerical values is also possible here.

```
# Create four tables from the two tables
train_data, test_data, train_col, test_col = train_test_split(data,
 col, test_size=0.2)

xgb = XGBClassifier()
```
Listing 4.15 Division into Four Tables and Instantiation of "XGBClassifier"

Let's go through the remaining lines of the source code one by one.

This is where the training and structuring of the trees takes place:

```
xgb.fit(train_data, train_col)
```

A column with forecast values is created:

```
predicted_col = xgb.predict(test_data)
```

The prediction column and the target column are compared and evaluated. With these few lines of source code, you can achieve a correct classification rate of over 93%:

```
score = metrics.accuracy_score(test_col, predicted_col)
print(score)
```

Let's now test XGBoost with a larger amount of data (*Winequality-red.csv*). You can find the source code in *K4_winequality-1.ipynb*. We've already used this dataset for the ANN. The aim is to determine the quality of the wine on the basis of chemical properties.

```
import pandas as pd
from sklearn.model_selection import train_test_split
from sklearn import metrics
from sklearn.preprocessing import LabelEncoder
from xgboost import XGBClassifier

path = "../Data/winequality-red.csv"
data = pd.read_csv(path, delimiter=';')
```

```
print(data.head())

le = LabelEncoder()
col = le.fit_transform(data['quality'])
data = data.drop(['quality'], axis = 1)

# Create four tables from the two tables
train_data, test_data, train_col, test_col = train_test_split(data,
 col, test_size=0.2, random_state=42)

xgb = XGBClassifier()

xgb.fit(train_data, train_col)
predicted_col = xgb.predict(test_data)

score = metrics.accuracy_score(test_col, predicted_col)
print(score)
```
Listing 4.16 Complete Source Code for the Classification of Red Wine

This simple setup has a correct classification rate of over 67% (we had achieved 50% with the ANN). But should you really use an AI that "only" achieves this correct classification rate? Now, if we had to guess or roll the dice as to how good the red wine is, we would only have a correct classification rate of 9%. Furthermore, if the quality in a row is rated 8 in the dataset, and the AI predicts 7, this is also an incorrect prediction, even though this value is close to the target value. You could divide the quality values into "poor," "good," and "very good," as we've done with the ANN. This could increase the correct classification rate.

We can also take a different approach and question the classification for this dataset altogether. Why not use regression? Then, the prediction for the quality has a value of 7.3 instead of 8, for example. Just because the quality is stored in integers in the dataset, you don't have to realize the prediction in integers. Compare the following scenarios:

- The AI for classification determines a quality of 8 for a bottle of wine. However, this AI has a correct classification rate of 67%.
- The AI for regression determines a quality of 7.3 for a bottle of wine. This AI has an average absolute deviation of 1.2.

Which of these AIs would you use if you were a wine merchant?

However, let's now set some hyperparameters (*K4_winequality-2.ipynb*) and try to increase the correct classification rate:

```
xgb = XGBClassifier(n_estimators=400, learning_rate=0.08, max_depth=6)
```

These three settings (number of trees, learning rate, and maximum depth of branching) increase the correct classification rate by 2%. Information on other hyperparameters can be found at *https://xgboost.readthedocs.io/en/stable/parameter.html*.

An ML solution for the *adult.csv* file should illustrate the performance once again. You'll also see how much simpler this solution is compared to the ANN variant (*K4_adult-1.ipynb*).

```
import pandas as pd
from sklearn.model_selection import train_test_split
from sklearn import metrics
from sklearn.preprocessing import LabelEncoder
from xgboost import XGBClassifier

path = "../Data/adult.csv"
data = pd.read_csv(path, delimiter=';', header=None)
print(data.head())
```

Listing 4.17 Loading Modules and the CSV File

The CSV file has no column headers (header=None). Missing values are marked with a question mark in this file. You can leave them as they are because XGBoost is able to handle missing values.

```
# convert these columns into numerical values 0...n
conv_num = [1, 3, 5, 6, 7]
data[conv_num] = data[conv_num].astype('category')
data[conv_num] = data[conv_num].apply(lambda x: x.cat.codes)

col = data[14]
```

Listing 4.18 Converting Data Types

As there are no column headings, the columns are accessed using indexes. Categorical data gets converted.

```
data.drop([8,9,13,14], axis = 1, inplace=True)
```

Listing 4.19 Removing Questionable Columns and the Target Column

Columns on ethnic origin and others will be removed. The same applies to the target column whose values will be predicted.

```
col = pd.get_dummies(col)

# Create four tables from the two tables
train_data, test_data, train_col, test_col = train_test_split(data,
```

```
col, test_size=0.2, random_state=42)

xgb = XGBClassifier()

xgb.fit(train_data, train_col)
predicted_col = xgb.predict(test_data)

score = metrics.accuracy_score(test_col, predicted_col)
print(score)
```

Listing 4.20 Dividing the Data, Followed by Training and Testing

With these default settings, we get a correct classification rate of over 87%. For the variant with the ANN, we achieved 83%, but only after standardizing the data and taking the correlations into account. None of this is necessary here, and the source code is clearer than with the ANN variant.

4.2.3 Automatic Hyperparameter Setting Using GridSearchCV

You've certainly noticed how time-consuming it is to optimize the hyperparameters. The *K4_adult-2.ipynb* program introduces you to the GridSearchCV library, which allows you to automatically test a predefined set of hyperparameters. At the end, you'll obtain the best result and the best hyperparameter settings. To avoid overfitting, we use *cross validation*, which will be explained using an example. You have a total of 100 rows of data, which are divided into groups of 20. With each iteration, a different group of 20 rows is always taken; the AI is trained with the remaining 80 rows and then evaluated using the withheld group of 20 rows. Each group is used once for evaluation. The final result is calculated from the averaged results of the evaluation groups. This type of cross-validation is called *k-fold cross-validation*. The result can then be checked with a test dataset.

Iteration	1	2	3	4	5
1st Division	Validation	Trainings	Trainings	Trainings	Trainings
2nd Division	Trainings	Validation	Trainings	Trainings	Trainings
3rd Division	Trainings	Trainings	Validation	Trainings	Trainings
4th Division	Trainings	Trainings	Trainings	Validation	Trainings
5th Division	Trainings	Trainings	Trainings	Trainings	Validation

Table 4.5 k-Fold Cross-Validation with k = 5

Let's take a look at the relevant sections in the source code.

4 Decision Trees

```
# Create four tables from the two tables
train_data, test_data, train_col, test_col = train_test_split(data,
 col, test_size=0.2, random_state=42)

xgb = XGBClassifier(seed=42)
```

Listing 4.21 Dividing the Table and Instantiating the "XGBClassifier"

The XGBClassifier is created in such a way that the procedure for creating the trees is reproducible. This ensures that it wasn't chance that led to an improvement, but the respective setting of the hyperparameters.

Dictionary Data Type

You can save key-value pairs of different data types in a dictionary, as shown in these examples:

```
tel_no = { 'john':555123, 'jane':5553234 }
my_dict = { 1: 'Newark', 2: 'Houston' }
```

"Range" Function

You can use the range function to generate a sequence of numbers. To do this, you need to enter the start and stop values as well as the increment. The start value is inclusive; the stop value is exclusive.

Forms the sequence 2, 3, 4 ... 9:

range (2, 10, 1)

Forms the sequence 0, 3, 6, 9:

range (0, 10, 3)

Forms the sequence 5, 4, 3, 2, 1:

```
range (5, 0, -1)
parameters = {
    'max_depth': range (2, 10, 1),
    'n_estimators': range(60, 200, 20),
    'learning_rate': [0.008, 0.01, 0.05, 0.1]
}
```

Listing 4.22 Creating Parameters for the Test

The parameters dictionary contains three keys: max_depth, n_estimators, and learning_rate. The parameters to be tested for max_depth and n_estimators were specified using the range function. This saves some typing work. Using learning_rate, the values to be tested are explicitly specified in the list. Let's go through the rest of the source code line by line.

The grid_search object of type GridSearchCV is created. In the constructor, we specify the classifier, the dictionary with the parameters, and the number of jobs to be run in parallel. Depending on the processor architecture, you can enter a positive number (e.g., 8 or 16). Using -1, the maximum possible number is determined by the library:

```
grid_search = GridSearchCV(estimator=xgb, param_grid=parameters, n_jobs=-1)
```

This is where the training and construction of the forest takes place. This may take a while, depending on your computer equipment. Many possible combinations need to be tested:

```
grid_search.fit(train_data, train_col)
```

This allows you to output the best value that has been obtained. In this case, it's again over 87%:

```
print(grid_search.best_score_)
```

The hyperparameter settings to achieve the best result are displayed. Here is the output:

```
print(grid_search.best_params_)
```

Determining good hyperparameters using the library relieves you of the tedious procedure of setting parameters, training, testing, setting parameters again, training again, and so on:

```
{'learning_rate': 0.1, 'max_depth': 4, 'n_estimators': 180}
```

4.3 XGBoost Regressor

For the sake of completeness, we also want to use XGBoost to determine a concrete numerical value using regressors for irises (*K4_iris-8.ipynb*).

```
import pandas as pd
from sklearn.model_selection import train_test_split
from sklearn import metrics
from xgboost import XGBRegressor

path = "../Data/iris.csv"
data = pd.read_csv(path, delimiter=',')
print(data.head())
```

4 Decision Trees

```
# What should be predicted? Save column name in variable.
col_name = 'sepal.length'

# convert these columns into numerical values 0...n
conv_num = ['species']
data[conv_num] = data[conv_num].astype('category')
data[conv_num] = data[conv_num].apply(lambda x: x.cat.codes)


# Here the division into two tables takes place (Input=data and Output=col).
col = data['sepal.length']
data = data.drop(['sepal.length'], axis = 1)
# Create four tables from the two tables
train_data, test_data, train_col, test_col = train_test_split(data,
 col, test_size=0.2)

xgb = XGBRegressor()

xgb.fit(train_data, train_col)

predicted_col = xgb.predict(test_data)

score = metrics.mean_absolute_error(test_col, predicted_col)
print(score)
```

Listing 4.23 Complete Source Code for Regression Using XGBoost

Nothing in these lines of source code should seem unfamiliar to you. This AI can predict the length of the sepal with an accuracy of approximately 2.4 mm (0.09"). The ANN variant was 3 mm (0.12"), but we had to pay attention to the standardization of the data.

4.4 Deployment

In the next example, we'll look at how you can first train and test the AI and then apply it to new, unknown data. The *K4_iris-8a.ipynb* program hardly differs from the preceding example. Only one code cell at the end is new:

```
# save the model
xgb.save_model("xgb.json")
```

Here, the model (the forest) is saved in a JavaScript Object Notation (JSON) file.

> **JSON File Format**
>
> In a JSON file, you can save related data in a hierarchical structure. JSON files are often used for the exchange between different programs. You can open the *xgb.json* file in the browser and view the contents. There, you'll find information on the attributes, hyperparameters, and so on.

We use the *K4_iris-8b.ipynb* program to open the JSON file, load the AI, and apply it to a new data set.

```
import pandas as pd
from sklearn.model_selection import train_test_split
from sklearn import metrics
from xgboost import XGBRegressor

path = "../Data/iris_test2.csv"
data = pd.read_csv(path, delimiter=',')
print(data.head())

conv_num = ['species']
data[conv_num] = data[conv_num].astype('category')
data[conv_num] = data[conv_num].apply(lambda x: x.cat.codes)

xgb = XGBRegressor()
xgb.load_model("xgb.json")

predicted_col = xgb.predict(data)
print(predicted_col)
```

Listing 4.24 Using a Pretrained XGB Regressor Model

The length of the sepal is given as 5.1 cm (2"). The source code is also very clear here.

For table data, you should consider using XGBoost, as preparing the data and building the model is relatively simple. However, there are other boosting libraries available such as AdaBoost or CatBoost that you may want to take a look at. The use of such libraries is always quite similar. Using `GridSearchCV`, you can now also automatically optimize the hyperparameters and get really good results with little effort.

4.5 Decision Trees Using Orange

Multiple tools are available that allow you to program decision trees graphically, and KNIME is one of them. But there are also leaner solutions. We'll use the open-source

4 Decision Trees

tool called Orange (*https://orangedatamining.com*) to train and test a simple decision tree without any source code. You can easily install Orange via Anaconda.

> **Installing Orange**
> On the Orange homepage, you'll also find installation files for Windows, macOS, and Linux if you don't want to install Orange via Anaconda.

Launch Anaconda, switch to the **Home** view, and click on **Channels**. Use the **Add** button to enter the "conda-forge" channel, and confirm this by pressing `Enter` (see Figure 4.2). This channel gives you access to many open-source tools.

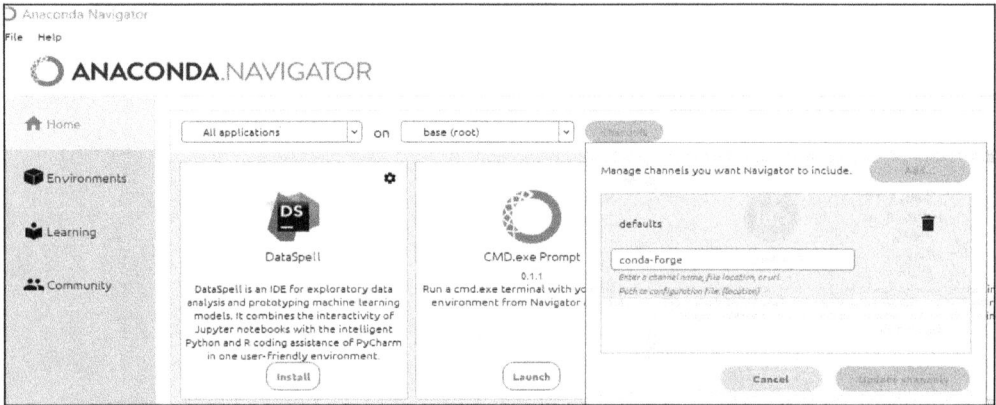

Figure 4.2 Activation of the conda-forge Channel

Switch to the **Environments** view, and install the **orange3** package (see Figure 4.3).

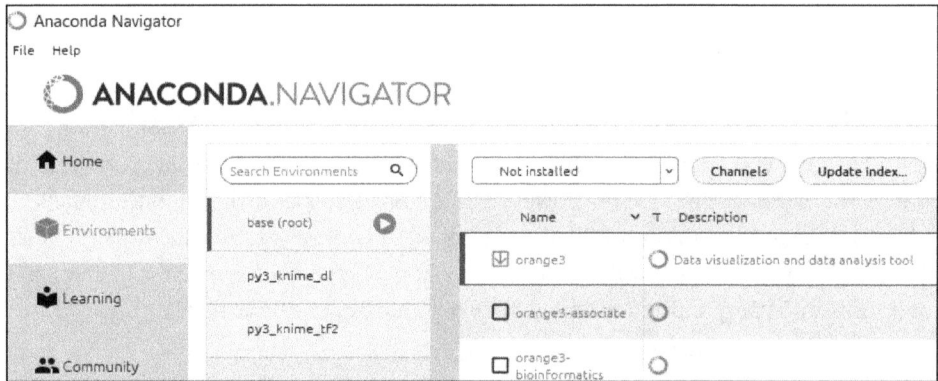

Figure 4.3 Installation of the orange3 Package

You can now start Orange via the terminal. Click on the **Play** button next to **base (root)**. Then, select **Open Terminal** (see Figure 4.4).

4.5 Decision Trees Using Orange

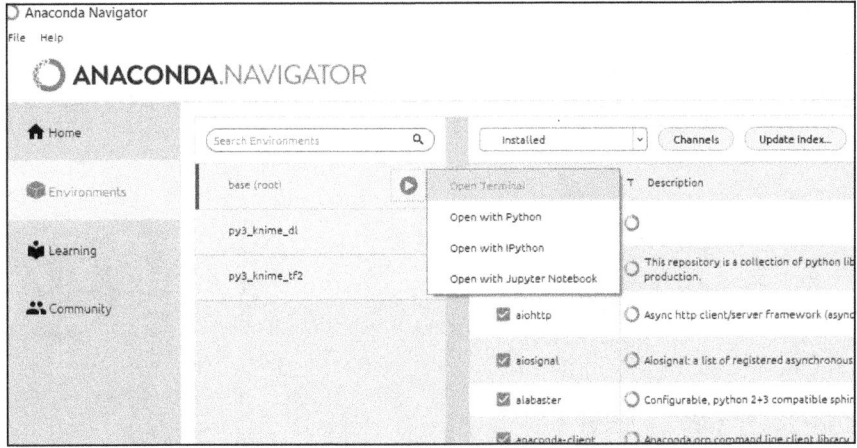

Figure 4.4 Opening the Terminal

Enter the "orange-canvas" command in the terminal, and confirm with ⟨Enter⟩. On the **Welcome** screen, click on **New**. Now let's create a decision tree and look at the results. From the **Data** section, drag the **CSV File Import** widget into the workspace, and select the *iris.csv* file. Then, select the **Select Columns** widget from the **Transform** section, and link the elements together. If you now double-click on **Select Columns**, you can set which column is the target column and must be predicted (see Figure 4.5).

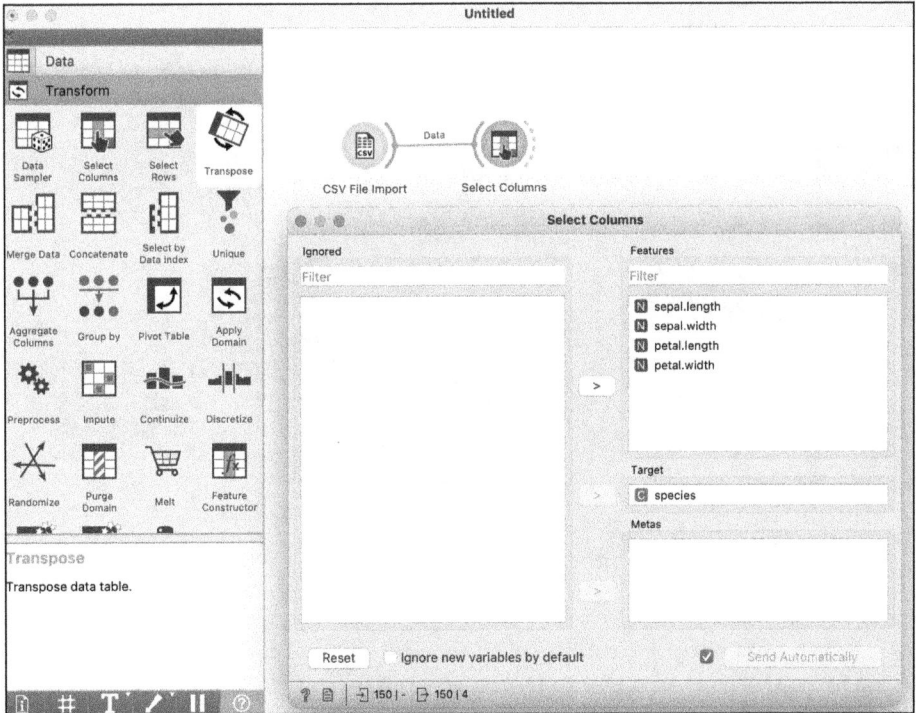

Figure 4.5 Configure the Columns Correctly

113

4 Decision Trees

Use the **Data Sampler** widget to select 80% of the data, connect the output to a **Tree**, and then connect the tree in turn to a **Tree Viewer** (see Figure 4.6).

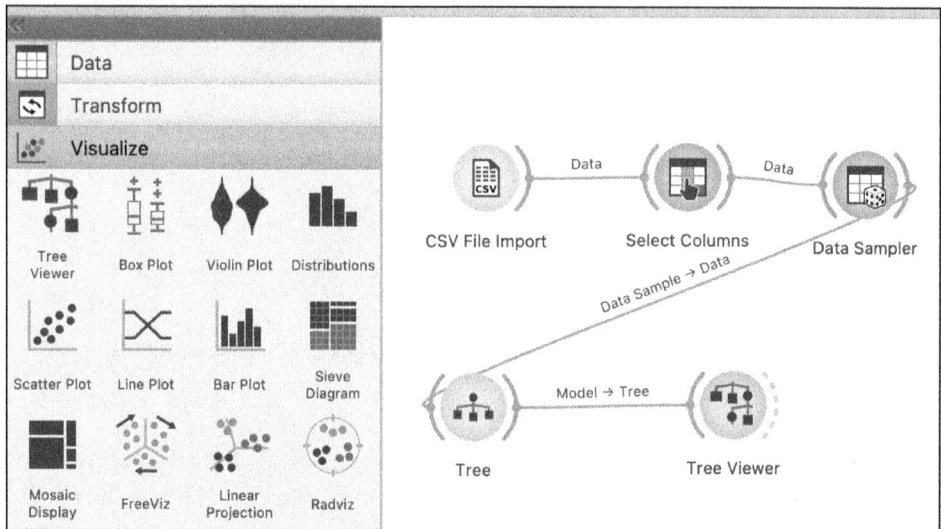

Figure 4.6 Structure of a Tree

Double-click on the **Tree Viewer** to see the decision tree (see Figure 4.7).

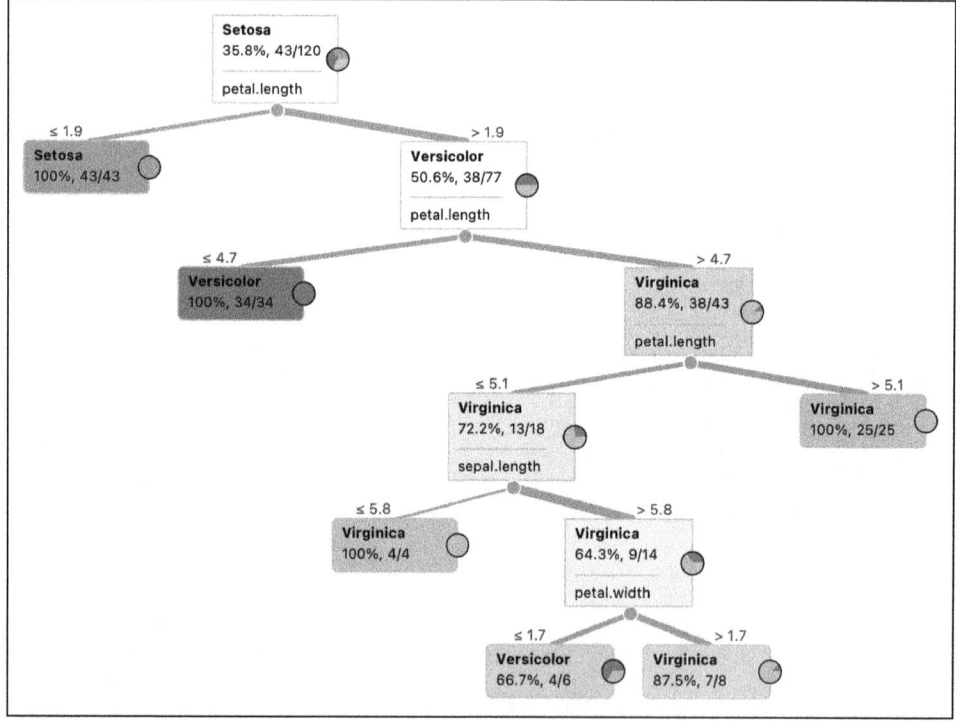

Figure 4.7 Tree Viewer

114

You can now add the **Test and Score** module and connect it to **Tree** and **Data Sampler** to test the generated tree using the test data. When connecting the **Tree** to the **Data Sampler**, make sure that you access the 20% of the data (**Remaining Data**).

The results can be viewed by double-clicking on **Test and Score**. The finished *K4_iris.ows* program is made available to you in the corresponding folder (see Figure 4.8).

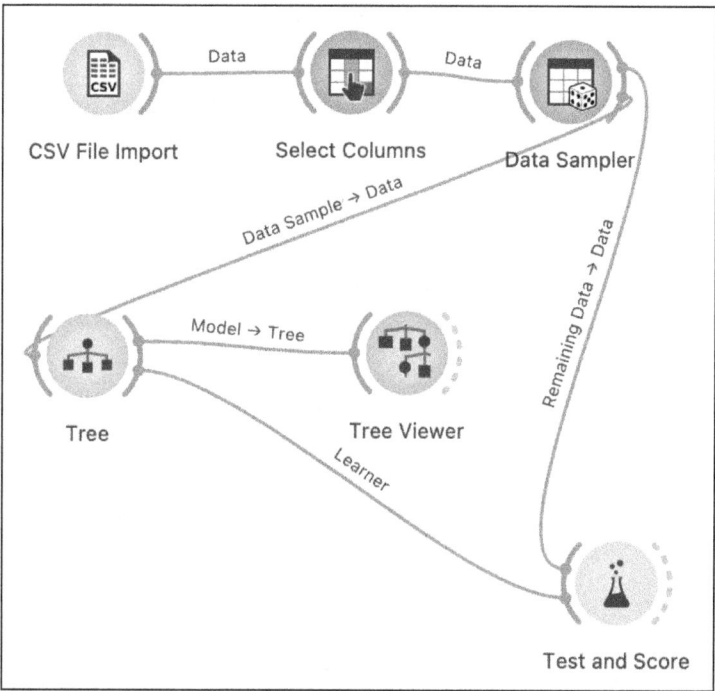

Figure 4.8 Finished Program

You can also create programs for other CSV files for practice. Not only is the Orange tool easy to install, but it can also be used to quickly create AIs using visual programming. The resulting tree gives you or other interested parties a complete insight into the inner workings of the AI. All decisions are comprehensible. We'll take a closer look at the topic of visual programming using KNIME in Chapter 11.

4.6 Exercises

4.6.1 Exercise 1: XGBoost for Classification

Program an AI using XGBoost to classify mushrooms into "edible" and "poisonous" categories. Use the *mushrooms.csv* file. Information on the attributes can be found at *www.kaggle.com/datasets/uciml/mushroom-classification*. You can also search for information on the data on other pages on the internet. Follow these steps:

4 Decision Trees

1. Output the first rows. Pay attention to the correct separator.
2. After splitting into two tables, you can convert all columns into numerical values in a single step:

   ```
   data = data.astype('category')
   data = data.apply(lambda x: x.cat.codes)
   ```

   It's not necessary to determine correlations when you use XGBoost.
3. Create, train, and test the model.
4. Try to improve the correct classification rate by adjusting the hyperparameters.

4.6.2 Exercise 2: XGBoost for Regression

Program an AI with ANN to determine the price of diamonds. Use the *diamonds.csv* file. Information on the attributes can be found at *www.kaggle.com/datasets/shivam2503/diamonds*, for example. Follow these steps:

1. Output the first rows. Pay attention to the correct separator.
2. Divide the data into two tables. In addition, remove the *Unnamed: 0* column from the data.
3. Convert nonnumeric values to numeric values.
4. Create, train, and test the model.
5. Try to minimize the deviation by adjusting the hyperparameters.

4.6.3 Exercise 3: Automatic Hyperparameter Optimization

Extend the preceding exercise so that `GridSearchCV` determines the best combination from a certain set of parameters.

Chapter 5
Convolutional Layers and Images

You can already develop AIs for table-based data. Now it's time to work with more complex data, such as images.

> **What This Chapter Is About**
> - Decision support for the use of XGBoost and artificial neural networks (ANNs) or methods based on them
> - Simple image classification with ANNs
> - Image classification with a convolutional neural network (CNN)
> - Use of pretrained networks

XGBoost is a software library that allows you to develop high-performance AI applications for table data relatively easily. Why should you bother with ANNs? As a matter of fact, many ML competitions have been won with XGBoost or other boosting libraries. To better assess which technology should be used for a particular dataset, it's advisable to follow current developments in the AI field. One website you should definitely check out is Kaggle: *www.kaggle.com/*.

> **Kaggle**
>
> The Kaggle platform (*www.kaggle.com/*) was founded in 2010 by economist Anthony Goldbloom. On this website, you'll find many freely available datasets. Users can download this data or program it directly in a *kernel* (similar to Google Colaboratory) on the site and make these applications available to other interested parties. You can view, copy, modify, and run the kernels of other users.
>
> The first competitions were held shortly after the platform was founded: datasets were made available and users were asked to program an AI application for them. There are now many companies that provide their own data and pay out bonuses to the first-placed development teams (or even individuals). This gives companies access to AI models that they can use for their projects at low cost.
>
> The platform was acquired by Google in 2017 and is very popular with more than 12 million users a year. You'll also find a large number of free courses on data analysis and AI on the platform.

Which approach should you use for table data—ANNs or XGBoost? Two Intel researchers, Ravid Shwartz-Ziv and Amitai Armon, addressed this question in "Tabulator Data: Deep Learning Is Not All You Need" (*https://arxiv.org/abs/2106.03253*). They came to the conclusion that with table data, XGBoost often achieves better results with less effort. However, the best results were obtained when ANN and XGBoost were combined in one dataset. But this combination of techniques leads to many hyperparameters that need to be optimized.

For example, you'll develop an AI based on ANNs or XGBoost for the Iris dataset. The predictions of these models are averaged and form the overall result. However, you have to optimize and test the hyperparameters of both models, which takes a lot of time.

But let's get back to the question of which specific approach you should take. As you're still at the beginning of your AI career, I recommend the following procedure:

- Use XGBoost for table data.
- For more complex data, such as images, you should use ANN and technologies based on it.

You'll also learn about options and modules that test and evaluate different approaches and hyperparameters for a given data set. You can then use the model with the best result for your application. We'll also take a look at how you can use pretrained models to find a good solution very quickly. Finally, once you've developed your knowledge in this area, you can view publications by other users and adapt them for your application or combine several models. However, it's important that you understand the basic relationships to be able to optimize your AI applications in a goal-oriented manner.

5.1 Simple Image Classification

There is a large, freely available database with images of handwritten digits called the Modified National Institute of Standards and Technology (MNIST) database. This database contains an extensive collection of handwritten digits from American high school students and public service employees. The data is processed in such a way that the digits are in the center of the image. The images are all 28 × 28 pixels in size and in black and white (monochrome). Thanks to the standardized, simple images, this database is well suited as an introduction to the technique of image classification. The dataset is also divided into test and training data, so that we have 60,000 images available for training and 10,000 images for testing. The MNIST database has long served as a benchmark in the development and evaluation of algorithms in the field of machine learning (ML).

Today, other more complex datasets are used to evaluate algorithms, but MNIST's simple structure makes access to image-recognition techniques particularly easy for beginners. We'll therefore deal with this dataset in our first program. The goal is to develop an AI that recognizes handwritten digits and assigns them correctly. The MNIST database is conveniently located in the TensorFlow module or in the Keras submodule, so we don't need to download and prepare files from the internet ourselves.

> **Grayscale**
> Monochrome images are often encoded in grayscale (grayscale). The value 0 means black, and 255 represents white. The values in between describe the gradations.

Let's take a look at the first simple example of image classification using the *K5_mnist_digits-1.ipynb* program.

```
import pandas as pd
import tensorflow as tf
from sklearn.model_selection import train_test_split
import matplotlib.pyplot as plt

(train_data, train_col), (test_data, test_col) = tf.keras.datasets.mnist.load_data()
```

Listing 5.1 Loading Required Modules and Data

The required modules are imported as usual. The images are also loaded using tf.keras.datasets.mnist.load_data(). The method call returns two tuples. The first tuple stores the image data (28 × 28 pixels) and the corresponding number in the image (e.g., 5). Accordingly, the test data (images and digits) are stored in the second tuple. Do you understand the structure of the dataset? The training and test data consist of images and digits, with the digits representing the target column. The images are supposed to be fed to the AI black box, which then ideally outputs the correct digit.

Let's output an image to illustrate the structure.

```
index = 111
plt.imshow(train_data[index], cmap='Greys')
plt.xlabel(train_col[index])
plt.show()
```

Listing 5.2 Output of Image and Digit with Index 111

Both the images and the digits are stored in arrays. The element with index 111 is to be output from each of these arrays with 60,000 entries. We use plt.imshow to output the train_data[111] image. Below the image, we use plt.xlabel(train_col[111]) to output the digit on the image. If you want to output a different image, simply change the index. By saving the index in a variable (here, index), you can specify at a central point

5 Convolutional Layers and Images

which image and which image information you want to output. For 60,000 images, you can set values between 0 and 59999.

If you want to prepare images for AIs yourself, you need to take them, edit them (here in grayscale), and cut them (e.g., 28 × 28 pixels). You must also enter the information about what is contained in the picture. This process is referred to as *labeling*. Fortunately, we don't need to implement this ourselves, as the publicly accessible data we use here is already labeled. In Figure 5.1, the value 3 (label) is displayed below the image for information purposes.

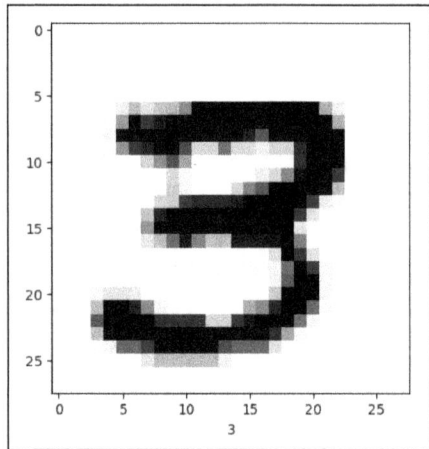

Figure 5.1 The Image with Index 111 Representing the Number 3

```
# Build ANN
model = tf.keras.Sequential([
    tf.keras.layers.Flatten(input_shape=(28, 28)),
    tf.keras.layers.Dense(256, activation=tf.nn.relu),
    tf.keras.layers.Dense(10, activation=tf.nn.softmax)
])
# Configuration of the training process
model.compile(optimizer='adam', loss='sparse_categorical_crossentropy',
 metrics=['accuracy'])
```

Listing 5.3 Building the ANN and Configuring the Training Process

Only the Flatten layer is new here. This layer does nothing other than line up the pixels by stringing all the lines together to form a single line (see Figure 5.2).

The result is an input layer with 784 (28 × 28) nodes. This call will provide you with information on the structure of the ANN:

```
model.summary()
```

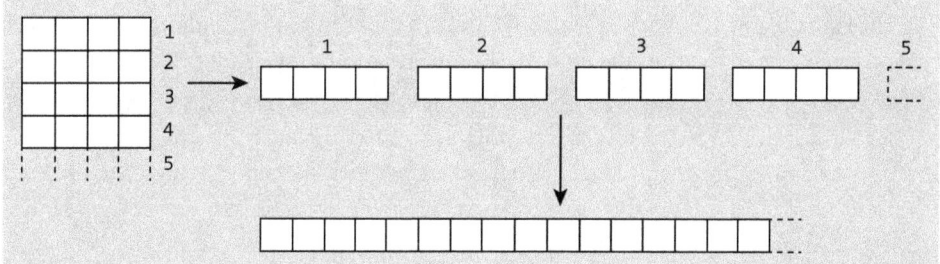

Figure 5.2 All Lines Strung Together

You can see the output in Figure 5.3.

```
Model: "sequential_1"
_____
Layer (type)                 Output Shape              Param #
=================================================================
flatten_1 (Flatten)          (None, 784)               0

dense_2 (Dense)              (None, 256)               200960

dense_3 (Dense)              (None, 10)                2570

=================================================================
Total params: 203,530
Trainable params: 203,530
Non-trainable params: 0
_____
```

Figure 5.3 Information on the Structure of the ANN

The layer named **flatten_1** has 784 nodes, but no parameters. These are simply passed through from the input layer. The next layer contains 256 nodes. The parameter **200960** is displayed. How does this value come about? Each node of one layer is connected to all nodes of the previous layer, whereby the connections are weighted. If you multiply the number of nodes of the input layer, 784, by the number of nodes of the first hidden layer, 256, you get 200,704 as a result, so it can't be the number of weights alone, as we have a difference of 256. The reason can be found in the structure of the ANN with Keras. As you've already learned, the individual connections are multiplied by the weights at the input of each node and added up (except for input layers). Keras and other libraries have extended the procedure and add a number to each node (*bias*, positive or negative) to make better use of the activation function. In addition to the weights, there are also summands (256, corresponding to the number of hidden layers), which together represent the parameters (200,960) of the first hidden layer.

> **Bias**
> Bias is the total of the nodes of an ANN, which is determined and optimized during the training process. However, this designation is also used for negative biases of AIs. This

is the case when an AI model is trained with data that contains an undesirable correlation, for example, and thus the model's statement becomes sexist or racist.

The addition of one summand to each ANN will be illustrated by means of an example: For values above 5, the sigmoid function always returns a value of approximately 1 (see Figure 5.4). If we add a bias of 5 to x, we have a better distinction for values around 5 (see Figure 5.5). The bias values are updated in many ANN libraries using the gradient method, similar to the weights.

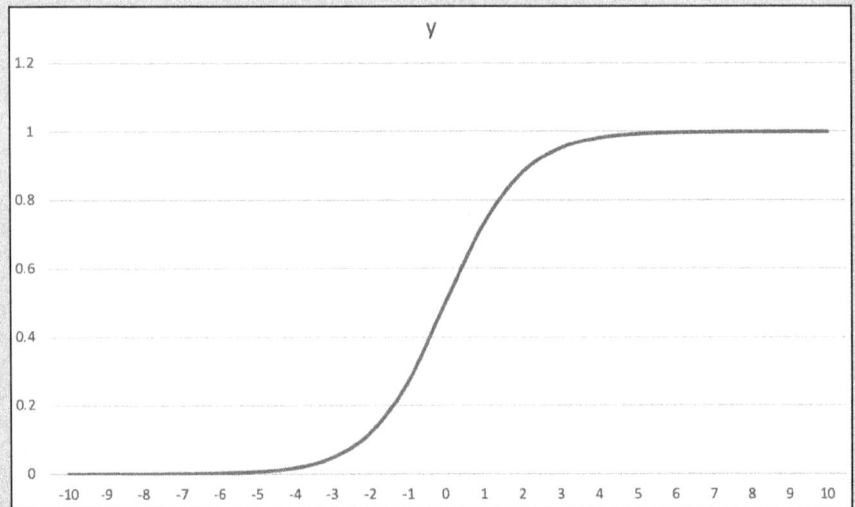

Figure 5.4 Sigmoid Function

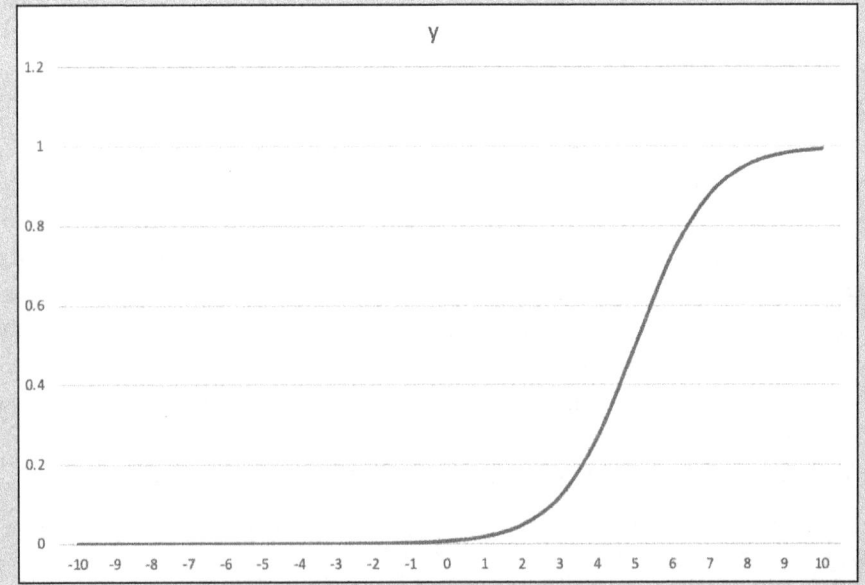

Figure 5.5 Sigmoid Function with a Bias of 5

With this new information, we can better interpret the values for parameters. The parameters are calculated here as follows:

$$Parameter = (number\ of\ inputs + 1) \times number\ of\ outputs$$

We've learned something new again. Let's take a look at the rest of the source code.

```
# 7 runs
model.fit(train_data, train_col, epochs=7)

test_loss, test_acc = model.evaluate(test_data, test_col)
print('Test accuracy:', test_acc)
```

Listing 5.4 Training and Testing Process

In contrast to the classification of table data, only the pixels were lined up, and, with this simple trick, we get a correct classification rate of 95%.

5.2 Hyperparameter Optimization Using Early Stopping and KerasTuner

For decision trees and forests, we've become acquainted with `GridSearchCV`, a module which allows us to automatically set hyperparameters. Let's take a look at the corresponding options for ANN. Let's start with the file named *K5_mnist_digits-2.ipynb* and again only look at the relevant parts.

```
cb_early = tf.keras.callbacks.EarlyStopping(monitor='val_loss', patience=3)

model.fit(train_data, train_col, epochs=100, validation_data=(test_data,
 test_col),callbacks=[cb_early])
```

Listing 5.5 Training Process with Callback Function

That's it for the changes. The `model.fit` method is called with additional parameters. We provide the test data so that tests are carried out with this data after each epoch. The `callbacks=[cb_early]` parameter refers to a variable that is defined in the line above and points to a special method: `tf.keras.callbacks.EarlyStopping`. What is exactly happening here? We've entered an epoch number of 100, but if there is no improvement for the `val_loss` output (result of the loss function for test data) for 3 epochs (an improvement in this case is a decreasing value), the process will be stopped. This allows us to determine the number of epochs without overfitting. In this example, the training procedure stops after 9 epochs. This is followed by an analysis with somewhat more complex data.

5 Convolutional Layers and Images

The Zalando company provides images of items of clothing for teaching and learning purposes, which are used in the *K5_mnist_fashion-1.ipynb* program.

```
import pandas as pd
import tensorflow as tf
from sklearn.model_selection import train_test_split
import matplotlib.pyplot as plt

(train_data, train_col), (test_data, test_col) = 
 tf.keras.datasets.fashion_mnist.load_data()

class_names = ['T-shirt/top', 'Trouser', 'Pullover', 'Dress', 'Coat',
 'Sandal', 'Shirt', 'Sneaker', 'Bag', 'Ankle boot']

index = 111
plt.imshow(train_data[index], cmap='binary')
plt.xlabel(class_names[train_col[index]])
plt.show()
```

Listing 5.6 Loading the Modules and Data and Outputting an Image

The labels are coded as numbers: 0 stands for T-shirt/top, 5 for sandals, and so on. We therefore define the class_names list for this assignment. With train_col[index], we get the number for the garment back. If we use this number as an index for class_names, we can access the character string in the list at the corresponding position. This nested call ensures that we don't need to struggle with numbers during output, but we can draw on specific descriptions.

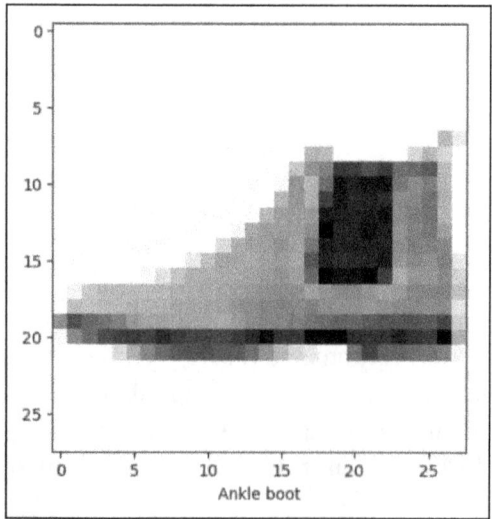

Figure 5.6 An Item of Clothing from the Training Data and Its Designation

5.2 Hyperparameter Optimization Using Early Stopping and KerasTuner

The remaining lines of source code can be used unchanged.

```
cb_early = tf.keras.callbacks.EarlyStopping(monitor='val_loss', patience=3)

model.fit(train_data, train_col, epochs=100, validation_data=(
 test_data, test_col),callbacks=[cb_early])
```

Listing 5.7 Loading the Modules and Data

After 10 epochs, the training procedure is completed with a correct classification rate of approximately 82%.

The KerasTuner module provides options for automatically testing additional hyperparameters. In the next example (*K5_mnist_fashion-2.ipynb*), we want to determine the best combination from a given set of values for the number of nodes and the learning rate.

> **Install the KerasTuner Module**
>
> You must install the KerasTuner module via Anaconda.

Again, we only look at the relevant lines of code. At this point, we need to define our first function.

> **Functions in Python**
>
> Here's a concrete example: You're supposed to define a function that can be used to form the square of a number. The result could look as follows:
>
> ```
> def myFunction1(param):
>     result = param*param
>     return result
> ```
>
> The name of the function is `myFunction1`. A parameter is passed to this function. What the function is supposed to do is described in the following indented lines. The number passed as a parameter is multiplied by itself, and the result is returned:
>
> ```
> res = myFunction1(3)
> print(res)
> ```
>
> You can call this function with different numbers and always get the square back as the result, which is stored in the `res` variable.
>
> You can also pass multiple (or no) parameters to a function and name the parameters as you wish. The name must not begin with a number or contain spaces. Only use letters from the English alphabet.
>
> ```
> def myFunction2(param1, param2):
>     result = param1*param2
>     return result
> ```

5 Convolutional Layers and Images

```
res = myFunction2(2,4)
print(res)
```

Listing 5.8 Definition and Call of a Function with One Return Value

In this case, you get back the product of the two numbers. If you want to return more than one value, you can use a tuple as the return value, for example.

```
def myFunction3(param1, param2):
    sum = param1+param2
    product = param1*param2
    return (sum, product)

(res1, res2) = myFunction3(2,4)
print(res1)
print(res2)
```

Listing 5.9 Definition and Call of a Function with Many Return Values

The sum is stored in res1, and the product is stored in res2.

We've learned something new once again. Now we can look at the source code (make sure to import the KerasTuner module).

```
def build_model(hp):
    model = tf.keras.Sequential()
    model.add(tf.keras.layers.Flatten(input_shape=(28, 28)))
    model.add(tf.keras.layers.Dense(
      hp.Choice('units', values=[100, 300, 500]), activation='relu'))
    model.add(tf.keras.layers.Dense(10, activation='softmax'))
    learning_rate = hp.Choice('learning_rate', values=[0.01, 0.001, 0.0001])

model.compile(optimizer=tf.keras.optimizers.Adam(learning_rate=learning_rate),
loss='sparse_categorical_crossentropy', metrics=['accuracy'])
    return model
```

Listing 5.10 ANN Model Built in a Function

The build_model function is called using the hp parameter. An ANN is built in this function. Pay attention to the places where the hidden layer is built and the learning rate is defined. The hp parameter makes the Choice method available, and the places with Choice are replaced one after the other with the values in the values list. Different values for the number of nodes in the hidden layer (100, 300, 500) and for the learning rate (0.01, 0.001, 0.0001) are therefore used one after the other.

5.2 Hyperparameter Optimization Using Early Stopping and KerasTuner

```
tuner = keras_tuner.GridSearch(
   build_model,
   objective='val_accuracy',
   project_name='mnistFashion',
   overwrite=True)
```

Listing 5.11 Instantiating "GridSearch"

The `tuner` variable is an object for a search procedure. The first parameter in the constructor is the name of our function, followed by the criterion for the evaluation—in this case, the correct classification rate with the test data. A folder with the project name is created, and the intermediate values are saved in JavaScript Object Notation (JSON) files. If you execute the cell again, the folder contents will be overwritten.

```
cb_early = tf.keras.callbacks.EarlyStopping(monitor='val_loss', patience=3)
tuner.search(train_data, train_col, epochs=100, validation_data=(
 test_data, test_col), callbacks=[cb_early])
```

Listing 5.12 Training Process with Various Parameters

Calling the `tuner.search` method is similar to calling `model.fit`, except that different parameters are tested one after the other. This can take a while (up to an hour depending on the computer equipment).

```
best_params_list = tuner.get_best_hyperparameters()
best_params = best_params_list[0]
print(best_params.get('units'))
print(best_params.get('learning_rate'))
```

Listing 5.13 Determining the Best Hyperparameters

We save a list of the best parameters in the `best_params_list` variable. The first element in the list is also the best, so we save it in the `best_params` variable. The values determined are displayed in the two lines below it.

```
model = tuner.hypermodel.build(best_params)
model.fit(train_data, train_col, epochs=100, validation_data=
 (test_data, test_col),callbacks=[cb_early])
```

Listing 5.14 Instantiating and Training a New Model with the Best Hyperparameters

It would also be possible to use the best model straight away, but here we build a new ANN with the determined parameters and then train and test the ANN model. The correct classification rate increases to approximately 88% with the automatic hyperparameter setting. With a little programming knowledge, you can have KerasTuner set many more parameters. You can find some examples on the Keras website (*https://keras.io/keras_tuner*).

5.3 Convolutional Neural Network

A convolutional neural network (CNN) is a special multilayer network consisting of a detection part and an identification part (see Figure 5.7).

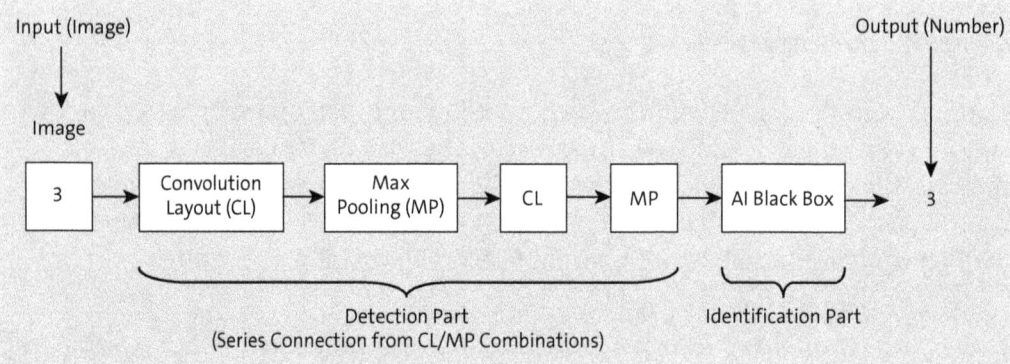

Figure 5.7 CNN with Detection and Identification Parts

Let's look at the detection part first. An image consists of many individual pixel values, and each pixel can have values between 0 and 255. A filter (e.g., 3 × 3 matrix) systematically scans the entire image and offsets the values of the filter and the image against each other (this is known as convolution). The result is written to a new layer, the *convolutional layer*. Depending on how the filter matrix is structured, certain features such as lines, edges, points, corners, and so on can be extracted. Typically, the image is convolved with several filters in succession to extract multiple features and use them as input data for the AI black box. For an AI model, it's not the sweet smile of a child in the picture that is important for classification, but only the features determined from the picture.

In the following, we'll take a closer look at the methods used to determine the characteristics. But here too, you don't have to carry out any calculations yourself later when developing your own AI models. However, once you've understood the concept, you can optimize the hyperparameters in a targeted manner. There are some new hyperparameters in addition to those you've already become familiar with in ANN models, such as size, type, and number of filters.

In Figure 5.8, you can see Bambam the family dog. This image demonstrates what the result of filtering can look like. The goal is to extract features from the image. These characteristics will then be used in the subsequent processes. Convolution reduces the image information and restricts the content to corners, edges, circles, and so on. This reduces the input data for the AI model to pixel patterns.

5.3 Convolutional Neural Network

Figure 5.8 Bambam before Applying the Filter

Special filters will be used for the pattern recognition. Filters with this matrix structure are referred to as kernels. The first kernel we want to look at is used for edge recognition:

$$Kernel\_1 = \begin{bmatrix} -1 & -1 & -1 \\ -1 & 8 & -1 \\ -1 & -1 & -1 \end{bmatrix}$$

The values in the matrix are empirical values that are also used in image-processing programs. Another kernel, which sharpens the image, has the following structure:

$$Kernel\_2 = \begin{bmatrix} 0 & -1 & 0 \\ -1 & 5 & -1 \\ 0 & -1 & 0 \end{bmatrix}$$

The last example is a kernel that creates a 3D effect:

$$Kernel\_3 = \begin{bmatrix} -2 & -1 & 0 \\ -1 & 1 & 1 \\ 0 & 1 & 2 \end{bmatrix}$$

A filter moves across the image from left to right and line by line from top to bottom. Values in the kernel are multiplied by the pixel values and added up. The results are then transferred to a new layer (convolutional layer) (see Figure 5.9).

OpenCV Python

OpenCV (Open Source Computer Vision Library) is an open-source library that provides many methods for image processing. Install the OpenCV Python module via Anaconda.

5   Convolutional Layers and Images

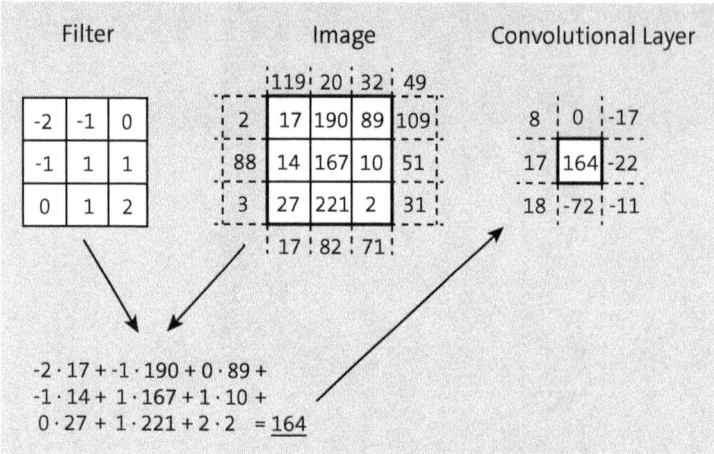

$-2 \cdot 17 + -1 \cdot 190 + 0 \cdot 89 +$
$-1 \cdot 14 + 1 \cdot 167 + 1 \cdot 10 +$
$0 \cdot 27 + 1 \cdot 221 + 2 \cdot 2 \;\; = \underline{164}$

Figure 5.9 Calculation of a Convolutional Layer

It's relatively easy to implement this procedure yourself with a few lines of Python source code.

```
import cv2
import numpy as np
img = cv2.imread('bambam.png')
kernel = np.array([[0, -1, 0],
                   [-1, 5, -1],
                   [0, -1, 0]])
img_final = cv2.filter2D(img, -1, kernel)
cv2.imwrite("bambam-sharp.png", img_final)
```

Listing 5.15 Applying Filters and Saving the Result

As usual, the required modules and then the initial image are loaded. A kernel in the form of a two-dimensional array (arrays in an array) is then defined and applied to the image using the `filter2D` method. When called, this method receives the initial image as the first parameter, followed by the information about the image depth (number of bits per pixel; at -1, the depth of the initial image is retained) and the kernel to be used as the last parameter. The result is saved as *bambam-sharp.png*.

Figure 5.10 shows the results. The first image in the top left is the original, on the right you can see the sharpened image. At the bottom left you can see a 3D effect, while at the bottom right the edges have been extracted.

Experiment with Kernel Parameters

Be sure to experiment with other kernel parameters. The result varies depending on the structure and numerical value.

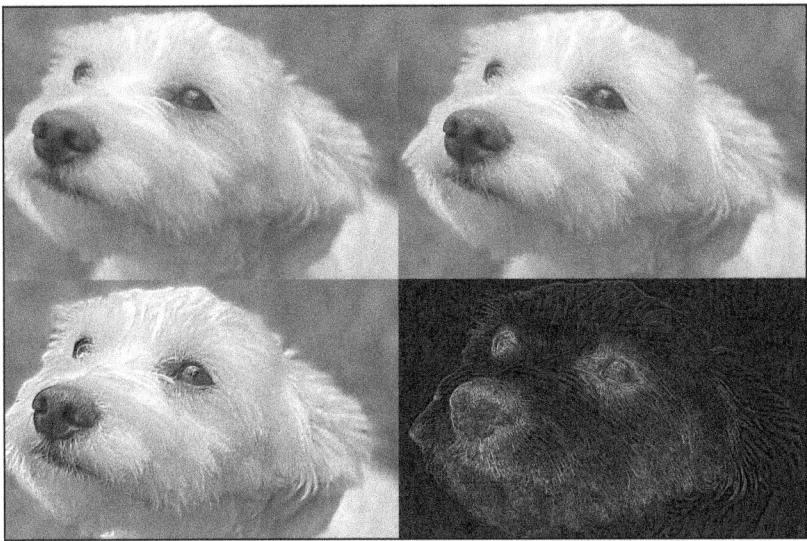

Figure 5.10 Images of Bambam: Original and after Applying Filters

What happens to the edge pixels when the kernel is processed depends on which padding you choose. With *same padding*, border areas are filled with zeros. With *valid padding*, the filter isn't filled, but only moves in the image area. The dimension of the output layer is therefore smaller than with the same padding filter.

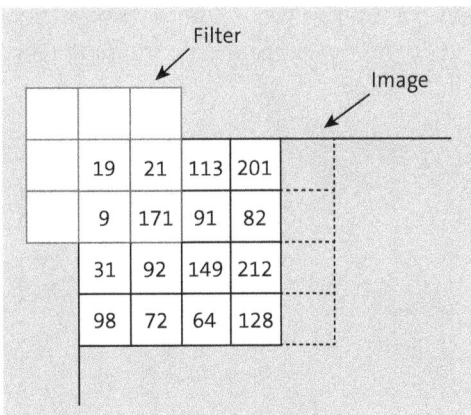

Figure 5.11 Padding Method

You're now familiar with the convolution process. You also know that the kernel can move over the edge or remain in the image. The number of pixels by which the filter moves can be specified using *stride*. The filter moves from left to right and line by line from top to bottom. A typical value for stride is 2, which is also an empirical value. In many examples on the internet and also in the literature, you'll find CNNs with a kernel size of 3 × 3 and a stride of 2. You can, of course, experiment with these values later.

As mentioned earlier, the result of the convolution is stored in the new convolutional layer. The convolutional layer is followed by the pooling layer, which reduces the dimension. In this context, *max pooling* is widely used. A filter (e.g., 2 × 2 matrix, also an empirical value) runs through the convolutional layer and copies only the largest number in the filter area into a new layer, the *pooling layer*. This reduces the dimensions of the image again and further simplifies the input data for the AI black box.

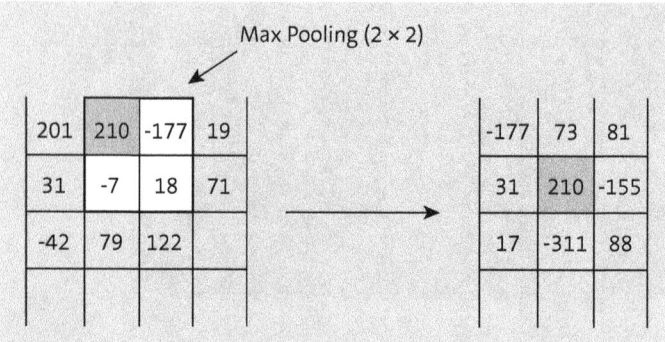

Figure 5.12 Max Pooling Method

The filters are then applied, and the data is further reduced by the pooling process. How should the data be fed into the AI model? As the last step in the detection part, the final pooling layers (as a result of several convolutions) are converted into a one-dimensional vector (*flatten*, see Figure 5.13). From here on, we can apply everything we've already learned about ANNs. The number of nodes in the last layer again depends on the task and must correspond to the number of possible classes.

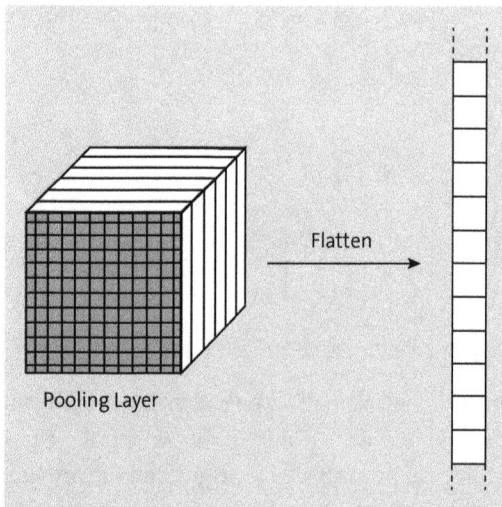

Figure 5.13 Flatten Method

> **RGB Color Model**
>
> The RGB color model is established in the IT world and means that the colors red, green, and blue can be used to mix all other colors. The information for any color is saved as follows: A total of 3 bytes are required, 1 byte for each basic color. The numbers 0 to 255, which represent the intensity of the respective basic color, are stored in each byte.
>
> Another connection between the basic colors is that the higher the individual intensities, the brighter the result (additive color mixing). So, if you select the highest intensity for each basic color (255), you get the color white. If, on the other hand, you select the intensity 0 for all basic colors, you get the color black. All other colors are somewhere in between. The connection can be verified as follows: if you shine different flashlights (with different colors) on one spot on the wall, the result becomes brighter and brighter.

You've already learned that black-and-white images created in grayscale require 1 byte per pixel. In color images, each pixel has 3 bytes, 1 byte each for red, green, and blue.

But why do we need a detection section? If the image objects are more difficult to distinguish from each other, you'll achieve better results with a CNN than with a simple ANN alone. The same applies to objects that aren't positioned nicely in the center of the image. Let's assume you want to recognize cat pictures. The cat can be at the top left or bottom right of the picture or on a person's lap. Here, too, better results are achieved if the original images are processed with filters and the features are extracted.

Let's apply what we've learned to a program (*K5_mnist_fashion-3.ipynb*). Here, we want to classify garments again, but this time with a detection and identification part. The only change is in the structure of the model.

```
model = tf.keras.Sequential([
    tf.keras.layers.Conv2D(32, (3,3), padding='same',
    activation=tf.nn.relu, input_shape=(28, 28, 1)),
    tf.keras.layers.MaxPooling2D((2, 2), strides=2),
    tf.keras.layers.Conv2D(64, (3,3), padding='same', activation=tf.nn.relu),
    tf.keras.layers.MaxPooling2D((2, 2), strides=2),
    tf.keras.layers.Flatten(),
    tf.keras.layers.Dense(128, activation=tf.nn.relu),
    tf.keras.layers.Dense(128, activation=tf.nn.relu),
    tf.keras.layers.Dense(10,  activation=tf.nn.softmax)
])
```

Listing 5.16 Structure of the CNN

Conv2d is a convolutional layer. The first parameter specifies the number of filters to be applied. This is followed by the size of the filter in parentheses. You can also see that the same padding layer is used here. The `MaxPooling2D` matrix is 2 × 2 in size and moves 2 pixels to the right each time. The combination of a convolutional and a max pooling layer is repeated in this example. The second time, 64 filters are applied. This example has a correct classification rate of 90%, but, in return, you have many additional hyperparameters:

- Number of filters
- Padding variant
- Size of the filters
- Stride
- Number of repetitions or combinations in the detection part

For your own applications, you should again copy a similar program and adapt the source code to the new task. However, this program was only intended as an introduction to the topic of CNNs. With large amounts of data and more complex images, the identification part is much more extensive, for example, 16 convolutional layers and 5 pooling layers. There are also models with even more layers. In Section 5.5, we'll use such a pretrained model and adapt it to our task.

5.4 Image Classification Using CIFAR-10

The algorithms for ML are now so advanced that a very high correct classification rate (almost 100%) can be achieved with the MNIST database. For this reason, other datasets are used to evaluate the performance of AI models. One such database is the CIFAR-10 database (Canadian Institute for Advanced Research). It comprises 10 classes of objects and animals. The images in this database are also all the same size (32 × 32 pixels), but unlike MNIST, they are stored in color. The dataset is divided into training and test data and comprises 60,000 training images and 10,000 test images. The aim is to correctly classify the objects in the images. The classes are airplane, car, bird, cat, dog, frog, horse, deer, ship, and truck.

Even beginners can write a program for this data and test their knowledge. It's fair to say that the dataset is somewhere in the middle in terms of complexity. It's more complex than the MNIST database, but simpler than the ImageNet database, for example, which you'll get to know in the next section.

Over time, when you've gained more experience in the field of AI, you can expand the source code presented here. The *K5_CIFAR_10.ipynb* program already provides a correct classification rate of approximately 85%.

I recommend that you run the program in Google Colab. CPUs are optimized for a serial execution of programs, whereas GPUs are optimized for a parallel execution. If you want your AI programs to use the GPU, you need to prepare your computer for this.

Depending on the graphics card and operating system, you may need to download drivers and configure the system accordingly. You can find instructions on the internet for your combination of graphics card and operating system.

If you use Colab, you can simply set the hardware accelerator via **Edit • Notebook Settings**. Using GPU instead of CPU can save you several hours of time with this program.

```
import pandas as pd
import tensorflow as tf
from sklearn.model_selection import train_test_split
import matplotlib.pyplot as plt

(train_data, train_col), (test_data, test_col) =
 tf.keras.datasets.cifar10.load_data()

class_names = ['airplane','automobile','bird','cat','deer','dog','frog',
 'horse','ship','truck']

index = 111
plt.imshow(train_data[index])
plt.xlabel(class_names[int(train_col[index])])
plt.show()

# Build ANN
model = tf.keras.Sequential([
    tf.keras.layers.Conv2D(32, (3,3), padding='same',
 activation=tf.nn.relu, input_shape=(32, 32, 3)),
    tf.keras.layers.Conv2D(32, (3,3), padding='same', activation=tf.nn.relu),
    tf.keras.layers.MaxPooling2D((2, 2), strides=2),
    tf.keras.layers.BatchNormalization(),
    tf.keras.layers.Dropout(0.4),

    tf.keras.layers.Conv2D(64, (3,3), padding='same', activation=tf.nn.relu),
    tf.keras.layers.Conv2D(64, (3,3), padding='same', activation=tf.nn.relu),
    tf.keras.layers.MaxPooling2D((2, 2), strides=2),
    tf.keras.layers.BatchNormalization(),
    tf.keras.layers.Dropout(0.4),

    tf.keras.layers.Conv2D(128, (3,3), padding='same', activation=tf.nn.relu),
    tf.keras.layers.Conv2D(128, (3,3), padding='same', activation=tf.nn.relu),
    tf.keras.layers.MaxPooling2D((2, 2), strides=2),
    tf.keras.layers.BatchNormalization(),
    tf.keras.layers.Dropout(0.4),
```

```
    tf.keras.layers.Flatten(),
    tf.keras.layers.Dense(128, activation=tf.nn.relu),
    tf.keras.layers.Dense(10, activation=tf.nn.softmax)
])
# Configuration of the training process
model.compile(optimizer=tf.keras.optimizers.Adam(learning_rate=0.0001),
 loss='sparse_categorical_crossentropy', metrics=['accuracy'])

cb_early = tf.keras.callbacks.EarlyStopping(monitor='loss', patience=3)

model.fit(train_data, train_col, epochs=100, validation_data=(test_data,
 test_col),callbacks=[cb_early])
```

Listing 5.17 CNN for CIFAR-10

Only three things are new: Batchnormalization, Dropout, and batch_size. The rest of the source code should be easy to interpret by now. This will also show you how good your knowledge has become. The three new items are described here:

- **Dropout**
  To reduce the risk of overfitting, it's enforced that all connections between the nodes must "work." For each iteration, a certain number (e.g., 20%) of randomly selected connections are hidden so that the other connections become active. This procedure sometimes leads to an improvement in the overall result.

- **batch_size**
  The data is processed in batches. The weights are updated for the entire stack. Not only does this save time, but it may also ensure greater stability because averaged values are used.

- **Batchnormalization**
  If you remember, we standardized attributes to get better results, and we discovered that it can sometimes be advantageous to standardize the values between the layers. The values are calculated for complete stacks.

Don't be surprised at the blurred images when you output them. Because the images are very small and in color, they look pixelated.

The program now has a very large number of hyperparameters. With large computing capacities and a lot of time, you can test various combinations. Another option is to search for high-performance AIs on the internet. You can use these networks as a guide when setting up the model. However, if you want to achieve good results as quickly as possible, you can also use pretrained networks.

5.5 Using Pretrained Networks

When you need an AI for a project, why reinvent the wheel? There are a lot of pretrained models available that you can easily integrate into your project. For the next program (*K5_vgg19-1.ipynb*), we use the VGG19 model (Visual Geometry Group Oxford University). This model consists of 19 layers and was trained with more than 1.2 million images for 1,000 categories. The images come from ImageNet, a database with 14 million labeled images. This database was initiated by researchers at Stanford University to develop and evaluate high-performance AI models for complex data.

Let's assume you're given the task of developing an AI to recognize teddy bears. You need a lot of pictures with and without teddy bears from different perspectives in all sizes and colors. You would have to label these images accordingly ("teddy" or "no teddy"). Then you can build an AI and test numerous possible combinations of hyperparameters to get an acceptable result. This approach is very good for teaching and learning purposes, but if the client wants a quick, cost-effective solution, it's not feasible. In this case, you should consider using ready-made modules or pretrained models—but that doesn't mean you don't need any AI knowledge. On the contrary, the knowledge of ANN, CNN, and so on will be very useful when dealing with such solutions.

```
import pandas as pd
import tensorflow as tf
import matplotlib.pyplot as plt
import numpy as np
from tensorflow.keras.applications import vgg19

model = vgg19.VGG19()
model.summary()
```

Listing 5.18 Importing the Modules and the Pretrained Model

The `model` variable already contains the entire network, so you don't need to do anything else and can apply it directly to new images. You can use `model.summary` to view the structure of the network. It contains 19 layers and more than 143 million parameters.

The model is quite large, but with the knowledge gained in this chapter, you can interpret the output. You should also now have an idea of how time-consuming it is to set hyperparameters for such a model if you want to develop it yourself. In addition, you would be at a loss with a PC as a development machine, as the training and test procedure would take forever after each setting.

```
Layer (type)                 Output Shape              Param #
=================================================================
input_1 (InputLayer)         [(None, 224, 224, 3)]     0
input_1 (InputLayer)         [(None, 224, 224, 3)]     0
block1_conv1 (Conv2D)        (None, 224, 224, 64)      1792
block1_conv2 (Conv2D)        (None, 224, 224, 64)      36928
block1_pool (MaxPooling2D)   (None, 112, 112, 64)      0
block2_conv1 (Conv2D)        (None, 112, 112, 128)     73856
block2_conv2 (Conv2D)        (None, 112, 112, 128)     147584
block2_pool (MaxPooling2D)   (None, 56, 56, 128)       0
block3_conv1 (Conv2D)        (None, 56, 56, 256)       295168
block3_conv2 (Conv2D)        (None, 56, 56, 256)       590080
block3_conv3 (Conv2D)        (None, 56, 56, 256)       590080
block3_conv4 (Conv2D)        (None, 56, 56, 256)       590080
block3_pool (MaxPooling2D)   (None, 28, 28, 256)       0
block4_conv1 (Conv2D)        (None, 28, 28, 512)       1180160
block4_conv2 (Conv2D)        (None, 28, 28, 512)       2359808
block4_conv3 (Conv2D)        (None, 28, 28, 512)       2359808
block4_conv4 (Conv2D)        (None, 28, 28, 512)       2359808
block4_pool (MaxPooling2D)   (None, 14, 14, 512)       0
block5_conv1 (Conv2D)        (None, 14, 14, 512)       2359808
block5_conv2 (Conv2D)        (None, 14, 14, 512)       2359808
block5_conv3 (Conv2D)        (None, 14, 14, 512)       2359808
block5_conv4 (Conv2D)        (None, 14, 14, 512)       2359808
block5_pool (MaxPooling2D)   (None, 7, 7, 512)         0
flatten (Flatten)            (None, 25088)             0
fc1 (Dense)                  (None, 4096)              102764544
fc2 (Dense)                  (None, 4096)              16781312
predictions (Dense)          (None, 1000)              4097000
=================================================================
Total params: 143,667,240
Trainable params: 143,667,240
Non-trainable params: 0
```

Figure 5.14 Structure of the VGG16 Model

```
file = "../Data/teddy.png"
img = tf.keras.utils.load_img(file,target_size=(224, 224,3))
img = np.array(img)

plt.figure()
plt.imshow(img)
plt.show()
```

Listing 5.19 Loading and Displaying Your Own Image

5.5 Using Pretrained Networks

The image of the teddy bear should be loaded and output, whereby the image is loaded with 224 × 224 pixels. Each pixel has 3 bytes, as this is a color photo. A conversion to the array data type is still necessary for further processing. Make sure that your own pictures are square; otherwise, the images will be very distorted, and the AI won't be able to classify them correctly.

Figure 5.15 Teddy Bear as a Test Object

The image still has to be fed to the model to classify it.

```
img = tf.keras.applications.vgg19.preprocess_input(img)
pred = model.predict(img.reshape(1, 224, 224, 3))

# Output the best 5 hits
top_five_predict = vgg19.decode_predictions(pred, top=5)
print(top_five_predict)
```

Listing 5.20 Classification of the Image with a Pretrained Model

The image can be prepared for VGG19 by using `preprocess_input`. The sequence for RGB bytes is ordered, and the pixel values are scaled. The image must have the same structure as the training images of the model with which it was trained. The large amount of data in the `img` array then needs to be "shaped" so that it's bundled image by image. The data contains one image with 244 × 244 pixels and 3 bytes per pixel. Then follows the classification. The pred variable now contains the result as the output of a softmax function with 1,000 digits. Just output it for checking, as you can do very little with this result.

```
top_five_predict = vgg19.decode_predictions(pred, top=5)
print(top_five_predict)
```

Listing 5.21 Output of the Best Hits

The result gets decoded, and the best five matches are displayed. The output is as follows:

```
[[('n04399382', 'teddy', 0.7304068), ('n03775071', 'mitten', 0.021119094),
('n04462240', 'toyshop', 0.020079762), ('n04548362', 'wallet', 0.018961832),
('n03908618', 'pencil_box', 0.017840609)]]
```

Each hit has a unique identifier, a name, and the hit probability (calculated using the softmax function). In first place, you see the teddy bear with about 73%, far ahead of all others. The program therefore works satisfactorily. Now you should test it with other images.

There are some other pretrained models available on the internet. However, you'll also find complete libraries with which you can achieve excellent results. But you'll need a good knowledge of the Python programming language. OpenCV allows you to implement face recognition, gesture recognition, and much more. You'll find many examples and tutorials on the homepage. If you want to label your own images, you can use the *labelme* or *labelImg* tools, for example. Using *Teachable Machine* from Google, you can "click together" an AI for object, gesture, or sound recognition in the browser within a few minutes.

5.6 Exercises

5.6.1 Exercise 1: Hyperparameter Optimization for CIFAR-10

In Google Colab, try to increase the correct classification rate for CIFAR-10. Research other models on the internet, and try to transfer the settings to your model. There is no sample solution for this exercise.

5.6.2 Exercise 2: Pretrained VGG19 Model

You'll hardly ever create, train, and test AIs for images from scratch for your own projects. Have a self-created image classified by VGG19. You can copy the *K5_vgg19-1.ipynb* file for this purpose. Don't forget to crop the picture into a square. You can also research which classes can be predicted at all (*https://image-net.org/challenges/LSVRC/2014/browse-synsets*). There is no sample solution for this exercise.

Chapter 6
Transfer Learning

What do you do if there are no pretrained networks for your use case, and you don't have the time or inclination to develop an AI model from scratch? You'll find out in this chapter.

What This Chapter Is About
- Downloading and using labeled images from the internet
- Adapting a pretrained model to new datasets
- Changing hyperparameters to increase the correct classification rate

You've already learned a few ways to program AI applications. In addition, you've also already used a pretrained network to classify images. Now let's assume you have to program an AI that recognizes from images whether the hand represents a rock, paper, or scissors. You probably know the popular game. How do you proceed?

You should research whether there are already pretrained AI models for your use case. Such models have generally been tested and optimized many times and have proven themselves in practice. But what if there are no models, or the models are chargeable or too expensive? Then, you could develop an AI model yourself.

First, you need to collect data, that is, many images of many different hands showing corresponding gestures. As far as how many images there should be, the more, the better, so maybe 500 to start with. These pictures must be of many different people and should include many ages and ethnicities, so you have images of hands in many colors and shapes. You must also pay attention to various backgrounds when collecting data. Otherwise, your model will only classify hands with a certain background correctly, as it (unintentionally) optimizes the parameters for this background information.

But it's not enough to simply shoot and edit images because they also need to be labeled. If, for example, scissors are shown in image 1, then this assignment must be recorded in a file, that is, image 1 = scissors. This is very time-consuming work.

If there is no pretrained model for your use case, perhaps you can find labeled images that you can use to train and test your model. Fortunately, there are many public datasets that you can draw on. The TensorFlow module even has some integrated datasets.

6 Transfer Learning

Know Your Data

Google provides many labeled images free of charge on the following website: *https://knowyourdata.withgoogle.com/* (see Figure 6.1). You can download the data and use it to develop your own AI models. However, it's also possible to download this data directly via the TensorFlow module. You'll find a lot of data, for example, on plants, cars, celebrities, and so on, and you can use all of this data for AI development.

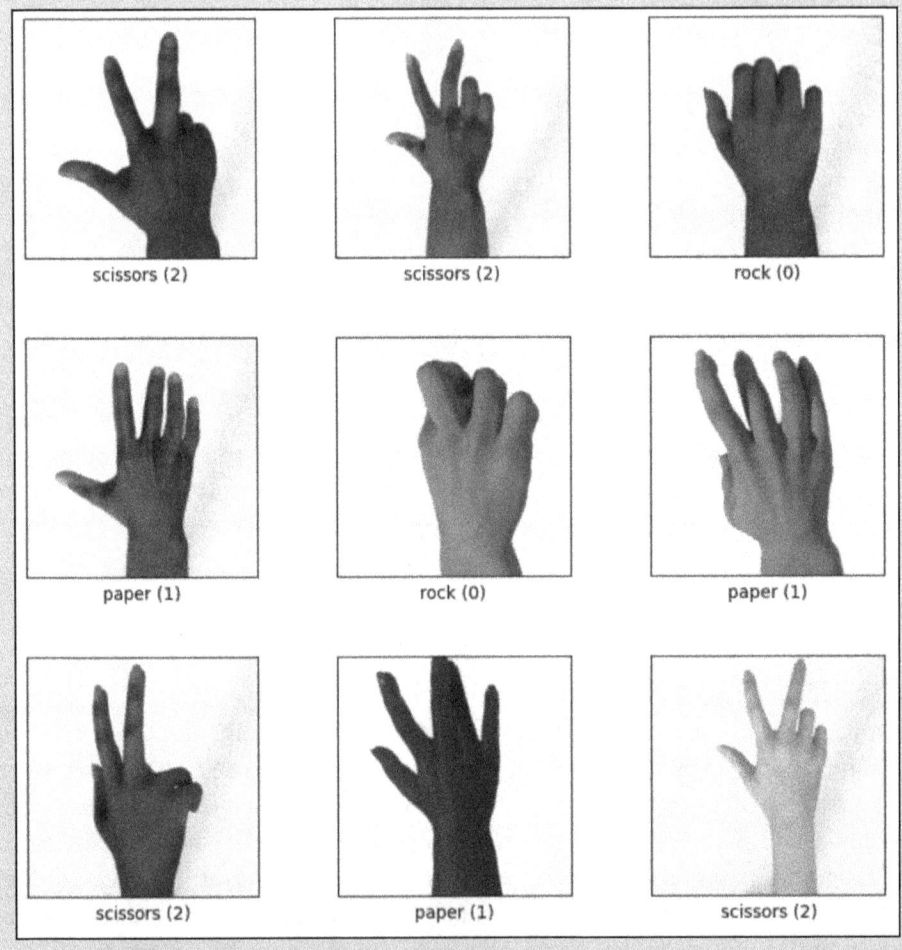

Figure 6.1 Some Images from the Rock, Paper, Scissors Dataset

Fortunately, we'll be spared the extensive work of collecting all the relevant data. What happens next? You could develop an AI based on convolutional layers with an identification and a detection part, in which case you would have to set many hyperparameters. There really are a lot of them:

- Selection and number of convolutional and pooling layers
- Repeating combinations

- Size of the filters
- Setting for stride
- Number and size of hidden layers
- With or without dropout
- Various activation functions
- Learning rate

Note that these aren't all the possible settings. You could use libraries such as Grid Search to find the best combination from a set of parameters. For such a large number of parameters, however, you need good programming skills, a high-performance computer, and a lot of time. Once your AI model has been trained and tested, you can evaluate it using data that is as yet unknown.

In this chapter, I want to introduce you to a different approach that requires significantly fewer resources: we download a pretrained model (not originally developed for the data in your use case) and adapt it to the new data. This works surprisingly well. This adaptation also costs a little computing time, but this is nothing compared to a new development.

In the following example, I recommend that you program the model in Google Colab with GPU support. You can then download this model, open it locally, and test it with your own images. You can also proceed in the same way if you're developing an AI for an embedded system such as Raspberry Pi: The AI is modeled, trained, and tested on a powerful computer. The model is then downloaded and used on the respective target platform.

So let's briefly summarize once again: We want to develop an AI model that recognizes the hand gestures for rock, paper, and scissors in an image. The labeled image data for training and testing processes comes from TensorFlow and the Know Your Data website. A pretrained model, which wasn't explicitly developed for this image data, is adapted to this very image data. This saves you having to completely redevelop the AI model.

6.1 How It Works

Again, we use the VGG19 model for our program. The detection part has already seen many objects and extracted and stored information such as edges, corners, and so on. As a result, the input layer is adapted to the new images, followed by the detection part (these weights are "frozen"), and finally an information part modeled by us (these weights must be trained). That's it. You'll be surprised how good the results of this approach are.

We use VGG19 because with the knowledge we've conveyed to you up to this point, you should find it easy to understand the structure of the model. There are a number of other pretrained models available (see *https://keras.io/api/applications*), but many of these models have an unfamiliar structure for beginners. For example, the layers of some models fork in places and are later merged again. There is nothing unusual for you about VGG19. As you'll see later, the model also delivers very good results.

Upload the *K6_vgg19-1a.ipynb* file to Google Colab (**File • Upload Notebook**), and activate GPU support under **Edit • Notebook Settings**.

```
import pandas as pd
import tensorflow as tf
from keras.models import Model
import matplotlib.pyplot as plt
import numpy as np
from tensorflow.keras.applications import vgg19
import tensorflow_datasets as tfds

base_model = vgg19.VGG19(include_top=False,input_shape=[300,300,3])
base_model.trainable = False
base_model.summary()
```

Listing 6.1 Downloading the Required Modules and the Pretrained Model

An object of type `VGG19` is stored in the `base_model` variable. The constructor is called with two parameters:

- `include_top=False` only loads the detection part of the model.
- `input_shape=[300,300,3]` adjusts the input (input layer) to the training images.

We also indicate that the weights of this model should not be updated. The purpose of our procedure is to use the trained detection part.

We use the `summary` method to look at the structure of the model. The first and last lines of the output are interesting.

```
input_1 (InputLayer)         [(None, 300, 300, 3)]     0
block1_conv1 (Conv2D)        (None, 300, 300, 64)      1792
block1_conv2 (Conv2D)        (None, 300, 300, 64)      36928
block1_pool (MaxPooling2D)   (None, 150, 150, 64)      0
...
block5_conv4 (Conv2D)        (None, 18, 18, 512)       2359808
block5_pool (MaxPooling2D)   (None, 9, 9, 512)         0
```

Now the input layer gets designed for our training images, which have 300 × 300 pixels and 3 bytes per pixel (RGB). At the end of the output, you'll see that the identification part is missing, so the model needs to be adjusted.

```
x = tf.keras.layers.Flatten()(base_model.output)
x = tf.keras.layers.Dense(256,activation='relu')(x)
x = tf.keras.layers.Dense(3,activation='softmax')(x)
```

Listing 6.2 Building the ANN as the Identification Part

These lines for building an artificial neural network (ANN) may seem unfamiliar to you. The Flatten method generates a partially evaluated function, which receives the output of our pretrained model as input, and the result is saved in the x variable. The entire process is repeated with the Dense methods, and the result is saved in x again. The output layer has three nodes, as there are three image types (rock, paper, and scissors) to classify.

```
model = Model(inputs=base_model.input, outputs=x)
model.summary()
```

Listing 6.3 Structure of the Model with Detection and Identification Parts

The preceding lines of code are used to build the final model. The input of the pretrained model serves as input, while the output of our new identification part serves as output. You can also add more hidden layers later and see if the result improves.

To verify that the model is structured as we imagined, we output it again. I think we can work with this result. There are more than 40 million parameters in total; fortunately, we only have to set half of them during the training process. These are the parameters of the identification part we added ourselves.

Now it's time to download the data to train the model later. The method call is shown in the following listing.

```
(train_data, train_col), (test_data, test_col) = tfds.as_numpy(tfds.load(
    'rock_paper_scissors',
    split=['train', 'test'],
    batch_size=-1,
    as_supervised=True
))
```

Listing 6.4 Downloading the Rock, Paper, Scissors Dataset

Let's take a closer look at the listing:
- The data is downloaded as NumPy arrays.
- split indicates that the data is split into training and test data.
- The next parameter, batch_size=-1, ensures that the entire dataset is available as one dataset.
- The last parameter, as_supervised=True, saves the images and labels in the tuple data type.

6 Transfer Learning

In the documentation (*www.tensorflow.org/datasets/overview*), you'll find further examples and descriptions of how you can handle datasets of this type.

The result consists of the four tables we need. Google Colab also provides us with sufficient working memory for this. Otherwise, we would first have to download the data into a directory and unpack it, which would result in a few more lines of code.

```
class_names = ['rock', 'paper', 'scissor']
index = 111
plt.imshow(train_data[index], cmap='binary')
plt.xlabel(class_names[train_col[index]])
plt.show()
```

Listing 6.5 Output of an Image from the Dataset

You should always output an image with the corresponding label (as in Figure 6.2) to verify that the data is available because you'll need it later to train and test the AI model.

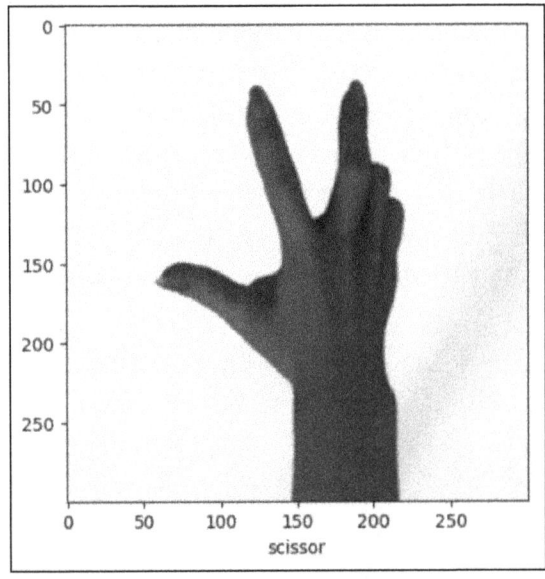

Figure 6.2 Image with Index 111

Let's take a closer look at the structure of the training data—in this case, the image data:

```
print(train_data.shape)
```

This line provides the following output:

```
(2520, 300, 300, 3)
```

This means that 2,520 images, each with 300 × 300 pixels and 3 bytes per pixel, are stored in the train_data variable. These images still need to be converted and adapted to the VGG19 model:

```
train_data = tf.keras.applications.vgg19.preprocess_input(train_data)
test_data = tf.keras.applications.vgg19.preprocess_input(test_data)
model.compile(loss='sparse_categorical_crossentropy', optimizer='adam',
 metrics=['accuracy'])
model.fit(train_data, train_col, epochs=5, validation_data=(test_data,
 test_col), batch_size=32)
```

Listing 6.6 Preparing Images and Training and Testing the ANN

The preceding code is used to train and test the model. In the first epoch, we achieve a correct classification rate of 100% with the training data, but around 80% with the test data. That's a typical sign of overfitting. As you can see, you have to test a lot to get an acceptable result. So let's test the whole thing again with the following configuration, namely with dropout.

```
x = tf.keras.layers.Dropout(0.3)(base_model.output)
x = tf.keras.layers.Flatten()(x)
x = tf.keras.layers.Dense(256,activation='relu')(x)
x = tf.keras.layers.Dropout(0.3)(x)
x = tf.keras.layers.Dense(3,activation='softmax')(x)

model = Model(inputs=base_model.input, outputs=x)
model.summary()
```

Listing 6.7 Structure of the AI Model

Then, reset everything via **Runtime • Restart and Run All Cells**, and run it again. But overfitting is also evident here. All right, for the next attempt, we also turn down the learning rate, mix the data, and set the number of epochs to 100. The optimum number, however, is determined automatically via callback.

```
cb_early = tf.keras.callbacks.EarlyStopping(monitor='val_loss', patience=3)

model.compile(loss='sparse_categorical_crossentropy', optimizer=
 tf.keras.optimizers.Adam(learning_rate=0.0001), metrics=['accuracy'])
model.fit(train_data, train_col, epochs=100, validation_data=
 (test_data, test_col), batch_size=32, shuffle=True, callbacks=[cb_early])
```

Listing 6.8 Training and Testing the AI Model Using Callback

Finally, with this setting, we achieve a correct classification rate of around 93%.

6 Transfer Learning

```
model.save('transfer.h5')
```

Listing 6.9 Saving the Trained Model

Subsequently, the model is saved and you can then download the file. You can find it in the File Explorer on the left side (see Figure 6.3).

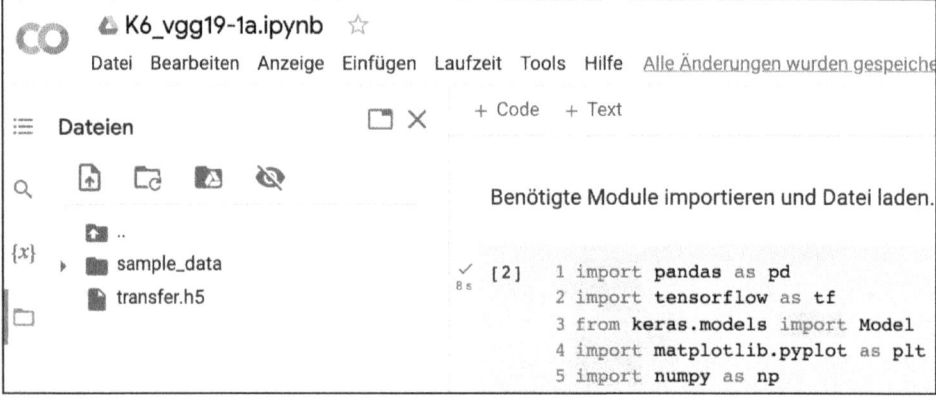

Figure 6.3 File Explorer

Note that all files you create or upload in Colab will be deleted after the session. Colab also provides you with a folder containing datasets that you can use for your own exercises, for example, *sample_data*.

We use the *K6_vgg19_1b.ipynb* file to open the model data locally (JupyterLab or Jupyter Notebook), load a new image of a gesture (from the test data or TensorFlow homepage), and have it classified. You can also plot the image and take a screenshot. The background of the images has a color cast so that your own images are poorly classified. You need the same background for this, or the training images have to be taken against different backgrounds.

```
import pandas as pd
import tensorflow as tf
from keras.models import Model
import matplotlib.pyplot as plt
import numpy as np

model = tf.keras.models.load_model('transfer.h5', compile=False)
file = "../Data/rock.png"

img = tf.keras.utils.load_img(file,target_size=(300, 300,3))
img = np.array(img)

plt.figure()
```

```
plt.imshow(img)
plt.show()
```

Listing 6.10 Loading a Pretrained Model and a Test Image

The model and the image file are loaded, and the image file is also output (see Figure 6.4). If you want to classify your own images, make sure that they have a square format.

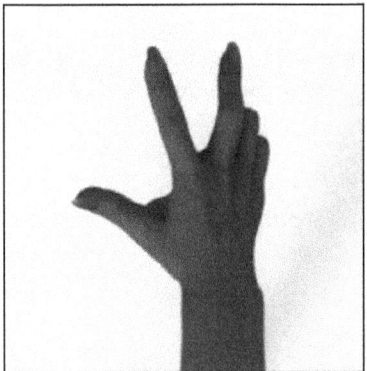

Figure 6.4 Hand Gesture (Scissors)

Now you need to determine the information that is important for the classification and output it.

```
class_names = ['rock', 'paper', 'scissor']

img = tf.keras.applications.vgg19.preprocess_input(img)
pred = model.predict(img.reshape(1, 300, 300, 3))

print("prediction softmax:",pred)
print("prediction max:",pred.argmax())
print("prediction string:",class_names[pred.argmax()]
```

Listing 6.11 Classification of the Test Image

The output of the classification is:

```
prediction softmax: [[0. 0. 1.]]
prediction max: 2
prediction string: scissors
```

So, the image is classified correctly, and we can be content. If you want to program an AI for large amounts of data, you'll need to have Python knowledge. In this case, the data must first be saved and prepared locally. An alternative to this is visual programming using KNIME, which you'll learn about in Chapter 11.

6.2 Exercises

6.2.1 Exercise 1: Rock-Paper-Scissors

Try to classify the hand gestures with using the *K6_vgg19_1b.ipynb* program and other images. There is no sample solution for this exercise.

6.2.2 Exercise 2: Human or Horse

Change the *K6_vgg19_1a.ipynb* and *K6_vgg19_1b.ipynb* files in such a way that the "horses_or_humans" data collection is now used (*https://knowyourdata-tfds.withgoogle.com*). This allows you to assign images to these two categories:

- When training, make sure that you only need two outlets.
- You can download images from the internet for testing purposes. Crop the pictures into square format.

Chapter 7
Anomaly Detection

Up to this point, we haven't considered whether data is balanced or not. When searching for anomalies, however, we have to make up for that.

> **What This Chapter Is About**
> - Classifying unevenly distributed data using XGBoost and resampling
> - Classifying unevenly distributed data using autoencoders (based on artificial neural networks [ANNs])
> - Analyzing results graphically using the confusion matrix

When searching for anomalies, you're dealing with unbalanced data. The dataset contains a lot of "normal" data but very little "abnormal" data. The aim of anomaly detection is to develop an AI that can distinguish between these highly unevenly distributed data groups.

Take, for example, measurement data on heart rhythms. To train an AI model that detects conspicuous heartbeats, you would have to record the heartbeats of many people. In the resulting dataset, the data of normal heartbeats would be significantly overrepresented, and only a small proportion of the data would contain abnormalities such as arrhythmias. Of course, you still need a person with specialist knowledge to classify the records in this case into normal and abnormal.

In Figure 7.1, you can see two plots of ECG recordings (electrocardiogram); the dataset comes from the Time Series Machine Learning website at *www.timeseriesclassification.com/description.php?Dataset=ECG500*.

The challenge in anomaly detection for data distributions of this type is therefore to develop an AI model that can divide a dataset into two categories, even if one category only accounts for 0.1% of the dataset, for example.

In this chapter, we'll look at data on credit card transactions. The data was collected, processed, anonymized, and made available to the public by the University of Brussels. It includes European credit card transactions from two days in 2013.

The corresponding comma-separated values (CSV) file contains 284,807 transactions, but only 492 of them are fraud cases, which is 0.173% of the data. Even if you program an AI that has a correct classification rate of 99%, this alone isn't a quality criterion. If your AI were to incorrectly classify all data records as normal, you would also achieve a correct classification rate of 99.8%. This means that we have to find other or additional evaluation criteria.

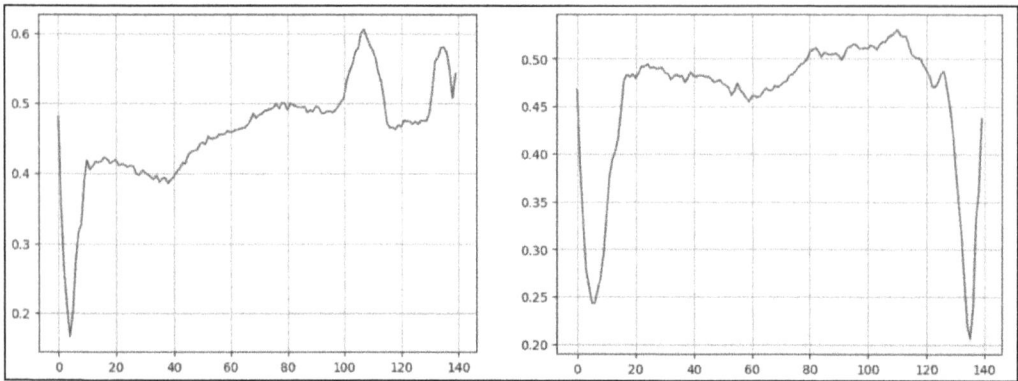

Figure 7.1 Two Examples from an ECG Recording: Normal on the Left and Abnormal on the Right

7.1 Unbalanced Data

As a first approach, let's try to simply apply what we've learned so far. Table-based data is available, which must be classified (there are two categories). You could build an ANN and try to optimize the model for this dataset. In Chapter 4, we discussed that models based on the gradient boosting method often return better results. In any case, developing an AI for this type of dataset is easier with these models. By using XGBoost in particular, you can achieve good results very quickly with tabular data. So let's start with this and analyze how good the result is in this example with unbalanced data (see *K7_credit-card-1.ipynb*).

```
import pandas as pd
from sklearn.model_selection import train_test_split
from sklearn import metrics
from xgboost import XGBClassifier

path = "../Data/creditcard.csv"
data = pd.read_csv(path, delimiter=',')
print(data.head())
```

Listing 7.1 Loading Modules and the Dataset

As always, it's recommended to output the first rows of the dataset, as shown in Figure 7.2, to get an overview.

```
   Time        V1        V2        V3        V4        V5        V6        V7  \
0   0.0 -1.359807 -0.072781  2.536347  1.378155 -0.338321  0.462388  0.239599
1   0.0  1.191857  0.266151  0.166480  0.448154  0.060018 -0.082361 -0.078803
2   1.0 -1.358354 -1.340163  1.773209  0.379780 -0.503198  1.800499  0.791461
3   1.0 -0.966272 -0.185226  1.792993 -0.863291 -0.010309  1.247203  0.237609
4   2.0 -1.158233  0.877737  1.548718  0.403034 -0.407193  0.095921  0.592941

         V8        V9   ...       V21       V22       V23       V24       V25  \
0  0.098698  0.363787   ... -0.018307  0.277838 -0.110474  0.066928  0.128539
1  0.085102 -0.255425   ... -0.225775 -0.638672  0.101288 -0.339846  0.167170
2  0.247676 -1.514654   ...  0.247998  0.771679  0.909412 -0.689281 -0.327642
3  0.377436 -1.387024   ... -0.108300  0.005274 -0.190321 -1.175575  0.647376
4 -0.270533  0.817739   ... -0.009431  0.798278 -0.137458  0.141267 -0.206010

        V26       V27       V28  Amount  Class
0 -0.189115  0.133558 -0.021053  149.62      0
1  0.125895 -0.008983  0.014724    2.69      0
2 -0.139097 -0.055353 -0.059752  378.66      0
3 -0.221929  0.062723  0.061458  123.50      0
4  0.502292  0.219422  0.215153   69.99      0
```

Figure 7.2 Output of the First Five Rows

The file or dataset has a total of 31 columns. The **Time** column indicates the seconds between the respective transaction and the first transaction, and **Amount** shows the amount. Then, the **V1** to **V28** columns contain anonymized and already standardized data. Unfortunately, there is no information about what these columns or attributes represent in the dataset. Finally, we have the **Class** column, which indicates whether there is a case of fraud (Class = 1) or not (Class = 0). The goal is to design an AI model that recognizes cases of fraud itself based on the attributes (without **Class**). For the development, however, we need a dataset that contains this target column.

The division of the dataset into four tables for the training and testing process is also decisive here. If you're still unsure about how to divide the data, take another look at Chapter 3.

```
Col = data['Class']
data.drop(['Class'], axis = 1, inplace=True)

# Output of the distribution
print(col.value_counts())

# Create four tables from the two tables
train_data, test_data, train_col, test_col = train_test_split(data,
 col, test_size=0.2, random_state=42)

xgb = XGBClassifier()

xgb.fit(train_data, train_col)
predicted_col = xgb.predict(test_data)
```

7 Anomaly Detection

```
score = metrics.accuracy_score(test_col, predicted_col)
print(score)
```

Listing 7.2 Division of the Dataset, Training, and Testing of the AI Model

This simple model delivers a correct classification rate of just under 100%. But you already know that this result must be viewed critically. Therefore, we'll look at the result using a *confusion matrix*, which allows more detailed analyses.

> **Confusion Matrix**
>
> In a confusion matrix, you can see clearly to which categories data belongs (according to the dataset) and to which categories it has been assigned, for example, by an AI model.
>
> In Figure 7.3, you can see that in the example for the classification of irises, six predictions of the AI model for *Iris versicolor* are incorrect (middle row, right-hand column). All other predictions correspond to the test data and are therefore correct.
>
>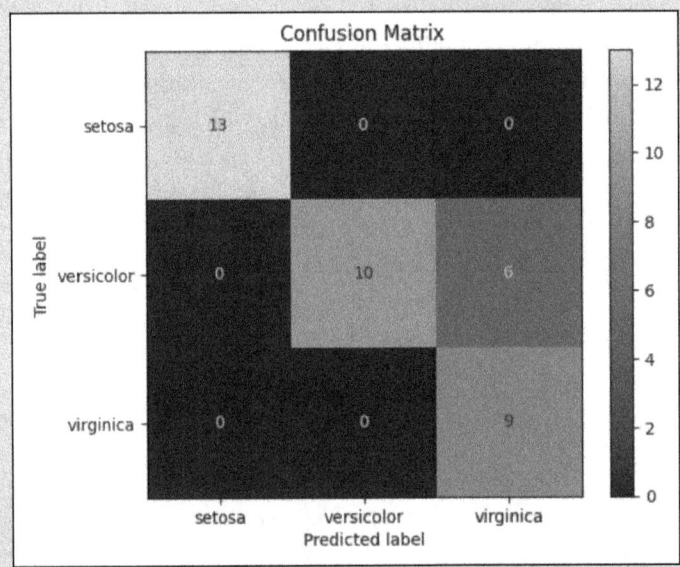
>
> **Figure 7.3** Confusion Matrix for the Graphical Evaluation of an AI Model for Irises
>
> The `ConfusionMatrixDisplay` class from the scikit-learn module enables you to easily display the confusion matrix of a classification. The values of the target column from the dataset are displayed on the y-axis, and the predictions for these values appear on the x-axis.

Let's create a confusion matrix to compare the predictions of the model with the actual values from the dataset (test data).

```
From sklearn.metrics import confusion_matrix, ConfusionMatrixDisplay

cm = confusion_matrix(test_col, predicted_col)
disp = ConfusionMatrixDisplay(cm)
disp.plot()
```

Listing 7.3 Output of the Confusion Matrix

The values are determined using the `confusion_matrix` function and saved in the `cm` variable. For this purpose, you must pass the actual values and the predictions as parameters. You can then output the content of the variable, but the result is easier to interpret graphically. For this reason, we create a `disp` object, and the constructor of `ConfusionMatrixDisplay` is called using the `cm` variable. Finally, we plot the confusion matrix (see Figure 7.4).

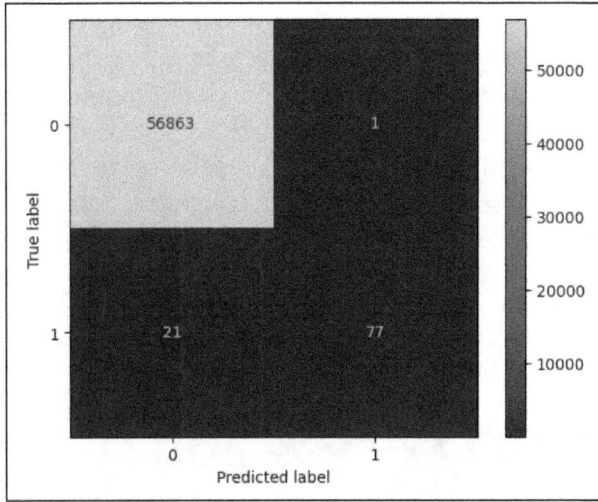

Figure 7.4 Confusion Matrix of the Results

With this graphical result, the AI can be evaluated much better, as the incorrect predictions that fall by the wayside in percentage terms are now more visible and listed here:

- 56,863 true negative results, that is, normal transactions that were recognized as such
- 77 true positive, that is, correctly identified cases of fraud
- 21 false negative transactions, that is, cases of fraud that have been assessed as normal transactions
- 1 false positive transaction, that is, incorrectly classified as a conspicuous transaction

You must always interpret the results of a confusion matrix for the respective application. In this instance, 79% of fraud cases were correctly identified. In one case, a normal transaction was assessed as a case of fraud. Given the amount of data, that's not bad. If account holders are notified of suspicious transactions via email, they can live with this result. But imagine a false statement in a cancer diagnosis or the swerving of an autonomously driven car. It's important that you always keep the respective application in mind. When AI models make serious predictions, there should always be (at least) one additional measure or assessment (e.g., medical staff, additional safety systems).

Let's get back to our example with the credit card transactions. One email more than necessary with the advice to check the account movements is really justifiable. However, 21% of fraud cases remain undetected. Let's see if we can do something about that.

7.2 Resampling

The unbalanced data doesn't make it easy for us. The simplest option would be to remove many data records with normal transactions so that the data is evenly distributed (*downsampling*). In doing so, however, we would forgo a large part of the data and reduce the quality of the AI. Another option is to generate more data records with conspicuous transactions so that the data is balanced again (*upsampling*); that is, there is roughly the same amount of normal as abnormal data. In the simplest case, you could copy the data records with the fraud cases as often as necessary until the balance is achieved. The scikit-learn module provides the `resample` method for this purpose.

Resampling

You can use the `resample` method from the scikit-learn module to adjust the amount of data in the dataset. This involves either copying and duplicating existing data (upsampling) or removing data (downsampling).

```
from sklearn.utils import resample

x = [1,2,3]
y = resample(x, n_samples=10, random_state=42)
print(y)
```

Listing 7.4 Example of Resampling (Here, Upsampling)

The x list is extended so that a total of 10 elements are included. The output looks as follows:

```
[3, 1, 3, 3, 1, 1, 3, 2, 3, 3]
```

In the next program, we want to apply this method to our dataset with the credit card transactions (*K7_credit-card-2.ipynb*). Only the relevant parts of the code are considered.

```
data_nfraud = data[data['Class']==0]
data_fraud = data[data['Class']==1]
```

Listing 7.5 One Table Each for Normal and Conspicuous Transactions

Two new tables are created, each containing normal and conspicuous transactions. You can see how easily this works with the pandas module.

```
print("data_fraud before Upsampling", data_fraud.shape)
print("datan_fraud before Upsampling", data_nfraud.shape)

data_fraud = resample(data_fraud, n_samples=data_nfraud.shape[0],
 random_state=42)

print("data_fraud after Upsampling", data_fraud.shape)
print("datan_fraud after Upsampling", data_nfraud.shape)

data = pd.concat([data_fraud, data_nfraud])
print("data after Upsampling", data.shape)
```

Listing 7.6 Tables Now Contain the Same Number of Entries

The table with conspicuous data is enlarged so that both tables are the same length. You can determine the length of the table using data_nfraud.shape[0]. Both tables are then merged and saved in the data variable. This allows the rest of the source code to remain unchanged. The output is shown in the following listing.

```
data_fraud before Upsampling (492, 31)
datan_fraud before Upsampling (284315, 31)
data_fraud after Upsampling (284315, 31)
datan_fraud after Upsampling (284315, 31)
data after Upsampling (568630, 31)
```

Although the correct classification rate has increased slightly, let's take a look at the confusion matrix, which provides more meaningful results (see Figure 7.5).

All conspicuous transactions were recognized as such. Only nine data records were incorrectly classified as conspicuous. It would be interesting to find out whether all columns are important. You can determine the correlation of the attributes, but with such unbalanced data, you have to remain critical. XGBoost can handle unimportant columns, but we should act on this distribution. Let's test our program after we've removed the **Time** column. The "time" attribute, that is, the time between the first and the respective transaction, could be unimportant. Indeed, after that, there are only seven data records that are incorrectly classified as conspicuous. Unfortunately, there is now a big "but"—the results are now impressive but not reliable. We've only copied data records and extended the table with them. There is a very high risk that our AI is only optimized for these 492 cases of fraud.

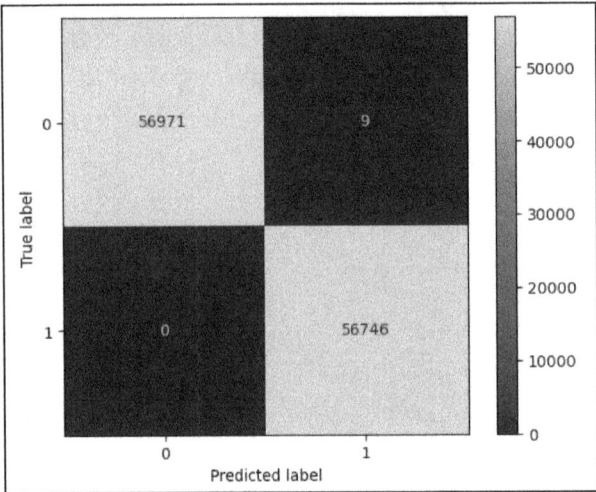

Figure 7.5 Confusion Matrix of the Results

It could therefore possibly be overfitting, which can't be ruled out with certainty. The program would have to be evaluated with new, unknown data to provide reliable information on the quality of the AI. With this distribution (only a few anomalies), it's difficult to obtain more data.

7.3 Autoencoders

We can also use ANN to detect anomalies. Autoencoders are ANNs with a special structure (see Figure 7.6).

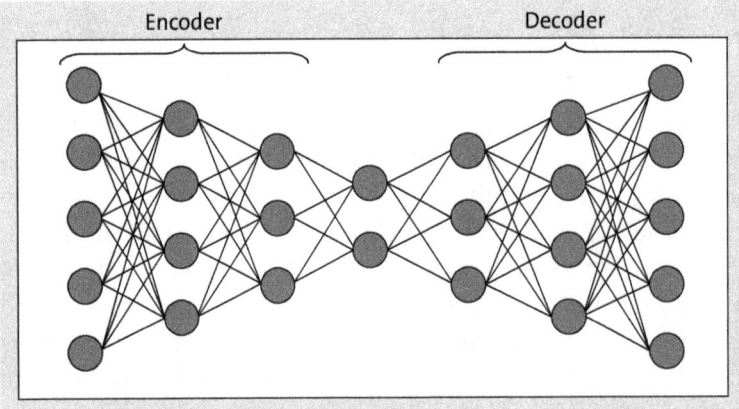

Figure 7.6 Structure of an Autoencoder

7.3 Autoencoders

The data is entered on the left in the *encoder*. A dimensional reduction takes place toward the center. From there, the dimension is expanded again so that the same number of nodes is available at the output of the *decoder* as at the input of the encoder.

Let's suppose we were to apply this structure to the iris data, and only to the *Setosa* subspecies. All other data would have to be removed. On the left, we enter the four variables (length and width of the leaves), and on the right, we compare the output with these variables. The identifier of the subspecies isn't necessary (the target column). The output is therefore not compared with the target column, but with the input data. The goal is to adapt the parameters in the ANN through training so that the same values are output at the output as are at the input. As no target column is required, autoencoders are assigned to the *unsupervised learning* area.

But why do we do this? The idea is that the ANN is precisely adapted to the data of the *Setosa iris* by this corset or bottleneck. Compression and decompression are designed for *Setosa*. If data for other subspecies is now entered at the input, the values at the output should not match the values at the input, or the differences should exceed a certain *threshold*). If this is the case, we know that the data entered doesn't belong to *Setosa*.

All right, let's get started with the *K7_credit-card-3.ipynb* file.

```
import tensorflow as tf
import pandas as pd
from sklearn.model_selection import train_test_split
from sklearn.utils import resample
from sklearn.metrics import mean_absolute_error
from keras.models import Model
from sklearn.preprocessing import StandardScaler
import numpy as np

path = "../Data/creditcard.csv"
data = pd.read_csv(path, delimiter=',')
print(data.head())

# Standardize Amount column
s_scaler = StandardScaler()
cols_to_s_scale=['Amount']
data[cols_to_s_scale] = s_scaler.fit_transform(data[cols_to_s_scale])

# remove Time
data.drop(['Time'], axis = 1, inplace=True)

# Split into two tables (both still contain the Class target column)
train_data, test_data = train_test_split(data, test_size=0.2, random_state=42)
```

7 Anomaly Detection

```
# training is only carried out with normal data
train_data = train_data[train_data['Class']==0]
# remove target column, is not required
train_data.drop(['Class'], axis = 1, inplace=True)

# The target column is removed from the test data and stored in a variable
test_col = test_data['Class']
test_data.drop(['Class'], axis = 1, inplace=True)

print(train_data.shape)
```
Listing 7.7 Distribution and Preparation of Data

In the preceding listing, the modules are imported, the first lines are output, and the data is prepared. The **Amount** column is standardized, and the target and **Time** columns are removed from the training data. The ANN later receives this data as input, and the same values are compared at the output. We also remove the target column from the test data, but save it in a variable. These values are required to evaluate the model.

```
encoder = tf.keras.Sequential(name='encoder')
encoder.add(layer=tf.keras.layers.Dense(units=20,
 activation=tf.nn.sigmoid, input_shape=[29]))
encoder.add(layer=tf.keras.layers.Dense(units=10, activation=tf.nn.sigmoid))
encoder.add(layer=tf.keras.layers.Dense(units=5, activation=tf.nn.sigmoid))

decoder = tf.keras.Sequential(name='decoder')
decoder.add(layer=tf.keras.layers.Dense(units=10,
 activation=tf.nn.sigmoid, input_shape=[5]))
decoder.add(layer=tf.keras.layers.Dense(units=20, activation=tf.nn.sigmoid))
decoder.add(layer=tf.keras.layers.Dense(units=29, activation=tf.nn.sigmoid))

autoencoder = tf.keras.Sequential([encoder, decoder], name='autoencoder')
autoencoder.compile(optimizer='adam', loss='mae', metrics=['mae'])
encoder.summary()
decoder.summary()
autoencoder.summary()
```
Listing 7.8 Building the ANN as an Autoencoder Consisting of an Encoder and a Decoder

The encoder and decoder are created, and these models in turn make up the autoencoder (see Figure 7.7). As you can see, you can also assign names to the individual models.

```
Model: "encoder"
_____
Layer (type)                 Output Shape              Param #
=================================================================
dense (Dense)                (None, 20)                600
dense_1 (Dense)              (None, 10)                210
dense_2 (Dense)              (None, 5)                 55
=================================================================
Total params: 865
Trainable params: 865
Non-trainable params: 0
_____
Model: "decoder"
_____
Layer (type)                 Output Shape              Param #
=================================================================
dense_3 (Dense)              (None, 10)                60
dense_4 (Dense)              (None, 20)                220
dense_5 (Dense)              (None, 29)                609
=================================================================
Total params: 889
Trainable params: 889
Non-trainable params: 0
_____
Model: "autoencoder"
_____
Layer (type)                 Output Shape              Param #
=================================================================
encoder (Sequential)         (None, 5)                 865
decoder (Sequential)         (None, 29)                889
=================================================================
Total params: 1,754
Trainable params: 1,754
Non-trainable params: 0
```

Figure 7.7 Structure of Encoder, Decoder, and Autoencoder

The ANN has 29 inputs and outputs. In between, the dimension is reduced to five nodes and then increased again:

```
autoencoder.fit(train_data, train_data, epochs=20, batch_size=16)
```

You'll see a significant difference from the previous models when you call the training program—the input and target data represent our training data without the target column:

```
train_pred = autoencoder.predict(train_data)
threshold = mean_absolute_error(train_pred, train_data)

print("Threshold:", threshold)
```

Listing 7.9 Determining the Threshold (First Draft)

7 Anomaly Detection

There are many ways to calculate the first draft for the threshold value. Here, we use the mean absolute error (MAE) between the training data and the prediction. It's important to note that the training data doesn't contain any cases of fraud; compression and decompression should be designed for normal data. The threshold is also a hyperparameter, but we still have to adjust this value (here, 0.59). Many books and tutorials add the standard deviation to the calculated error. You must then iteratively increase and decrease the value to find the optimum amount. But here, we only use the MAE as the initial value. Later, we'll gradually increase this value to achieve better results. The value therefore only needs to be changed in one direction.

```
new_threshold = 0.75

test_pred = autoencoder.predict(test_data)
maes = tf.keras.losses.mae(test_data, test_pred)
pred_col_bool = tf.math.greater(maes, new_threshold)
pred_col = tf.cast(pred_col_bool, dtype=tf.int32)
```

Listing 7.10 Generating Predictions

The new threshold value used here is 0.75. The model provides a prediction for the test data, which also contains cases of fraud. An entire column is created with the values for the MAE between the test data and the prediction. The system then checks line by line whether the calculated error is greater than the new threshold value. Accordingly, a column is created with true or false values (True or False of the Boolean data type). It's therefore checked whether the first parameter is greater than the second one. The values of pred_col_bool are then converted into integer values (called *type conversion* in computer science), namely 0 for False and 1 for True. In the case of fraud, this call therefore returns True (1), and the number 1 in test_col represents a case of fraud as well.

> **Boolean Data Type**
>
> Variables of the Boolean type can only have the values True or False; alternatively, the values 1 for True and 0 for False.

The following holds true for the autoencoder: if the MAE is greater than the threshold value, this is a special case; that is, maes > new_threshold. The tf.math.greater(maes, new_threshold) call then returns True (1). In this dataset, the number 1 also indicates an anomaly.

> **Select the "Greater" or "Less" Comparison**
>
> If the dataset contained the number 1 for the normal case, you could use the tf.math.less(maes, new_threshold) call. Is maes less than new_threshold? Normally, this is the case, and the call returns True (1).

This is followed by another graphical analysis to better evaluate the model.

```
from sklearn.metrics import confusion_matrix, ConfusionMatrixDisplay
cm = confusion_matrix(test_col, pred_col)
disp = ConfusionMatrixDisplay(cm)
disp.plot()
```

Listing 7.11 Output of the Confusion Matrix

A confusion matrix is now created again so that the quality of the AI can be evaluated (see Figure 7.8).

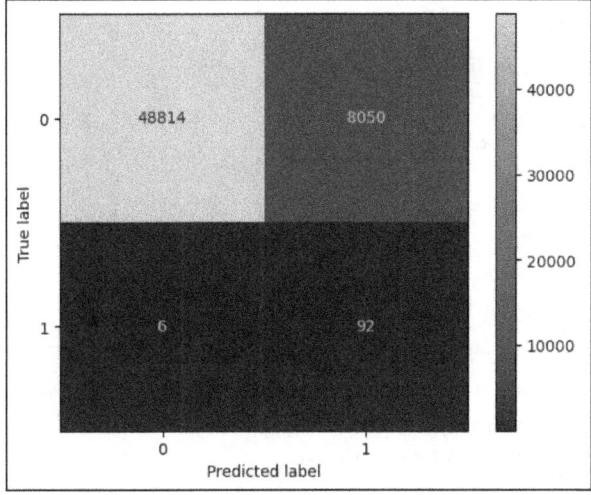

Figure 7.8 Confusion Matrix of the Results

As you can see, 94% of fraud cases are recognized. However, 8,050 transactions are incorrectly classified as cases of fraud. If, in the event of suspicion, only an email is sent with the instruction to check the transaction, you may be able to live with it. This would happen in 14% of normal transactions. The model alone isn't suitable for critical applications, and additional measures would have to be taken.

You can test the whole thing with other hyperparameters and check whether you get better results. I have made the following changes:

```
cb_early = tf.keras.callbacks.EarlyStopping(monitor='val_loss', patience=1)

autoencoder.fit(train_data, train_data, validation_data=(test_data, test_data),
 epochs=100, batch_size=16, callbacks=[cb_early])
```

In addition, I've made this change:

```
new_threshold = 0.82
```

You can see the confusion matrix that was created after these changes in Figure 7.9. For example, you can change the number of layers and nodes and compare the results.

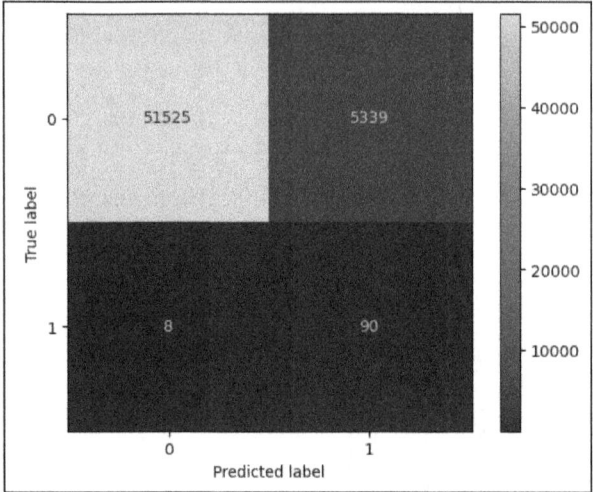

Figure 7.9 Confusion Matrix of the Results

Autoencoders with ANN and XGBoost with resampling are generally suitable for an anomaly detection. You should test both variants for your use cases and decide on the model with the better results. It's also worth researching at *www.kaggle.com*, for example, to get ideas for hyperparameter settings. Once you've gained more experience in this field, you can also analyze and evaluate other AI models.

7.4 Exercises

7.4.1 Exercise 1: Anomaly Detection Using XGBoost and Upsampling

Use the *ecg.csv* file to detect anomalies in the data (heart rhythms). The value of the last column must be predicted. Warning: in this data, the number 0 classifies the anomaly.

7.4.2 Exercise 2: Anomaly Detection Using an Autoencoder

The task is the same as for Exercise 1, but this time, you're supposed to use an autoencoder. I recommend using Google Colab with GPU support for the training process.

Chapter 8
Text Classification

An AI that interprets text content isn't much more complicated than the ones you've seen so far.

> **What This Chapter Is About**
> - Transforming words into integers using text vectorization
> - Transforming integers into vectors using the embedding layer
> - Averaging the vectors using the `GlobalAveragePooling1D` layer
> - Analyzing the functionality of individual layers
> - Performing text classification using ANN

What are the use cases for AIs that classify texts? You could train an AI with restaurant ratings (text) and a score (number). This AI would then be able to automatically calculate the score for a new review once it has been trained. All you would have to do is describe your impressions in a text, and the AI would generate a score accordingly. Literary texts could be automatically categorized into different genres, and emails could be marked as either spam or being important according to their content. AI applications of this type are summarized under the term *natural language processing* (*NLP*).

We'll use an artificial neural network (ANN) again to classify texts. However, we still lack the option of feeding entire texts to the ANN as input data. In this chapter, you'll learn the procedures for processing texts in such a way that the ANN can subsequently be trained and tested. We'll get familiar with this topic on the basis of real-life examples.

8.1 Embedding Layer

You already know that character strings can't be used for calculations, which means they must be transformed into numerical values. For example, you could replace the words "good" and "bad" in a classification with the numbers 0 and 1. We took a similar approach with the Iris dataset when we replaced the subspecies names with numbers. With this measure, we made sure that the ANN could be trained and tested because calculations have to be carried out to determine the deviation or error and to update the

8　Text Classification

weights. The character strings for text classification must also be transformed into numbers.

One possibility is to encode the individual words in the text using OHE (one hot encoding), but then there would be many long lists. The length of each list and the number of lists would correspond to the number of words in the text. The lists would only contain the number 1 in one place, and the rest would be filled with the number 0. For example, when rating restaurants, the ratings should be in text form, and the AI model should determine a score from the text. If 5,000 different words occur in the evaluations (training data and test data), and the words will be OHE-encoded, you would have a list with 5,000 elements for each word, 4,999 times the number 0 and one times the number 1, whereby the position of the number 1 uniquely identifies a word. The result would therefore be 5,000 lists with 5,000 elements each. This isn't a viable solution.

How about replacing each word with an integer? This could save a lot of storage space. In fact, this process is used as an intermediate step. The embedding layer extends this approach: each word is represented by a vector with n elements (n is also a hyperparameter). First, however, the words must be converted into integers. We'll discuss the advantages of this type of coding later. First of all, it's important that you understand how this layer works.

The order of the transformation is shown in Figure 8.1: first, the individual words are transformed into integers, and then the integers are transformed into vectors (here with three elements) using an embedding layer.

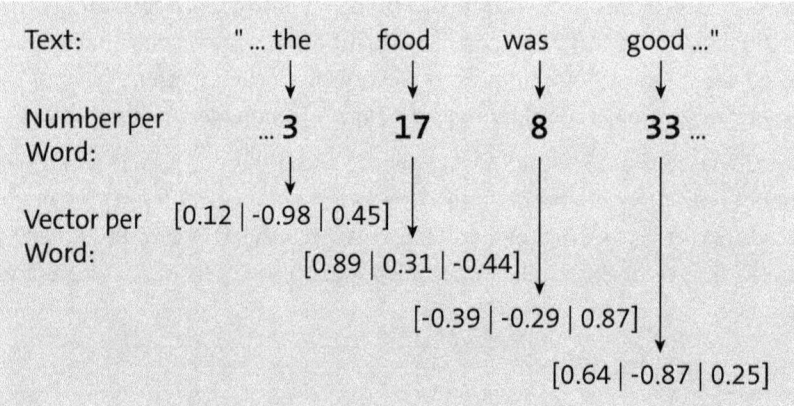

Figure 8.1 Transformation of the Words to Integers, Then to Vectors with Three Floats Each

---

Random Numbers with NumPy

The NumPy module provides several options for generating random numbers. In the following examples, we need integers (for the substitution of words), which can be realized using the `randint` method.

8.1 Embedding Layer

```
import numpy as np

# Generate random numbers 0-99
arr = np.random.randint(100, size=(2, 3))
print(arr)
```

The first parameter specifies up to which number (exclusive) the numbers are supposed to be generated (100 means 0 to 99), and the second parameter describes the size or dimensions of the array. A two-dimensional array (i.e., arrays within an array) is created here, whereby an array contains two further arrays with three elements each.

The output looks as follows:

```
[[18 53 62]
 [30 31 77]]
```

There are many ways to generate random numbers using NumPy. The NumPy website (*http://r-wrk.de/9763-numpy-random*) provides many examples and explanations.

The analysis of the relationships takes place in the *K8_emb-globAverage.ipynb* program.

```
import pandas as pd
import tensorflow as tf
from keras.models import Model
import numpy as np

# Generate random numbers 0-99
input_array = np.random.randint(100, size=(2, 4))
print(input_array)
```

Listing 8.1 Generating a Two-Dimensional Array with Random Numbers

We create two lists with four elements each. You can imagine these lists as two sentences with four words, where each word has been converted into an integer. The actual conversion of the words into integers will follow, but for now, we'll focus on how the embedding layer works. The output looks as follows:

```
[[10 48 39 81]
 [46 33 44 84]]
```

The embedding layer transforms these numbers into vectors with floats.

```
emb = tf.keras.layers.Embedding(100, 3, input_length=4)

model = tf.keras.Sequential()
model.add(emb)
```

8 Text Classification

```
output_array = model.predict(input_array)
print(output_array)
```

Listing 8.2 Transformation to Vectors Using the Embedding Layer

The embedding layer has four inputs, as the sentences each have four coded words. Each word can have 100 possible values (0 to 99), and each word is represented by a vector with three floats. The size of the vector is also a hyperparameter and depends on the task and the amount of data. For larger texts you can, for example, use 16. The result here is shown here.

```
[[[-0.01320171  0.03148221 -0.00856429]
  [-0.04194389  0.01545546  0.01220403]
  [-0.04243801  0.04818739  0.04436051]
  [ 0.02788809  0.00041498  0.01722099]]

 [[-0.00167985 -0.03003917  0.00516564]
  [-0.02093614  0.02270022 -0.0044214 ]
  [ 0.04464257  0.01409246  0.03318931]
  [ 0.02844888 -0.03360491 -0.00909815]]]
```

The two sentences in turn consist of four vectors, as the sentences also have four words. The transformation or coding was carried out successfully. Remember that the ANN isn't trained. These are randomly generated initialization values.

The embedding layer therefore ensures that integers are transformed into vectors. However, the individual words must first be transformed into integers, whereby the assignment is random. Complete examples follow, but you should first understand the process here.

The output of the embedding layer consists of vectors that can also be used for calculations. To reduce complexity, these vectors aren't fed directly into the ANN. Rather, the `GlobalAveragePooling1D` layer, which we'll look at in the next section, reduces the amount of input data for the ANN by averaging.

8.2 GlobalAveragePooling1D Layer

If an AI is to be programmed to classify texts, it must be remembered that the text usually consists of more than just one sentence. For a restaurant review, for example, you could start with a text of 500 words (per review). If the texts are longer, you can use the first 500 words and ignore the rest. However, if the text is shorter, the remainder can be filled with the number 0. Here's what it looks like: 500 words, each represented by a vector with 10 elements, for example. You need a dataset consisting of many texts to train the AI. As you can imagine, that's a lot of data. With regard to the output layer of

8.2 GlobalAveragePooling1D Layer

the ANN, we need a method or approach to reduce the dimension. With ratings of "good" and "bad," we can use the `softmax` function for the last layer, which only requires two outputs. But how can the number of parameters prior to this layer be reduced? The `GlobalAveragePooling1D` layer provides an option for reducing the dimension.

We proceed as follows to analyze the relationships: We create a model with an embedding layer and view the output. This output data is the input data of the second model with a `GlobalAveragePooling1D` layer. A comparison of the two models' output data should help you understand how these layers work.

```
emb = tf.keras.layers.Embedding(100, 3, input_length=4)

model = tf.keras.Sequential()
model.add(emb)

output_array = model.predict(input_array)
print(output_array)

model2 = tf.keras.Sequential()
model2.add(tf.keras.layers.GlobalAveragePooling1D())

output_array2 = model2.predict(output_array)
print(output_array2)
```

Listing 8.3 Two Models with One Layer Each

The output for `output_array` after the embedding layer is shown in the following listing.

```
[[[ 0.0303422   0.00916262 -0.02895613]
  [-0.04773426  0.00604304 -0.00436129]
  [-0.04543605 -0.04671688 -0.04545806]
  [ 0.03782098  0.02155227  0.02097246]]

 [[-0.00668789 -0.01421345  0.03394908]
  [ 0.0288301   0.03731045  0.04348401]
  [ 0.00281034  0.03294808  0.04089836]
  [-0.02651856  0.04532225 -0.0345437 ]]]
```

There are two sentences of four words each, so the result is two groups with four vectors each.

Here's the output for `output_array2` after the `GlobalAveragePooling1D` layer:

```
[[-0.00625178 -0.00248974 -0.01445076]
 [-0.0003915   0.02534183  0.02094693]]
```

A dimensional reduction takes place so that the sentence is only represented by a single vector and no longer one vector per word. The calculation is very simple: the arithmetic mean value is calculated for each column. Let's take the first column of the first word in the output as an example:

$$\frac{0.030422 - 0.04773426 - 0.04543605 + 0.03782098}{4} = -0.00625178$$

The result matches the first numerical value of the vector after the `GlobalAveragePooling1D` layer. The calculation is that simple. Remember that because the network hasn't yet been trained, these figures are initialization values.

But let's get back to the question of why the words are transformed into vectors. We've already established that a lot of memory is saved if the words are encoded in numbers instead of using OHE. And if you also write all the letters in lower case, you can save even more because "This" (e.g., at the beginning of a sentence) and "this" would be represented by the same number. The lowercase sentence "this is a nice building" could be represented by the numbers [17, 22, 31, 46], one number for each word. The sentence "this is a nice house" could be encoded with the numbers [17, 22, 31, 55]. Now a problem arises: the semantic relationship between building and house is completely neglected. How is learning progress supposed to take place here? The sentence "this is a nice car" would only differ from the other sentences in one place, but it means something completely different. We therefore need a solution that takes semantic relationships into account. And this is where vectors come into play. Each word is represented by a vector. The training updates the numerical values in the vectors and the weights between the layers in an ANN so that the sentences "this is a nice car" and "this is a nice vehicle" lead to the same output. Vector representation and the weights of the ANN can be used to encode simple semantic relationships between the words. This is the big advantage over other transformation options.

8.3 Text Vectorization

The words must therefore first be transformed into numerical values before the embedding layer transforms them into vector representations. Prior to that, the words are also converted to lowercase letters, and the punctuation is removed. TensorFlow or Keras provides an easy-to-use option for this, namely, *text vectorization*.

The *K8_textVectorization* program classifies teaching assessments.

```
import pandas as pd
import tensorflow as tf
from keras.models import Model
import numpy as np
```

8.3 Text Vectorization

```
# Evaluation texts
train_data =[
        'You are the best teacher!',
        'I will name my first-born after you.',
        'Your lessons are boring',
        'terrible',
        'excellent!',
        'I was just sleeping.',
        'The best lessons',
        'I have learned nothing.',
        'lame',
        'You should stop teaching.'
]
# 0 for bad, 1 for good
train_col = np.array([1,1,0,0,1,0,1,0,0,0])
```

Listing 8.4 Evaluation of the Lessons

The train_data variable contains the texts, and train_col contains the corresponding ratings. The individual words are now converted into integers.

```
transform = tf.keras.layers.TextVectorization(max_tokens=50,
 output_sequence_length=10)

transform.adapt(train_data)

tain_data_transformed = transform(train_data)

print(tain_data_transformed)
```

Listing 8.5 Transformation of Words to Integers

We specify that there is a maximum total of 50 words to be coded. If there are more words, only the most common 50 words are used. We also stipulate that the texts have a maximum of 10 words each. Using tain_data_transformed = transform(train_data), we create the table for the assignment, while tain_data_transformed = transform(train_data) finally transforms the texts into vectors. Pay attention to the differences: These vectors with the integers represent the texts, whereby each word is represented by an integer. The embedding layer, on the other hand, converts each individual word or integer into an n-dimensional vector with floats. The code that follows is stored in the tain_data_transformed variable.

```
tf.Tensor(
[[ 2  7  4  6 13  0  0  0  0  0]
 [ 3  9 18 19 24 27  2  0  0  0]
```

8 Text Classification

```
[ 8  5  7 26  0  0  0  0  0  0]
[11  0  0  0  0  0  0  0  0  0]
[25  0  0  0  0  0  0  0  0  0]
[ 3 10 22 15  0  0  0  0  0  0]
[ 4  6  5  0  0  0  0  0  0  0]
[ 3 23 20 17  0  0  0  0  0  0]
[21  0  0  0  0  0  0  0  0  0]
[ 2 16 14 12  0  0  0  0  0  0]], shape=(10, 10), dtype=int64)
```

As the values are generated randomly, the output may look different on your side. Compare the individual vectors with the texts. The first element in the first vector is the number 2. The `print(transform.get_vocabulary()[2])` call provides the output you, which is the first word in the first sentence. You should also call the method with other numbers in the vectors. You'll see that the words have been replaced by numbers in the appropriate places. Now let's build the model.

```
model = tf.keras.Sequential([
  tf.keras.layers.Embedding(50, 16),
  tf.keras.layers.GlobalAveragePooling1D(),
  tf.keras.layers.Dense(2, activation=tf.nn.softmax)
])
```

Listing 8.6 ANN Structure

The embedding layer is called with two parameters: there is a maximum of 50 words in total, and each individual word is to be transformed into a 16-dimensional vector.

```
model.compile(optimizer='adam', loss='sparse_categorical_crossentropy',
 metrics=['accuracy'])

model.fit(tain_data_transformed, train_col, epochs=200)
```

Listing 8.7 Training the Model

After 200 epochs, we get a correct classification rate of 90%. With this small dataset and this large number of epochs, overfitting certainly plays a major role, but that is beside the point here, as we only want to understand the principle.

```
reviews = [
    "terrible",
    "the best lessons",
    "please stop",
    "still asleep"
]
txt = transform(reviews)
```

```
pred = model.predict([txt])
print(pred)
```

Listing 8.8 Applying the Trained Model

Now the trained AI is supposed to classify new ratings. I've deliberately used similar wording in the text fragments. For everything else, the training data is too small and the number of epochs too high. The output is show in the following listing.

```
[[0.6918644  0.30813566]
 [0.48320082 0.5167992 ]
 [0.678091   0.32190904]
 [0.64080495 0.35919502]]
```

The second rating is assigned to one class (good = 1); all other ratings are assigned to the other class (bad = 0). This very simple example actually works.

8.4 Analysis of the Relationships

We can now program an AI that classifies larger texts. In other disciplines (e.g., electrical engineering), it's common to illustrate certain relationships by means of experiments, and we want to apply this procedure here as well: What values do the vectors have before and after training? Can you recognize relationships from this? You can read the Keras documentation on the layers, but you probably won't find any details on the structure. The documentation is designed to support you in using the modules. We now use the *K8_analysis.ipynb* file to try to unravel the secrets of the vectors.

```
import pandas as pd
import tensorflow as tf
from keras.models import Model
import numpy as np

# Texts for tests
train_data =[
        'eating a hamburger',
        'eating a steak',
        'eating a banana',
        'reading interesting books',
        'reading interesting magazines',
        'reading interesting texts',
]
# 1 for eating, 0 for reading
train_col = np.array([1,1,1,0,0,0])
```

Listing 8.9 Generating Simple Training Data

8 Text Classification

These simple texts are similar. The first three text fragments are about eating something, while the others are about reading interesting texts. Accordingly, these texts are classified as 0 or 1.

```
transform = tf.keras.layers.TextVectorization(max_tokens=20,
 output_sequence_length=3)

transform.adapt(train_data)

tain_data_transformed = transform(train_data)

print(tain_data_transformed)
```

Listing 8.10 Transforming Words into Integers

After the transformation, there are six vectors, one for each text. As the texts consist of three words, the vectors also have three elements. The content of the `tain_data_transformed` variable is given in the following:

```
tf.Tensor(
[[ 4  5  9]
 [ 4  5  7]
 [ 4  5 11]
 [ 2  3 10]
 [ 2  3  8]
 [ 2  3  6]], shape=(6, 3), dtype=int64
```

Each word has been replaced by a number. You can use `print(transform.get_vocabulary()[4])` to transform the value back (results in "eating").

```
emb = tf.keras.layers.Embedding(20, 4, input_length=3)
glob = tf.keras.layers.GlobalAveragePooling1D()
dense = tf.keras.layers.Dense(2, activation=tf.nn.softmax)
```

Listing 8.11 Saving Individual Layers in Variables

These layers are saved in variables so that they can be called repeatedly.

```
model = tf.keras.Sequential()
model.add(emb)

output_array = model.predict(tain_data_transformed)
print(output_array)

model2 = tf.keras.Sequential()
model2.add(glob)
```

```
output_array2 = model2.predict(output_array)
print(output_array2)
```

Listing 8.12 Two Models with One Layer Each

One model each is created with an embedding layer and a `GlobalAveragePooling1D` layer. The first model receives the transformed texts as input, and the second model receives the output of the first model. The output of the first model is shown in the following listing.

```
[[[ 0.01049133 -0.04971594 -0.00242941 -0.01039904]
  [-0.0146427  -0.03859086 -0.04132865 -0.04318663]
  [ 0.04486222  0.00906948  0.00579925  0.04466336]]

 [[ 0.01049133 -0.04971594 -0.00242941 -0.01039904]
  [-0.0146427  -0.03859086 -0.04132865 -0.04318663]
  [-0.00364053 -0.00436382  0.01563049  0.00180672]]

 [[ 0.01049133 -0.04971594 -0.00242941 -0.01039904]
  [-0.0146427  -0.03859086 -0.04132865 -0.04318663]
  [-0.03530319 -0.02164061  0.00330987  0.00134084]]

 [[ 0.0252908  -0.02948647  0.02983702  0.00434661]
  [-0.04164221  0.02052603  0.01361841 -0.04364735]
  [ 0.00749851  0.04306162  0.00157911  0.04272323]]

 [[ 0.0252908  -0.02948647  0.02983702  0.00434661]
  [-0.04164221  0.02052603  0.01361841 -0.04364735]
  [ 0.03996739 -0.04306407 -0.01306259  0.0460949 ]]

 [[ 0.0252908  -0.02948647  0.02983702  0.00434661]
  [-0.04164221  0.02052603  0.01361841 -0.04364735]
  [-0.01442187 -0.01923801  0.03613246  0.00200254]]]
```

Each word is represented by a vector with four floats, and each text consists of three vectors. Take a close look at the output, and compare it with the texts before the transformation. Try to recognize the assignment. The first vectors of the first three vector groups are the same, as they represent the word "eating". Try to find out more connections. The output of the second model is given in the following:

```
[[ 0.01357028 -0.02641244 -0.01265294 -0.00297411]
 [-0.0025973  -0.03089021 -0.00937586 -0.01725965]
 [-0.01315152 -0.03664914 -0.01348273 -0.01741494]]
```

```
[-0.00295097  0.01136706  0.01501151  0.00114083]
[ 0.00787199 -0.0173415   0.01013095  0.00226472]
[-0.01025776 -0.00939948  0.0265293  -0.01243273]]
```

The mean values are calculated column by column for the texts, so that at the end there is one vector with floats per text.

```
model3 = tf.keras.Sequential()
model3.add(emb)
model3.add(glob)
model3.add(dense)
model3.compile(optimizer='adam', loss='sparse_categorical_crossentropy',
 metrics=['accuracy'])
model3.fit(tain_data_transformed, train_col, epochs=100)
```

Listing 8.13 Creation and Training of a Third Model with the Previously Used Layers

We use these layers to build and train a third model. We want to see how the values of the vectors change. To do this, we create new models again, but with the same two layers used previously.

```
model4 = tf.keras.Sequential()
model4.add(emb)

output_array4 = model4.predict(tain_data_transformed)
print(output_array4)

model5 = tf.keras.Sequential()
model5.add(glob)

output_array5 = model5.predict(output_array4)
print(output_array5)
```

Listing 8.14 Creation of Two Models with One Layer Each to Analyze the Values

These simple models are again used to analyze the output vectors. The output of the first model (model4) is given in the following:

```
[[[-0.090047   -0.15439196 -0.10349681  0.08845942]
  [-0.11518102 -0.1432669  -0.14239605  0.05567183]
  [-0.05569917 -0.09562722 -0.09529078  0.14354557]]

 [[-0.090047   -0.15439196 -0.10349681  0.08845942]
  [-0.11518102 -0.1432669  -0.14239605  0.05567183]
  [-0.10420263 -0.10906133 -0.08546025  0.10068955]]
```

```
[[-0.090047   -0.15439196 -0.10349681  0.08845942]
 [-0.11518102 -0.1432669  -0.14239605  0.05567183]
 [-0.13578954 -0.12626342 -0.0977053   0.10014768]]

[[ 0.12583329  0.07519313  0.1309083  -0.09451628]
 [ 0.0589003   0.12520564  0.11468967 -0.1425102 ]
 [ 0.10802086  0.1477194   0.10262994 -0.05612   ]]

[[ 0.12583329  0.07519313  0.1309083  -0.09451628]
 [ 0.0589003   0.12520564  0.11468967 -0.1425102 ]
 [ 0.14055818  0.06166125  0.08805657 -0.05281706]]

[[ 0.12583329  0.07519313  0.1309083  -0.09451628]
 [ 0.0589003   0.12520564  0.11468967 -0.1425102 ]
 [ 0.08608709  0.08540642  0.1371699  -0.09682731]]]
```

You can see that the vectors of the first three texts are similar; in this example, they are all negative. The vectors of the next three texts are positive. The relationships become clearer after calculating the mean value.

```
[[-0.08697573 -0.13109536 -0.11372788  0.09589227]
 [-0.10314355 -0.1355734  -0.11045104  0.08160693]
 [-0.11367252 -0.14130743 -0.11453272  0.08142632]
 [ 0.09758481  0.1160394   0.11607596 -0.0977155 ]
 [ 0.10843059  0.08735334  0.11121818 -0.09661451]
 [ 0.09027356  0.0952684   0.12758929 -0.11128459]]
```

The similarity of the respective vectors can be seen particularly clearly here. For the last dense layer, it's no longer a problem to assign the vectors to the classes using the softmax function. You can use this experimental setup to better illustrate the processes surrounding the vectors. Now we can turn our attention to the classification of larger texts where the principle remains the same.

8.5 Classifying Large Amounts of Data

In this example, the AI is trained with the IMDb (Internet Movie Database). The *IMDB-Dataset.csv* file contains 50,000 movie reviews, where each review contains the information as to whether the text contains a positive or negative review.

```
Import tensorflow as tf
import pandas as pd
from sklearn.model_selection import train_test_split

path = "../Data/IMDB-Dataset.csv"
```

8 Text Classification

```
data = pd.read_csv(path, delimiter=', ')

MAX_FEATURES = 10000
SEQUENCE_LENGTH = 250

print(data.head())
print(data.shape)
```
Listing 8.15 Loading the Modules and Dataset

You can see that the MAX_FEATURES (maximum number of words) and SEQUENCE_LENGTH (maximum length of a review) variables have already been defined. If you want to adjust these values, you can do this here in a central location and don't need to search for the relevant places in the source code. These variables don't get changed in the source code, so they are constant variables. It has become established that constants are capitalized.

```
# Transform string to integer
# new column, set all values to 0
data['rating'] = 0

# if the content of the 'sentiment' column is equal to 'positive', set
# the entry in the 'rate' column to 1
data.loc[data['sentiment'] == 'positive', 'rating'] = 1

col = data['rating']

# remove the 'sentiment' and 'rate' columns
data.drop(['sentiment', 'rating'], axis = 1, inplace=True)
```
Listing 8.16 Preparing the Data

The values in the sentiment column are positive or negative. Here, you can see another simple way of transforming these values into numerical values. First, a new rating column is created and filled with the number 0. The transformation is then carried out using pandas as follows:

```
data.loc[data['sentiment'] == 'positive', 'rating'] = 1
```

If the sentiment column has the value positive, the value in the rating column in the same row gets set to 1. Then, the two tables are created.

```
train_data, test_data, train_col, test_col = train_test_split(data,
 col, test_size=0.2)
transform = tf.keras.layers.TextVectorization(max_tokens=MAX_FEATURES,
 output_sequence_length=SEQUENCE_LENGTH)
```

```
transform.adapt(train_data)

tain_data_transformed = transform(train_data)
test_data_transformed = transform(test_data)

print(tain_data_transformed)
```
Listing 8.17 Creation of the Four Tables and Transformation of the Words to Integers

The four tables are created. The mapping table for the transformation is only created for the training data (`transform.adapt(train_data)`). The test data simulates unknown data, which is why text vectorization must not know this data for the creation of the mapping table. Test data and training data are then transformed. If the test data contains new, unknown words, these will be replaced by a number that stands for "unknown word."

```
model = tf.keras.Sequential([
  tf.keras.layers.Embedding(MAX_FEATURES, 16),
  tf.keras.layers.GlobalAveragePooling1D(),
  tf.keras.layers.Dropout(0.2),
  tf.keras.layers.Dense(2, activation=tf.nn.softmax)
])
model.compile(optimizer='adam', loss='sparse_categorical_crossentropy',
 metrics=['accuracy'])

cb_early = tf.keras.callbacks.EarlyStopping(monitor='val_loss', patience=3)

model.fit(tain_data_transformed, train_col, validation_data=(
 test_data_transformed, test_col), epochs=100, callbacks=[cb_early])
```
Listing 8.18 Setting Up, Training, and Testing the Model

By now, you should be able to interpret these lines yourself. We achieve a correct classification rate of over 88%.

```
examples = [
    "The movie was great",
    "It was boring",
    "Don't waste your time",
]

txt = transform(examples)
pred = model.predict([txt])
print(pred)
```
Listing 8.19 Applying the Trained Model

8 Text Classification

The AI is now being tested with three new unknown texts. The output looks as follows:

```
[[0.4801555  0.5198445 ]
 [0.8410477  0.15895234]
 [0.91192335 0.08807658]]
```

You can see from the output that the positive text is assigned to one class and the negative texts to the other class. The AI can now read movie reviews and classify them itself, with a correct classification rate of 88%.

The text contains formatting characters (
 for line break), which you can remove via the following method call:

```
data['review'] = data['review'].str.replace("<br />"," ")
```

The formatting character is replaced by a space.

With a little programming knowledge, you can further improve the AI. *Stemming* and *lemmatization* can be used to transform the terms into basic forms. As a result, there are fewer variations in the transformation, and the dictionary is smaller. When you use stemming, some letters of the word may be removed. For example, the algorithm can convert the words "programming" and "programmed" to "program." Lemmatization goes one step further and recognizes word variations. The words "went," "go," and "gone" can be translated to "go."

8.6 Exercises

8.6.1 Exercise 1: Hyperparameter Optimization

Try to increase the correct classification rate of the *K8_imdb.ipynb* program by means of hyperparameter optimization. There is no sample solution for this task.

8.6.2 Exercise 2: Text Classification

Program an AI that reads the *spam.csv* file and classifies the texts as "ham" or "spam."

8.6.3 Exercise 3: Text Classification Using Upsampling

Expand the previous task to ensure a balance in the data.

Chapter 9
Cluster Analysis

The cluster analysis enables you to recognize similarities in the datasets and to form groups accordingly.

> **What This Chapter Is About**
> - Graphical analysis of the data using a scatterplot
> - How k-means clustering works
> - Grouping unlabeled data using k-means clustering

Let's take our iris data again as an example. How can such a collection of data come about? A biologist walks through the fields, measures the flowers of the various irises, and collects this data in a table. Thanks to the biologist's expertise, the name of the subspecies can be added to the table straight away.

You can now use this table to program an AI that will take over the classification into the subspecies in the future. Only the leaves need to be measured (this no longer requires a biologist), and the AI can then determine the subspecies itself.

Cluster analysis goes one step further: As a layman, I would like to measure the leaves of the irises and record the data in a table, without any knowledge of subspecies. The AI should then analyze whether there are similarities in the dataset. Accordingly, the data should be divided into groups, ideally into subspecies. This means that biologists are no longer needed to classify this data or determine subspecies.

But even with the simple Iris dataset, the generated groups don't exactly match the expected target column (i.e., the subspecies). The procedure presented here groups the data based on the available attributes; the result doesn't have to match your expectations. We'll look at the whole thing by means of examples.

What is the purpose of cluster analysis then? Now, let's say you have a large, unlabeled dataset of customer orders. You can use cluster analysis to group this data and try to recognize patterns. Perhaps you can deduce from the generated groups that, for example, customers from a certain area place similar orders.

For the presentation of the procedure, the simple Iris dataset is again used to better evaluate the results. You should regard cluster analysis as a method of recognizing patterns in a dataset and forming groups accordingly. The result or the target column that you expect doesn't necessarily come out.

9 Cluster Analysis

9.1 Graphical Analysis of the Data

The visualization of the data is very helpful for understanding the cluster analysis. We'll use the *K9_analysis.ipynb* program to analyze the data graphically.

```
import pandas as pd
import matplotlib.pyplot as plt
from sklearn.cluster import KMeans

path = "../Data/iris.csv"
data = pd.read_csv(path, delimiter=',')
print(data.head())
```

Listing 9.1 Loading the Modules and File

The required modules are imported, the file is loaded, and the first lines are output.

```
plt.scatter(data['sepal.length'], data['sepal.width'])
plt.xlabel('sepal.length')
plt.ylabel('sepal.width')
plt.show()
```

Listing 9.2 Output of the Scatterplot: Sepal

In the first plot, we look at the sepal length and width (see Figure 9.1). Here, you can see the first problem with the visualization of data: only the length and width of a leaf type (here, sepalum) can be displayed at the same time on the x-axis and y-axis, respectively. Three columns or attributes can also be output with a 3D plot, but you can no longer display four columns visually.

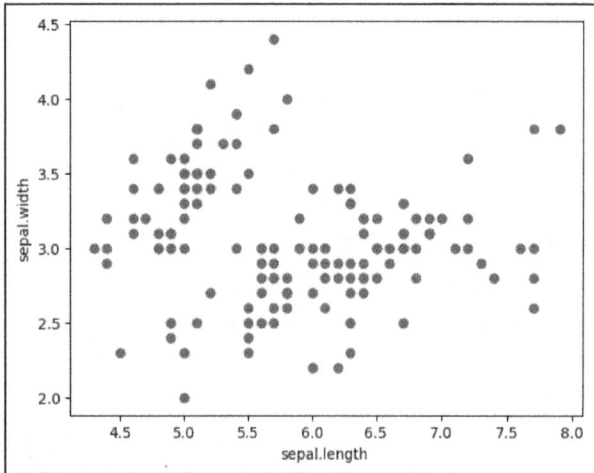

Figure 9.1 Scatterplot of the Sepal

The Matplotlib module has numerous options for visualizing data. We're only looking at the scatterplot here. Each point in this scatterplot corresponds to a row in the comma-separated values (CSV) file. For each point, you can (approximately) read off the length and width of this leaf.

```
plt.scatter(data['petal.length'], data['petal.width'])
plt.xlabel('petal.length')
plt.ylabel('petal.width')
plt.show()
```

Listing 9.3 Output of the Scatterplot: Petal

Accordingly, we can look at the length and width of the petal, see Figure 9.2. The rows remain similar, only the new data for the axes must be updated.

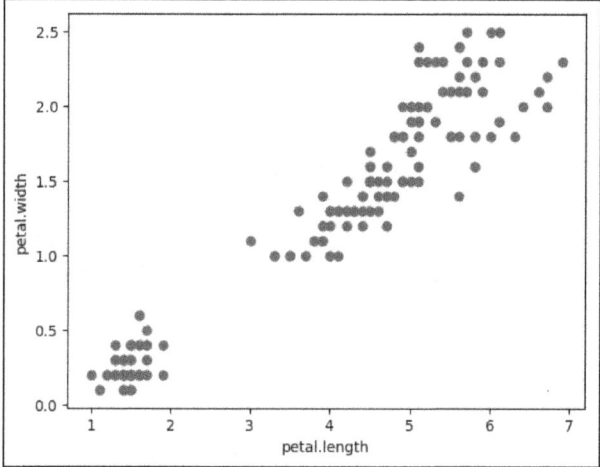

Figure 9.2 Scatterplot of the Petal

It's important that you select two meaningful columns whose context you want to look at. Here, it's the length and width of a leaf type. For example, it would probably make little sense to look at the relationship between sepal length (x-axis) and petal width (y-axis).

It's not very difficult to select the correct two columns for iris data. But what could you look at in the wine quality data, for example? There are many ways to do this, for example, the pH value (x-axis) and quality (y-axis) or the alcohol content and quality. A 3D plot would also be possible with three contiguous columns.

Up to now, we've output the data without reference to the grouping. The Iris dataset already contains the assignment to the subspecies. We can use pandas to divide this data into the different groups with one line each and then visualize this data using different colors and shapes.

9 Cluster Analysis

```python
setosa = data[data['species']=='Setosa']
versicolor = data[data['species']=='Versicolor']
virginica = data[data['species']=='Virginica']

plt.scatter(setosa['sepal.length'], setosa['sepal.width'], color = 'red',
marker='s')
plt.scatter(versicolor['sepal.length'], versicolor['sepal.width'], color =
'blue',
 marker='o')
plt.scatter(virginica['sepal.length'], virginica['sepal.width'], color =
'green',
 marker='^')

plt.xlabel('sepal.length')
plt.ylabel('sepal.width')
plt.show()
```

Listing 9.4 Scatterplot with Consideration of Grouping: Sepal

The three subspecies are output with red, blue, and green colors and square, circle, and triangle shapes. Take a look at the plot in Figure 9.3. If our table only contained the length and width of the sepal, a clear classification would not be possible. Many points in the scatterplot overlap or are distributed in such a way that they can't be divided into areas.

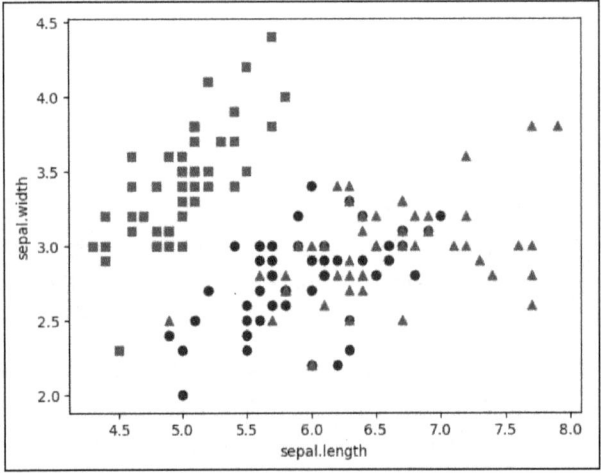

Figure 9.3 Scatterplot of the Sepal with Division into Different Subspecies

What should an algorithm be based on? No matter which areas you define for certain groups in this 2D illustration, there will always be points assigned to the wrong group.

9.1 Graphical Analysis of the Data

A procedure is therefore required to group ungrouped data (as in Figure 9.1 shown earlier) in such a way that the result corresponds to Figure 9.3.

Now, we can also look at the length and width of the petal when we update the data for the axes.

```
setosa = data[data['species']=='Setosa']
versicolor = data[data['species']=='Versicolor']
virginica = data[data['species']=='Virginica']

plt.scatter(setosa['petal.length'], setosa['petal.width'], color = 'red',
marker='s')
plt.scatter(versicolor['petal.length'], versicolor['petal.width'], color =
'blue',
 marker='o')
plt.scatter(virginica['petal.length'], virginica['petal.width'], color =
'green',
 marker='^')

plt.xlabel('petal.length')
plt.ylabel('petal.width')
plt.show()
```

Listing 9.5 Scatterplot with Consideration of Grouping: Petal

The distribution in Figure 9.4 looks more like areas could be defined here that contain the respective subspecies. There are overlaps here too but nowhere near as many as with the sepal.

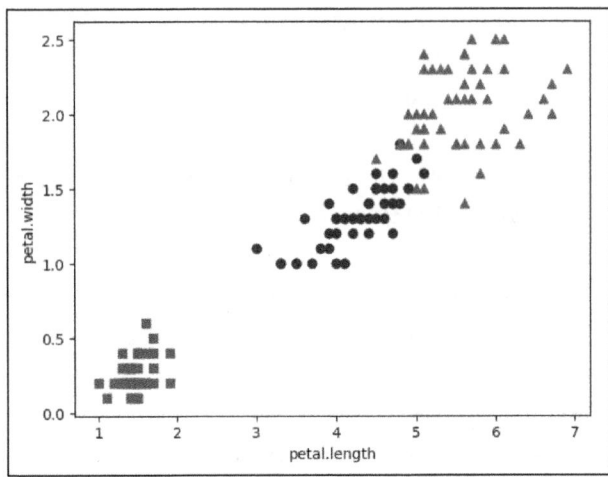

Figure 9.4 Scatterplot of the Petal with Division into Different Subspecies

Let's recap: a simple grouping of this data in 2D space isn't possible (or hardly possible). The comparison of the scatterplots of already grouped and ungrouped data is intended to illustrate this.

9.2 The k-Means Clustering Algorithm

Let's take a look at how the data can be assigned to specific groups. We won't program the algorithm ourselves but instead use modules with appropriate methods. However, it's helpful for the development of the model and the later evaluation of the result if you have an idea of how the process works.

Basically, k-means clustering is about randomly placing certain data centers and then iteratively moving them so that these data centers are always at the center of a group of data.

> **Yellowbrick Module**
>
> The Yellowbrick module in Python is based on Matplotlib and provides numerous options for graphically analyzing data. The module was developed with regard to important plots that facilitate or support development in the field of machine learning. It's worth taking a look at the module's website (*www.scikit-yb.org/en/latest*).
>
> Please install the module before you run the programs discussed in this chapter.

First, we need to determine how many groups there should be. Now you may be wondering how we're supposed to know that because we need a process that does this exact task for us. Well, we'll go into that later.

Let's use the Iris dataset again, but without the target column. To demonstrate how the process works, we provide three groups. Three points are selected at random from the dataset. Let's call these points "centers." In Figure 9.5, you can see the centers in the scatterplot.

Each data point calculates the distance to the three centers. The data points are assigned to the center from which they have the smallest distance. This way, the first imprecise (and temporary) groups are formed.

Then, the position of the three centers gets updated. For this purpose, the positions or distances of the data points to the centers in the temporary group are determined, and the mean value is calculated. These are the new positions of the centers.

After that, the procedure will be repeated. The temporary groups are dissolved, the distances are recalculated, new temporary groups are formed, the respective centers are formed, and the position of the centers is updated.

This procedure is repeated until there is no more change in position. The centers then form the focal points of the respective final groups (see Figure 9.6).

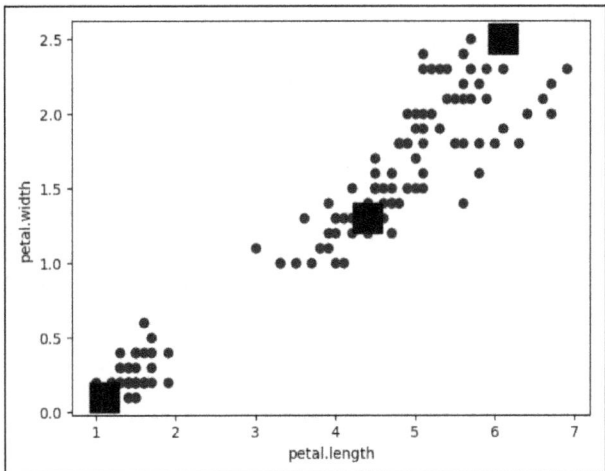

Figure 9.5 Three Randomly Selected Points in the Dataset

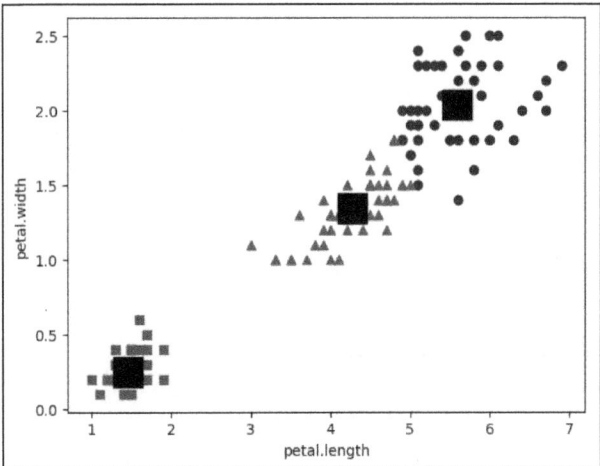

Figure 9.6 Three Centers as Focal Points of the Final Groups

Let's now get back to the question: How are we supposed to know in advance how many groups there will be? Well, we have no way of knowing.

We test the algorithm iteratively, for example, using 2 (k = 2) to 10 (k = 10) centers, and calculate a quality criterion in each case. The results are then compared with each other. One possible quality criterion is *sum of squares error* (*SSE*), also called *distortion score* or *inertia*:

$$\text{SSE} = \sum_{i=1}^{n} (y_i - \hat{y}_i)^2$$

y_i is the actual value (here, the position of the data point), and $\hat{y}_i$ is the target value (position of the center). The individual differences of all data points to the respective focal points are calculated, squared, and then added up. For our example with the three focal points, SSE = 31.37. If we choose two centers, SSE = 86.39, and if we choose four centers, SSE = 19.47.

Note that, in general, the distances become smaller when there are more centers. Compare this with the following situation: The more people there are in a room, the smaller the average distance between them. The interpretation of these distances alone is therefore not sufficient to calculate the recommended number of groups. Otherwise, the statement would be, the more centers (the higher the value for k), the better. Let's analyze the SSE from $k = 2$ to $k = 8$ in graphically.

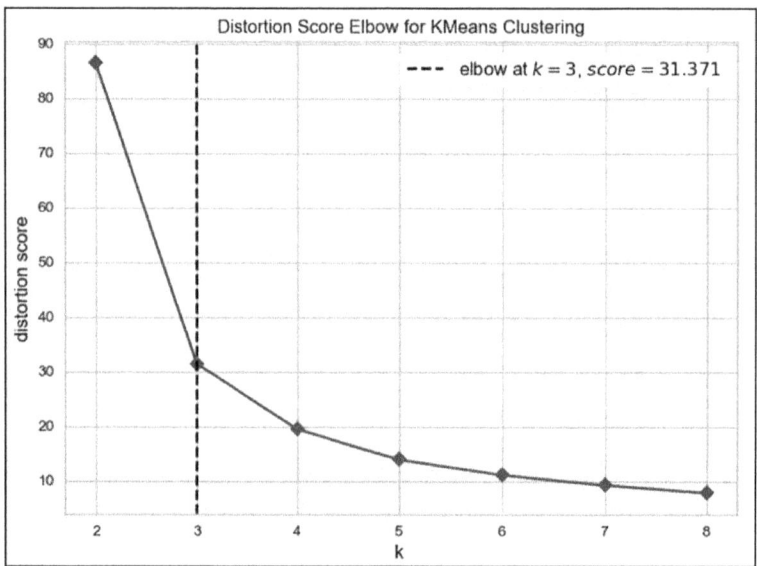

Figure 9.7 Distortion Score from k = 3 to k = 8

How do we determine the optimal number for k (number of centers)? We start with $k = 2$ and observe when the negative slope flattens out. In our example, that's the case with $k = 3$, after which the SSE no longer falls as rapidly. This is also known as the elbow method. We don't need to visually determine this inflection point ourselves, as it can be determined automatically using the Yellowbrick module. This inflection point (based on distances) implies the determined number of groups in the dataset.

Note that a satisfactory result can't always be obtained by means of k-means clustering. If the data points are very scattered or if there are many overlaps, groups may be formed incorrectly. On the other hand, the method can also be used with multidimensional data (in that case, more dimensions mean more columns or attributes), increasing the risk of incorrectly grouped data.

9.3 The Finished Program

We did a lot of analysis with the Iris dataset to understand the algorithm. Using the *K9_iris.ipynb* program, we'll briefly and concisely apply what we've learned to the Iris dataset so that you can adapt the program to other datasets with minor changes.

```
import pandas as pd
from sklearn.cluster import KMeans
from yellowbrick.cluster import KElbowVisualizer
from sklearn.preprocessing import StandardScaler

path = "../Data/iris.csv"

data = pd.read_csv(path, delimiter=',')

# Create data without target columns
data_unknown = data.drop(['species'], axis = 1)

print(data_unknown.head())
```
Listing 9.6 Loading Modules and the Dataset, and Removing the Target Column

The dataset is loaded, and the target column is immediately removed. We want to group the data ourselves or have it grouped by the algorithm.

```
s_scaler = StandardScaler()

data_unknown = pd.DataFrame(s_scaler.fit_transform(data_unknown),
 columns = data_unknown.columns)
```
Listing 9.7 Standardizing Data

Standardizing the data can lead to better results; otherwise (with some data), the positions of the focal points are distorted, especially if different columns have very different value ranges. The scaler discards the column headings, so we create a `DataFrame` with the result, whereby the old column headings are adopted.

```
model = KMeans()

visualizer = KElbowVisualizer(model, k=(2,9), timings=False)
visualizer.fit(data_unknown)
visualizer.show()
```
Listing 9.8 Visualization of the Results for k = 2–9

9 Cluster Analysis

A model is created and transferred to the visualizer. We also specify the numerical range for *k*. The elbow gets plotted, and the inflection point is determined. The output is shown in Figure 9.8. Here, you can also specify that the calculation time should not be output. You can set timings=True as a test.

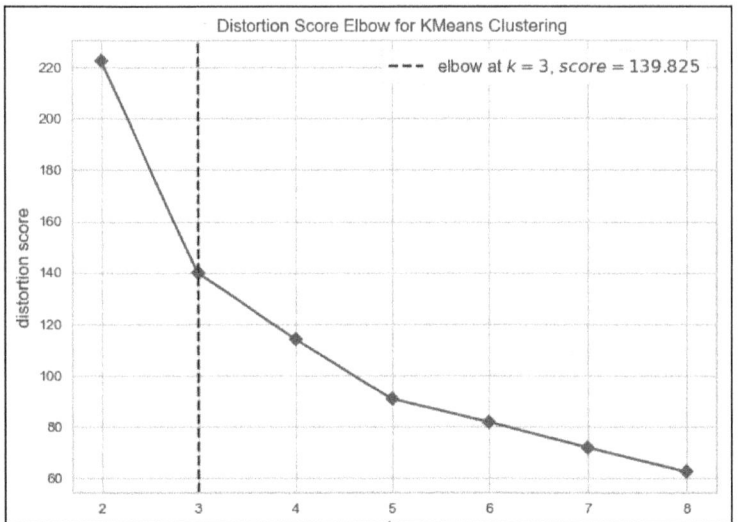

Figure 9.8 Distortion Score for k = 2 to k = 8 with Standardized Data

```
kmeans = KMeans(n_clusters=3)

pred = kmeans.fit_predict(data_unknown)
```
Listing 9.9 k Set to 3 and Data Grouped

Now that we've determined the number 3 for *k* graphically, we create a model with three centers and have the data grouped. The results are in pred.

```
data_new = pd.concat([data, pd.DataFrame(pred, columns=['label'])], axis=1)
print(data_new)
```
Listing 9.10 Dataset Expanded to Include the New Target Column

The new column is appended to the original table with the "label" heading. You'll see in the result that the data has been assigned to groups 0 to 2. The result is saved in a CSV file.

```
data_new.to_csv("./data_new.csv")
```

You can open that file from within Jupyter by double-clicking on it.

sepal.length	sepal.width	petal.length	petal.width	species	label
5.7	2.8	4.5	1.3	Versicolor	1
6.3	3.3	4.7	1.6	Versicolor	2
4.9	2.4	3.3	1.0	Versicolor	1
6.6	2.9	4.6	1.3	Versicolor	1
5.2	2.7	3.9	1.4	Versicolor	1
5.0	2.0	3.5	1.0	Versicolor	1
5.9	3.0	4.2	1.5	Versicolor	1
6.0	2.2	4.0	1.0	Versicolor	1
6.1	2.9	4.7	1.4	Versicolor	1
5.6	2.9	3.6	1.3	Versicolor	1
6.7	3.1	4.4	1.4	Versicolor	2
5.6	3.0	4.5	1.5	Versicolor	1
5.8	2.7	4.1	1.0	Versicolor	1
6.2	2.2	4.5	1.5	Versicolor	1
5.6	2.5	3.9	1.1	Versicolor	1
5.9	3.2	4.8	1.8	Versicolor	2
6.1	2.8	4.0	1.3	Versicolor	1

Figure 9.9 Extract from the Generated CSV File

Analyze the resulting CSV file, and look in particular at the **species** and **label** columns. There are some incorrect classifications or the results don't always match the original subcategories.

Remove the line that contains the standardization from the code (or comment it out) and look at the results again. The standardization of data with k-means clustering is particularly recommended if attributes have very different value ranges (e.g., see the discussion at *http://r-wrk.de/9763-scale-data*). If the datasets have nonnumeric columns, these must first be converted into numeric values.

The result of this procedure also depends on the available data. In the case of wine data, clustering in the search for groups in terms of quality can't be compared with the evaluation of people, as people don't always evaluate objectively or different people evaluate the same bottle of wine differently.

The procedure attempts to find groups in the given data. It doesn't have to be the "target column" you expect. You may need to group a dataset according to quality, but the result is a grouping according to chemical properties. It's desirable to have some test data (with the target column) to be able to evaluate the results. If, on the other hand, there is a lot of data that includes the target column, you can use a different method (e.g., a classification using XGBoost). The model can then be trained with this data and subsequently classify new data independently.

9.4 Exercises

9.4.1 Exercise 1: Grouping of Diamonds

Program an AI to group the data in the *diamonds.csv* file. To do this, remove the *cut* column and try to achieve this grouping again by way of clustering. You'll notice that the result differs from the target column. The question for such tasks should rather be the following: Can new insights be gained from the generated groups?

9.4.2 Exercise 2: Grouping of Mushrooms

Program an AI to group the data in the *mushrooms.csv* file. Use the complete dataset for the procedure.

> **Transform Columns into Numerical Values**
>
> You can use the following lines to transform all columns of a dataset into numerical values:
>
> ```
> data_unknown = data_unknown.astype('category')
> data_unknown = data_unknown.apply(lambda x: x.cat.codes)
> print(data_unknown)
> ```

Chapter 10
AutoKeras

AutoKeras builds the AI for you and prepares the data. You can achieve impressive results with just a few lines of source code.

> **What This Chapter Is About**
> - Classification and regression using table-based data
> - Image and text classification

AI models are developed in this chapter using AutoKeras for known problems that we've dealt with in previous chapters. The goal is to simplify the development of AI models even further.

When developing AI models, you've likely noticed that many steps repeat themselves, such as preparing data, and building, training, testing, and optimizing AI. AutoKeras automates many things for you, making the development of artificial neural networks (ANNs) even easier.

> **AutoKeras**
>
> François Chollet developed Keras, a library that allows you to easily program ANNs. The AutoKeras module can be regarded as the next evolutionary stage of this library. Haifeng Jin, François Chollet, Qingquan Song, and Xia "Ben" Hu have created a module that relieves you of many steps in the development of AI models. Several very clear programs enable you to develop AI models that relieve you of many work steps and deliver really good results.
>
> You can find more tutorials at *https://autokeras.com/* if you want to deepen your knowledge. Before we create our first program, you need to install the AutoKeras module via Anaconda.

In previous chapters, we've already programmed AI models for the datasets used in this book. In this chapter, we'll develop new AI models using AutoKeras. This way, you'll have a direct comparison of the development methods for the same datasets. Again, we use classification, regression, and image and text classification.

10 AutoKeras

10.1 Classification

Let's again classify our irises (*K10_iris-1.ipynb*). The goal is to classify an iris based on the length and width of its leaves. This time, the program for this dataset is even clearer than the models we developed in the previous chapters.

```
import pandas as pd
from sklearn.model_selection import train_test_split
import autokeras as ak

path = "../Data/iris.csv"
data = pd.read_csv(path, delimiter=',')
print(data.head())
print(data.info())
```

Listing 10.1 Loading Modules and the Dataset

The modules are imported, and the first rows are output (see Figure 10.1). The `info()` method provides you with further information on the data, such as the data types of the columns and information on whether empty cells are included.

```
   sepal.length  sepal.width  petal.length  petal.width species
0           5.1          3.5           1.4          0.2  Setosa
1           4.9          3.0           1.4          0.2  Setosa
2           4.7          3.2           1.3          0.2  Setosa
3           4.6          3.1           1.5          0.2  Setosa
4           5.0          3.6           1.4          0.2  Setosa
####################################################
<class 'pandas.core.frame.DataFrame'>
RangeIndex: 150 entries, 0 to 149
Data columns (total 5 columns):
 #   Column        Non-Null Count  Dtype
---  ------        --------------  -----
 0   sepal.length  150 non-null    float64
 1   sepal.width   150 non-null    float64
 2   petal.length  150 non-null    float64
 3   petal.width   150 non-null    float64
 4   species       150 non-null    object
dtypes: float64(4), object(1)
memory usage: 6.0+ KB
```

Figure 10.1 Output from Listing 10.1

As usual, the tables must be created.

```
# What should be predicted? Save column name in variable.
col_name = 'species'

# Here the division into two tables takes place (Input=data and Output=col).
col = data[col_name]
```

```
data = data.drop([col_name], axis = 1)

# Create four tables from the two tables
train_data, test_data, train_col, test_col = train_test_split(data,col,
 test_size=0.2, random_state=42)
```
Listing 10.2 Division into Four Tables

The required four tables have been created. Thanks to AutoKeras, we don't have to transform any data types here.

```
# Build and train the model.
model = ak.StructuredDataClassifier(max_trials=3, overwrite=True)
model.fit(train_data, train_col, validation_data=(test_data, test_col))
```
Listing 10.3 Creating, Training, and Testing the Model

Using `StructuredDataClassifier`, we create a model (actually an object that builds and tests various models) for table-based data and classification. Three different configurations are tested. If you have a fast computer and some time, you can increase the number or leave it out completely, in which case 100 models will be tested. Using `overwrite=True`, we specify that a new execution overwrites any existing models—and that's all. This enables us to achieve a correct classification rate of 100% with the test data.

```
best_model = model.export_model()
best_model.save('model.h5')
```
Listing 10.4 Saving the Model with the Best Result

If you wish, you can save the model with the best test values for reuse later.

So far, the recommendation in this book has been to use XGBoost for table data in AI models. The optimization of hyperparameters was often very time-consuming for ANNs, and you also need a lot of experience to test possible parameters with certain value ranges. Thanks to AutoKeras, you can now also test models based on ANNs for such datasets and then decide on an approach (XGBoost or ANN with AutoKeras).

10.2 Regression

For the regression (prediction of the sepal length), we can copy the last program and then adapt it minimally (*K10_iris-1.ipynb*).

```
import pandas as pd
from sklearn.model_selection import train_test_split
import autokeras as ak
```

```
path = "../Data/iris.csv"
data = pd.read_csv(path, delimiter=',')
print(data.head())
print(data.info())

# Division into Two Tables
col = data['sepal.length']
data = data.drop(['sepal.length'], axis = 1)

# Create four tables from the two tables
train_data, test_data, train_col, test_col = train_test_split(data,
 col, test_size=0.2, random_state=42)

# Build and train the model.
model = ak.StructuredDataRegressor(max_trials=3, overwrite=True)
model.fit(train_data, train_col, validation_data=(test_data, test_col))
```
Listing 10.5 Complete Program for the Regression

So, what has been changed? We have a new target column (sepal.length), and we now use StructuredDataRegressor. That's it for the changes. By default, AutoKeras uses the mean square error (MSE) for the calculation and not the mean absolute error (MAE). If you want to change this setting, you can call the constructor.

```
model = ak.StructuredDataRegressor(max_trials=3, overwrite=True,
 loss="mean_absolute_error", metrics="mean_absolute_error")
```
Listing 10.6 Using MAE instead of MSE

The MAE is 0.29 cm (0.11") for the best model. You can increase max_trials to generate and test even more models.

10.3 Image Classification

Using AutoKeras, it's also no problem to create your own AI models for image classification. In the next example, we want to program an AI that recognizes and classifies hand gestures for the rock, paper, scissors game (*K10_image_classification.ipynb*). We had already solved this task by means of transfer learning and achieved a correct classification rate of approximately 93%.

You can save the images used here on your computer. The *Rock-Paper-Scissor* folder contains the *train* and *test* subfolders. These folders have additional subfolders—one subfolder for each image class, that is, *paper*, *rock*, and *scissors*. Finally, these subfolders contain the corresponding images. The folder structure is shown in Figure 10.2.

10.3 Image Classification

```
▸ Rock-Paper-Scissor
  ▸ train
    ▸ paper
      ▸ pic1.png
      ▸ pic2.png
      ▸ ...
    ▸ rock
      ▸ ...
    ▸ scissors
      ▸ ...
  ▸ test
    ▸ ...
```

Figure 10.2 Labels Derived from the Folder Structure

The great thing is that you don't need to label the images. The labels (target column) are derived from the folder structure. You don't need to create a list of labels with information about which class is shown on which picture.

```
import autokeras as ak
import tensorflow as tf
import numpy as np
import matplotlib.pyplot as plt

IMG_HEIGHT = 180
IMG_WIDTH = 180

data_dir_train = "../Data/Rock-Paper-Scissor/train"
data_dir_test = "../Data/Rock-Paper-Scissor/train"

train_data = ak.image_dataset_from_directory(
    data_dir_train,
    image_size=(IMG_HEIGHT, IMG_WIDTH),
    batch_size=BATCH_SIZE
)

test_data = ak.image_dataset_from_directory(
    data_dir_test,
    image_size=(IMG_HEIGHT, IMG_WIDTH),
    batch_size=BATCH_SIZE
)
```

Listing 10.7 Loading the Modules and Images

We use constants to define some values right at the beginning. If these settings need to be changed, this can be done at a central location. The training and test data are also defined in the preceding listing.

```
model = ak.ImageClassifier(overwrite=True, max_trials=2)
model.fit(train_data, validation_data=test_data, epochs=2)
```

Listing 10.8 Creating, Training, and Testing a Model

Here, we define that only two settings are to be tested and two epochs are to be trained per setting so that the program doesn't take too long. If you have time, you can increase the number. You can see that we don't specify the target columns at all, as these are generated automatically from the folder structure.

And again, that's all it takes. With these few lines, I achieve a correct classification rate of approximately 100%. Let's test the best model with an image of a hand gesture.

```
file = "../Data/gesture.png"

img = tf.keras.utils.load_img(file,target_size=(IMG_HEIGHT, IMG_WIDTH,3))
img = np.array(img)

plt.figure()
plt.imshow(img)
plt.show()
```

Listing 10.9 Output of the Test Image

You can see the output of Figure 10.3.

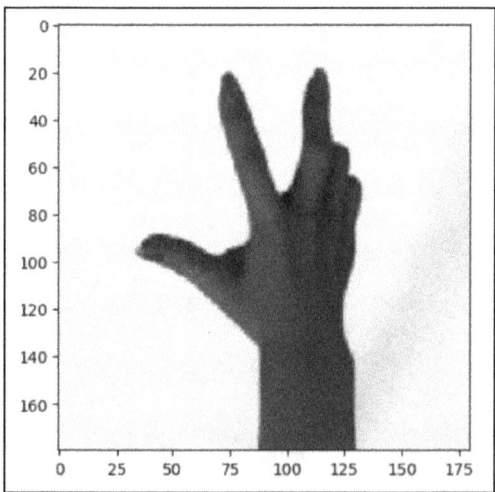

Figure 10.3 "Scissors" Test Image

The image is loaded and output. You should be familiar with these lines from Chapter 5, Section 5.5.

```
pred = model.predict(img.reshape(1, IMG_HEIGHT, IMG_WIDTH, 3))
print(pred)
```

Listing 10.10 Classification of the Test Image

The scissors are recognized correctly, as the output shows:

```
[['scissors']]
```

Now you can use pretrained networks for images in AI projects, adapt these pretrained networks to new data using transfer learning, or simply create the ANN from scratch using AutoKeras.

10.4 Text Classification

The *K10_text_classification.ipynb* program trains the AI in such a way that movie reviews can be read and automatically classified. You may remember that the reviews are available in text form, together with information as to whether the criticism is positive or negative. The goal is to develop an AI model that will automatically classify movies into positive and negative based on a movie rating in text form.

We've already developed an AI model for this dataset in Chapter 8. The procedure was as follows:

- The words of a text were transformed into integers using text vectorization.
- These integers were transformed into vectors with floats using the embedding layer.
- The data was dimensionally reduced by averaging the vectors using the `GlobalAveragePooling1D` layer.
- The result was fed to an ANN, which classified it as positive or negative.
- With this approach, we achieved a correct classification rate of 88%.

Here, too, the AutoKeras module will save us a lot of work. You should run the following program in Google Colab with GPU support; otherwise, it will take a very long time to run. To do this, upload the program and the *IMDB-Dataset.csv* file to Colab.

```
# for the execution in google colab,
# use the following line:
!pip install autokeras
```

Listing 10.11 Installation of "AutoKeras" in Colab

This installs `AutoKeras` in your Google Colab environment. Remember that the installation must be renewed when you end and restart the session.

```
import pandas as pd
from sklearn.model_selection import train_test_split
import autokeras as ak
import numpy as np

path = "/content/IMDB-Dataset.csv "

# to ensure that the file is imported correctly in Colab,
# we have to specify the character encoding and pandas using
# the Python engine
data = pd.read_csv(path, delimiter=',', error_bad_lines=False,
 encoding='utf-8', engine="python")
print(data.head())
print(data.info())

data['review'] = data['review'].str.replace("<br />"," ")

# Transform string to integer
# new column, set all values to 0
data['rating'] = 0

# if the content of the 'sentiment' column is equal to 'positive', set
# the entry in the 'rate' column to 1
data.loc[data['sentiment'] == 'positive', 'rating'] = 1

col = data['rating']

# remove the 'sentiment' and 'rate' columns
data.drop(['sentiment', 'rating'], axis = 1, inplace=True)

print(data)
print(col)

# Create four tables from the two tables
train_data, test_data, train_col, test_col = train_test_split(data,
 col, test_size=0.2, random_state=42)
```

Listing 10.12 Loading the Data and Creating the Four Tables

The data is loaded, and the values in the target column are transformed into numerical values. Up to this point, the lines largely correspond to the example in Chapter 8, Section 8.5.

```
# Transform data into arrays
train_data = np.array(train_data)
train_col = np.array(train_col)
test_data = np.array(test_data)
test_col = np.array(test_col)
```

Listing 10.13 Transformation into Arrays

The model expects data of the array data type, so we need to transform it.

```
# Build and train the model.
model = ak.TextClassifier(max_trials=3, overwrite=True)
model.fit(train_data, train_col, validation_data=(test_data, test_col), epochs=
2)
```

Listing 10.14 Building, Testing, and Training the Model

The best model has a correct classification rate of approximately 93%. You can increase the number of trials and the number of epochs to obtain better results.

Although only three models are tested with different settings, one of these models uses the high-performance and pretrained language model called *BERT*.

> **BERT (Bidirectional Encoder Representations from Transformers)**
>
> The BERT language model was developed by Google. Here, too, words or word components are transformed into numbers or vectors, but the model is then trained using a sophisticated algorithm. At the word or word component level, the procedures are as follows:
>
> - A word in a sentence is omitted.
> - A word in a sentence is replaced by a randomly selected word.
> - The sentence remains the same (i.e., with the original word).
>
> The model attempts to predict this word in each case.
>
> At the sentence level, two sentences are linked together. The model is trained to predict whether the connection makes sense or not.
>
> In the "BERT: Pre-Training of Deep Bidirectional Transformers for Language Understanding" paper, you'll find detailed information on the procedure (available at *https://arxiv.org/abs/1810.04805*). The model was trained with Wikipedia data, among other things, and can assign words contextually thanks to this training procedure. Google uses BERT, for example, to interpret user input in the search engine.

10.5 Exercises

10.5.1 Exercise 1: Classification

Program an ANN to classify mushrooms as edible and poisonous. Use the *mushrooms.csv* file.

10.5.2 Exercise 2: Regression

Program an ANN to determine the price of diamonds. Use the *diamonds.csv* file. Note: The execution can take a very long time. Try it with and without a prior standardization of the data.

10.5.3 Exercise 3: Image Classification

Download the Horses or Humans dataset and have it classified by the ANN. You can find the data at *www.kaggle.com/datasets/sanikamal/horses-or-humans-dataset*.

10.5.4 Exercise 4: Text Classification

Program an AI that reads the *spam.csv* file and classifies the texts as ham or spam. Use upsampling.

Chapter 11
Visual Programming Using KNIME

Can you program an AI without learning a programming language or optionally integrating source code? KNIME makes it possible.

What This Chapter Is About
- Introduction to visual programming
- Regression and classification using artificial neural networks (ANNs)
- Regression and classification using XGBoost
- Image classification using pretrained models
- Transfer learning
- Autoencoders
- Text classification
- Automatic creation of AI models using AutoML
- Cluster analysis
- Time series analysis
- Text generation

KNIME enables you to program a wide variety of AI models using graphical modules. One of the advantages of visual programming is its clarity. Even if the program was developed by someone else, you can familiarize yourself with it relatively quickly because the data flow is easy to understand. But KNIME is anything but a gimmick. Many companies and researchers use this software for professional tasks.

KNIME also makes it very easy to work with data from different sources. With programming languages such as Python, depending on the data source, you have to import various modules and use them to load the data. There are separate modules for file formats such as Excel or comma-separated values (CSV), each of which provides different functions. If the data is supposed to be loaded from a database, you must familiarize yourself with the SQL database language (Structured Query Language [SQL]). KNIME has different modules for the data sources, such as the **Excel Reader** or **CSV Reader** modules. There are even modules for databases so that you can import the data without any knowledge of SQL. Once the data has been imported from the respective source, it can be processed in a standardized way and used for the development of AI models. As an AI developer, you don't need to spend a lot of time connecting the data source or importing the data. You can also include Python source code where you think it makes sense.

> **Python Modules as an Alternative to Keras and TensorFlow Modules**
>
> In Chapter 2, we created Keras and TensorFlow environments for Python deep learning. If you have problems with the configuration on your computer, you should set all environments to Base, as all the required modules have been installed in this environment. Then, you can use a Python module with a few lines of code instead of Keras and TensorFlow modules. The loading and processing of data, on the other hand, can be realized with other modules, as you'll see from the examples in this chapter.

You can program the programs presented in this book yourself or import the programs provided. After the import, the path to the respective data source (CSV file) must be updated. You can import the program by right-clicking on the target location in the KNIME Explorer (*LOCAL* or a subfolder, which can also be created by right-clicking and selecting **New Workflow Group**). Let's get started.

11.1 Simple ANNs

Let's now apply what we've learned about artificial neural networks (ANN). We'll compare the KNIME programs with the Python programs that we've already created in previous chapters.

11.1.1 Classification

The first workflow is supposed to classify irises using an ANN. Based on the length and width of the leaves, we want to determine the subspecies of the iris. We've already developed AI models for this problem using ANN and XGBoost in the Python programming language. In this first program, we'll take a close look at the components used (*nodes*) (see Figure 11.1).

Data Preparation

Let's start with the data source. You can drag the **CSV Reader** module into the workflow editor and make the desired settings by right-clicking and selecting **Configure**. As an alternative, you can also drag and drop the CSV file into the workflow editor. The appropriate module is then AUTOMATICALLY inserted, and the settings window opens (Figure 11.2).

11.1 Simple ANNs

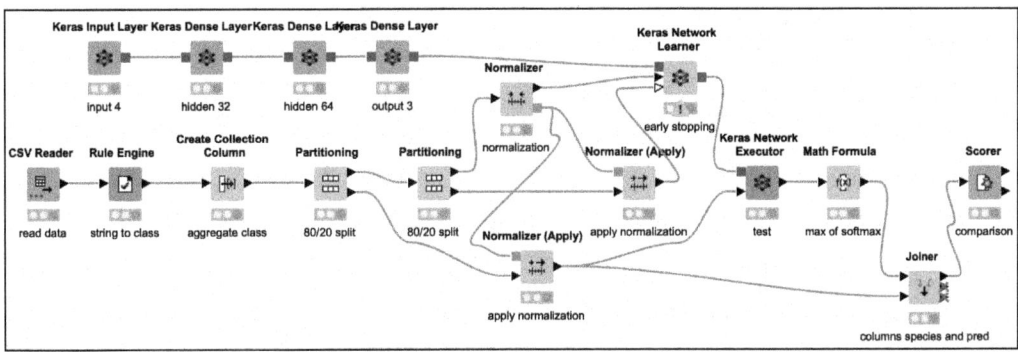

Figure 11.1 The 01-iris.knwf Program

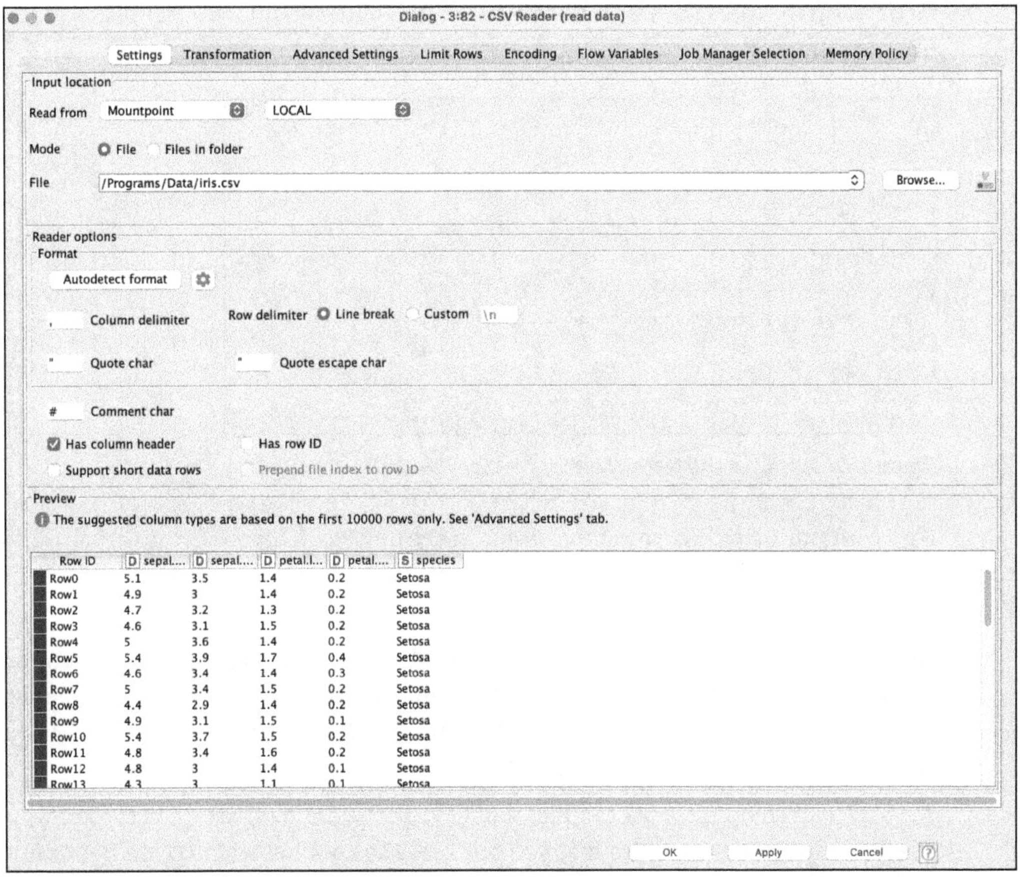

![Figure 11.2 CSV Reader settings dialog]

Figure 11.2 Settings Window for the CSV File

The most important settings you need to make are as follows:

- **File**
  Specify the path to the file.

11 Visual Programming Using KNIME

- **Column delimiter**
  Choose the delimiter used for the columns.

- **Has column headers**
  Specify whether column headers are used for a table.

- **Has row ID**
  Specify whether a table has a row ID. If the table itself has a column, for example, with ascending numbering, to clearly identify the rows, this column can be used. Otherwise, a new, temporary column with unique indexes will be created.

The status display of the module is yellow after the configuration. You can execute the module by right-clicking and selecting **Execute**; the status display then changes to green. Alternatively, you can select the module, and click on the **Play** button in the menu bar at the top. You can see the output values of the module (see Figure 11.3) in the **Node Monitor** screen.

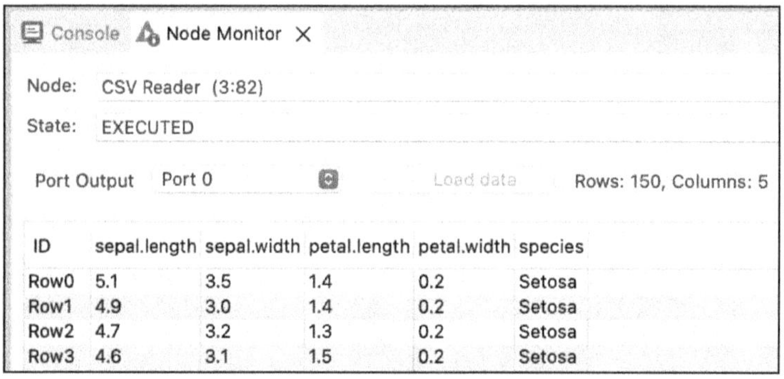

Figure 11.3 Output of the Node Monitor for "CSV Reader"

We'll continue using the data at the output of the **CSV Reader** module. From then on, it doesn't matter which data source the data comes from because it's always available in table format.

To enable the ANN to use the input data for calculations, all nonnumerical columns required for this must be transformed into numbers. This only applies to the species column. In this example, we'll use the **Rule Engine** module. The expressions that can be entered in this module have an unusual syntax, as you can see in Figure 11.4.

The first lines contain some examples that have been commented out. In the left window you can see the variables that you can use. The column headings are important here, especially **species**.

```
$species$ = "Setosa" => 0
$species$ = "Virginica" => 1
TRUE => 2
```

Listing 11.1 Transforming the Data Using "Rule Engine"

11.1 Simple ANNs

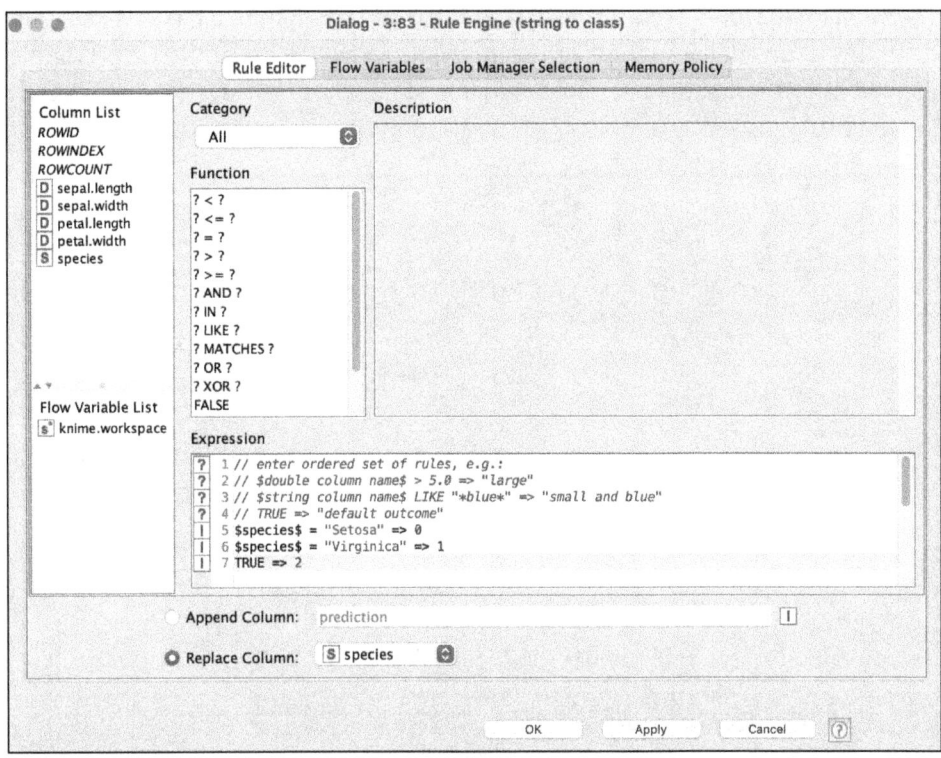

Figure 11.4 Settings Window of the Rule Engine Module

The first line means that if the content of the $species$ variable is "Setosa", the number 0 will be used. The second line must be interpreted in the same way for the next subspecies. The third line could also be formulated in the same way for "Versicolor", but here you can see another possibility: for all other variable contents, the values are replaced by the number 2. You can also set whether the content of the **species** column should be overwritten or a new column with the numerical values should be created. After executing the module, the **Node Monitor** shows that the transformation has been carried out successfully. Take a look at the **species** column in Figure 11.5.

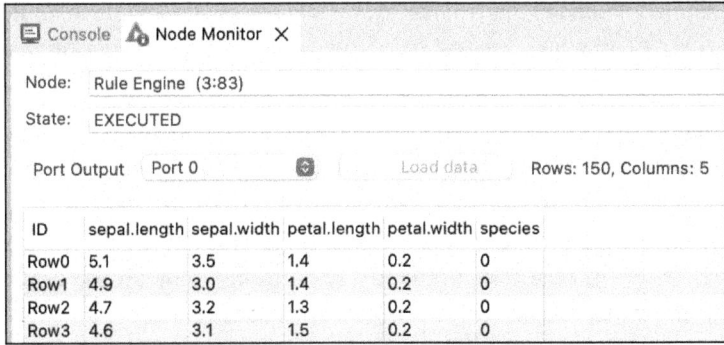

Figure 11.5 Output of the Node Monitor for Rule Engine

207

11 Visual Programming Using KNIME

The module for ANN expects lists with numerical values as output data (target column) if the softmax function is supposed to be used as an activation function. This gets implemented using the **Create Collection Column** module, whose settings window is shown in Figure 11.6.

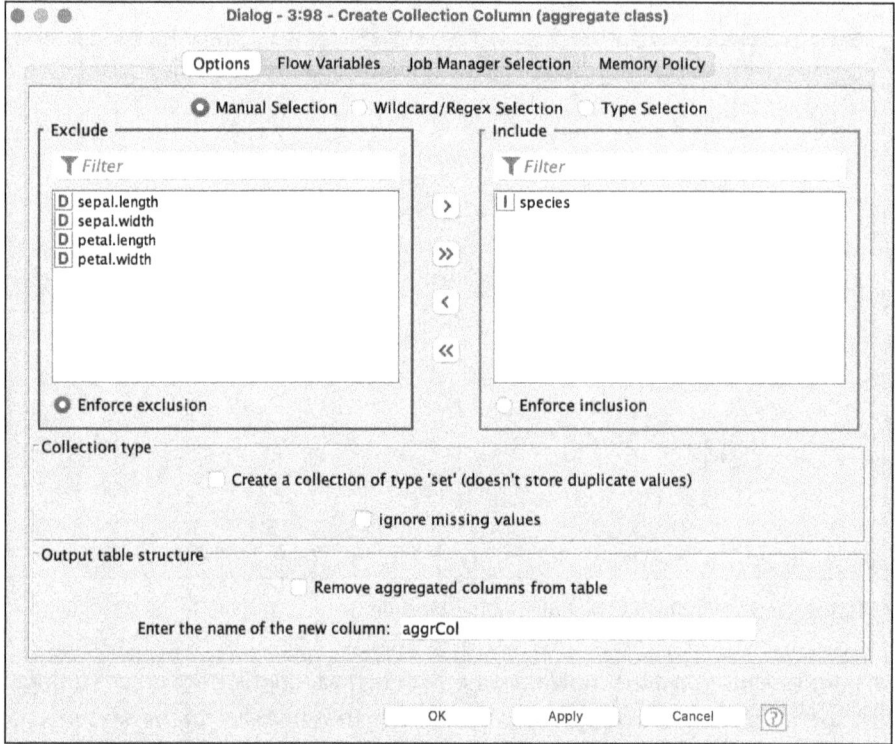

Figure 11.6 Settings Window of the Create Collection Column Module

This module can be used to merge multiple columns into one list. We only need to "pack" the individual elements into a list so that we can use the softmax function later. This step is therefore only necessary because the module for the ANN with softmax expects this data type as an activation function. The columns to be merged are listed in the **Include** section (here, only **species**). Furthermore, we add the **aggrCol** column to the table, as shown in Figure 11.7, in which the results of the data type transformation (list with one element each) are stored.

When you execute the module, you'll see in the **Node Monitor** that the numerical values of this new column are in square brackets (see Figure 11.7). For example, the number 0 becomes the list [0]. This means the data has now been imported and prepared. The data can now be divided into training, test, and validation data using the **Partitioning** modules (see Figure 11.8). In this case, 80% of the mixed data should be output at the top output node (training data), and the rest should be at the bottom. The data at the lower output is divided again into test and validation data (80% and 20%, respectively).

We also set that the data should be mixed in such a way that it can be reproduced (**Use random seed**).

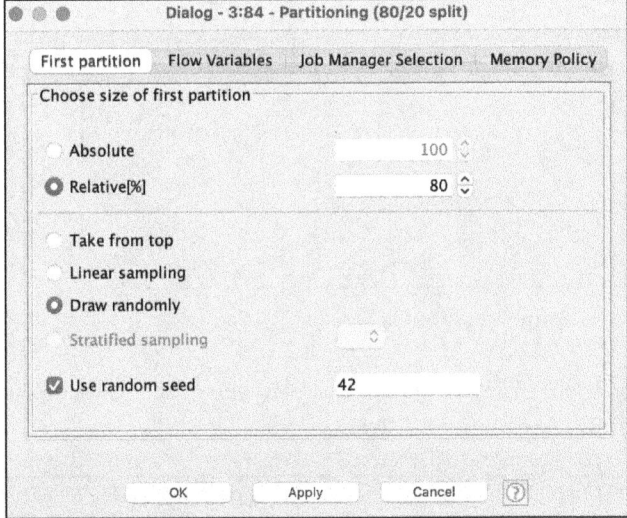

Figure 11.7 Output of the Node Monitor for the Create Collection Column Module

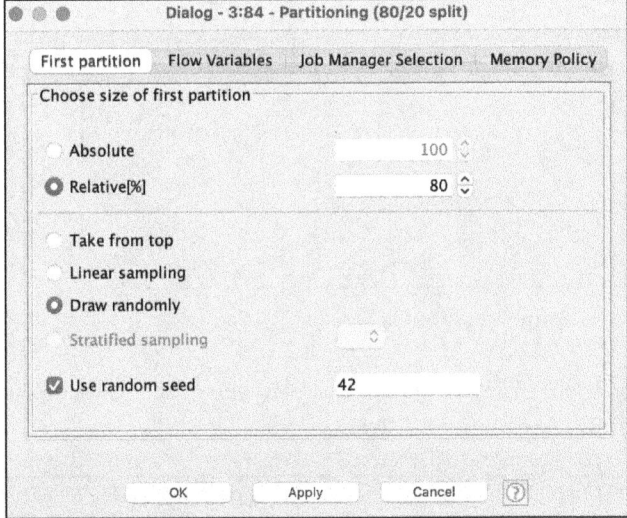

Figure 11.8 Settings Window of the Partitioning Module

For devices with multiple outputs, you can set which data should be displayed in the **Node Monitor (Port Output)**.

The training data is used to train the ANN and thus update the weights. The test data is used to determine the correct classification rate after each epoch during the training phase. In this program, the training data is used to detect whether stagnation occurs and further optimization of the weights isn't possible. In this case, the training procedure will be aborted to prevent overfitting. At the end, the new, unknown evaluation data is used to determine the correct classification rate.

As you already know from previous chapters, ANNs can make better predictions with standardized data, as otherwise the different attributes lead to a distortion when updating the weights due to very different value ranges of the data. The input data is

11 Visual Programming Using KNIME

standardized here using Z-score (see Figure 11.9, corresponds to the standard scaler from sklearn).

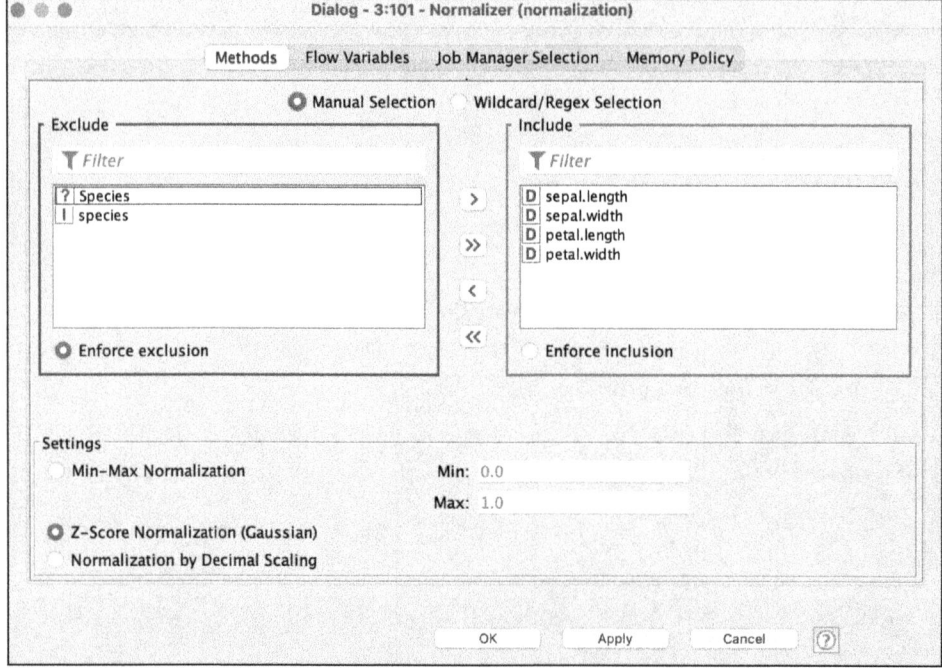

Figure 11.9 Settings Window of the Normalizer Module

The module is only applied to the training data. You can view the output data of the module again in the **Node Monitor** (see Figure 11.10).

ID	sepal.length	sepal.width	petal.length	petal.width	species	aggrCol
Row2	-1.336724926334261	0.3763136790200976	-1.3714143107808723	-1.2865140668853283	0	[0]
Row3	-1.4583605594414548	0.14621104726258682	-1.256078547456331	-1.2865140668853283	0	[0]
Row7	-0.9718180270126817	0.8365189425351183	-1.256078547456331	-1.2865140668853283	0	[0]
Row9	-1.0934536601198745	0.14621104726258682	-1.256078547456331	-1.4190305372941177	0	[0]

Figure 11.10 Output of the Node Monitor for Normalizer

You can see that the test and validation data in this program are each standardized using the **Normalizer (Apply)** module and that this module is linked to **Normalizer** (see Figure 11.1). The mean value and the standard deviation are only determined using the training data. The test and validation data are intended to simulate new, unknown data, so standardization is only applied to this data. This is the correct procedure for standardizing datasets when you program for practical applications. Test data is intended to simulate new, unknown data, so it's not advisable to use these values to calculate the mean and standard deviation.

Building the ANN

Now it's time to build the ANN. The connections of the modules used so far show the data flow symbolized by arrows. The connections of the layers of an ANN are simple lines, which only symbolize a connection without data flow. We'll make some central settings for the entire network in the **Keras Network Learner** module (see Figure 11.11). This module must first be connected to **Keras Input Layer** and three **Keras Dense Layer** modules that are connected in series. The first layer has four input nodes. This is followed by two hidden layers with 32 and 64 nodes, respectively; we select **ReLu** as the activation function in the settings of the hidden layers. The output layer has three nodes with **softmax** as the activation function.

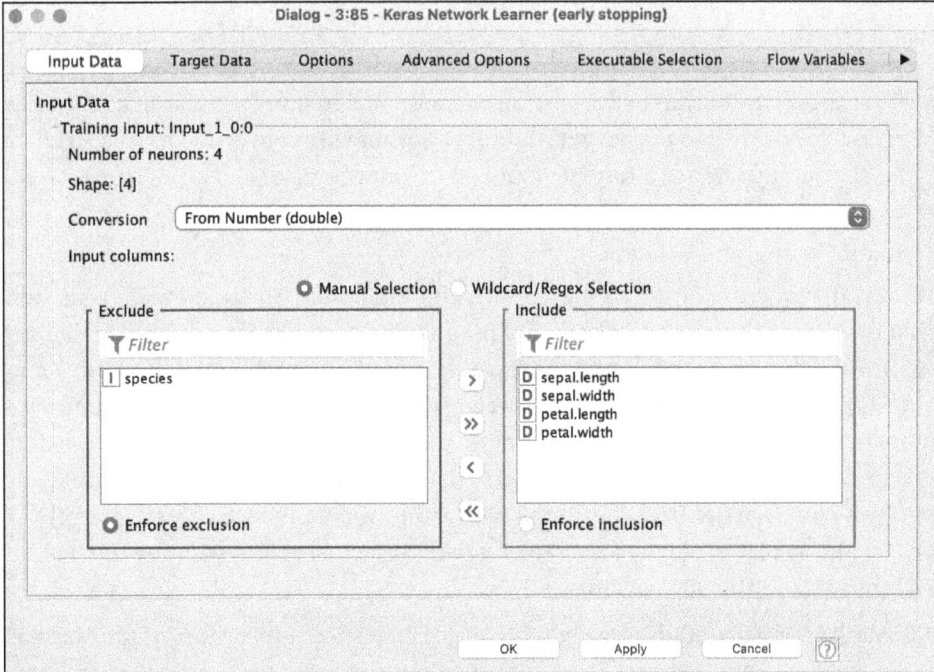

Figure 11.11 Settings Window of Keras Network Learner

The first four tabs of the **Keras Network Learner** settings window are as follows:

- **Input Data**
  The four input variables are selected, and **From Number (double)** is set under **Conversion**.

- **Target Data**

    The **col** column is selected as the input column, and **From Collection of Number (integer) to One-Hot Tensor** is set under **Conversion**. In addition, **Categorical cross entropy** is set as a loss function.

- **Options**

    The number of epochs is **40**, and **Batch Size** is **10**. Use **Shuffle training data before each epoch** to select that the shuffle should be reproducible. You must enter a number for **Use random seed** here, but the value is irrelevant.

- **Advanced Options**

    Here, you can set **Early Stopping** (via **Validation Loss**). If you decide to do this, you can set the number of epochs under **Options** to **100**, for example. In this case, the training process is terminated if there is no further improvement.

The **Keras Network Learner** module is connected to **Keras Network Executor**. There are also some settings to be made for this module, all under the **Options** tab (see Figure 11.12):

- The four input variables are selected, and **From Number (double)** is set under **Conversion**.

- You can add the output by clicking on **add output**. You must select which of the layers is to be defined as the output (**Output_1/Softmax:0**). **Conversion** must also be set for the output data as **To Number (double)** for our example.

Training, Testing, and Validation

What is the procedure for training, testing, and validation? The training data and test data is fed to the Keras Network Learner module. The training data is used for training (i.e., the weights are updated), and the test data is used after each iteration to check how good the prediction of the ANN is (the weights aren't updated). If there is no improvement in the prediction of the test data (early stopping), the training and test procedure is aborted. The validation data is then fed to the Keras Network Executor module. In the field, we would have to trust the predictions of the module. But we know the target values for the validation data, so we can compare the actual values with the predictions and thus validate the ANN again.

The **Math Formula** module (see Figure 11.13) compares the three values of the **Softmax** function of the Keras Network Executor module and determines the index of the maximum value. As a result, the numbers 0 to 2 are possible, which correspond to the coded subspecies.

```
colMax($Output_1/Softmax:0_0$, $Output_1/Softmax:0_1$, $Output_1/Softmax:0_2$)
```

Listing 11.2 Expression in the Math Formula Module

11.1 Simple ANNs

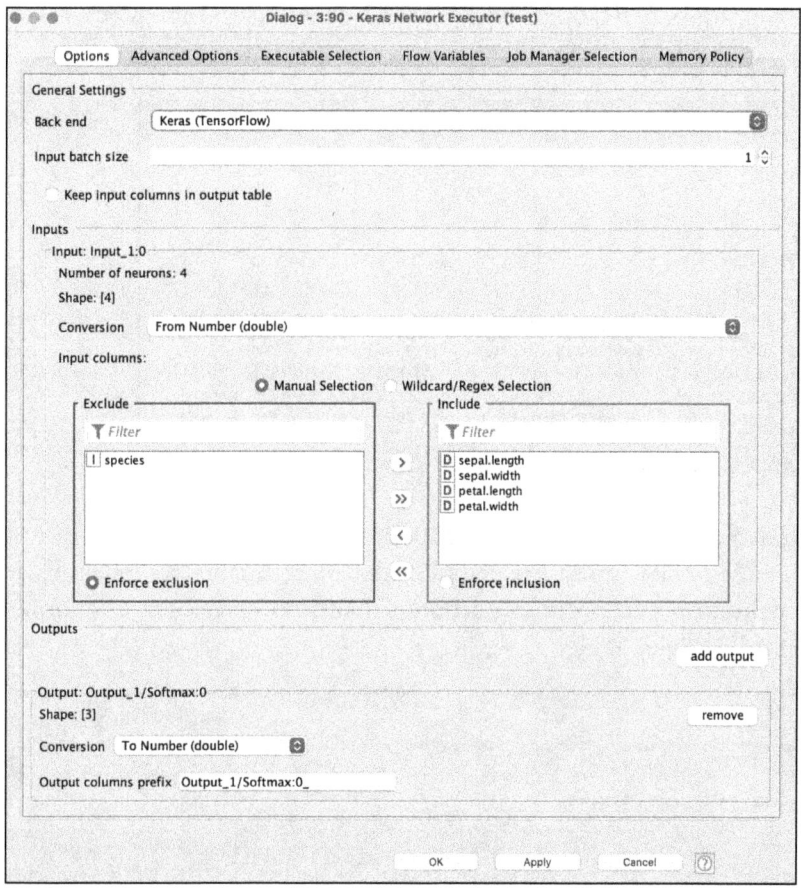

Figure 11.12 Settings Window of Keras Network Executor

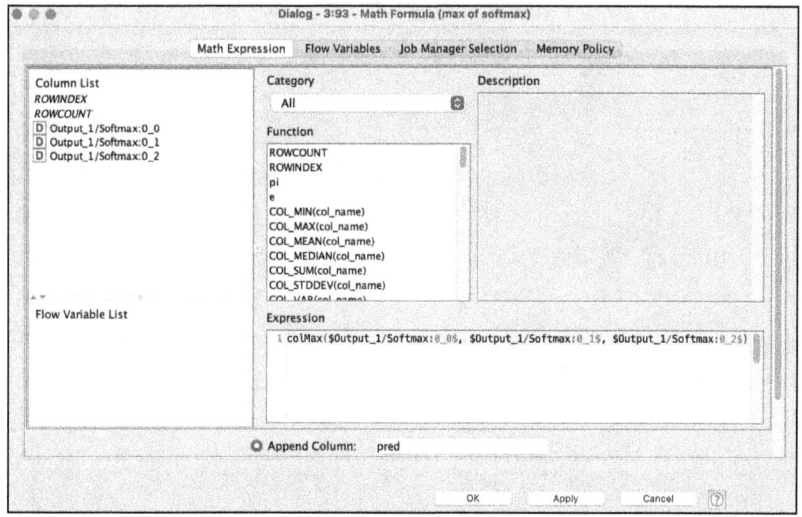

Figure 11.13 Settings Window of the Math Formula Module

The result gets transformed to the data type integer and saved in the new **pred** column (see Figure 11.14).

![Node Monitor output]

ID	Output_1/Softmax:0_0	Output_1/Softmax:0_1	Output_1/Softmax:0_2	pred
Row1	0.9997072815895081	1.4895428535055544E-7	2.9260184965096414E-4	0
Row6	0.9999563694000244	6.3045668241557E-8	4.3553736759349704E-5	0
Row30	0.9998106360435486	1.4558473537817918E-7	1.8926318443845958E-4	0
Row32	0.9999721050262451	2.786611830174479E-8	2.7860316549777053E-5	0

Figure 11.14 Output of the Node Monitor for Math Formula

This **pred** column must be compared with the **species** column of the validation data. A new table with these two columns is created using the **Joiner** module. We can then use the values from this table to perform calculations—in this case to calculate the correct classification rate (see Figure 11.15).

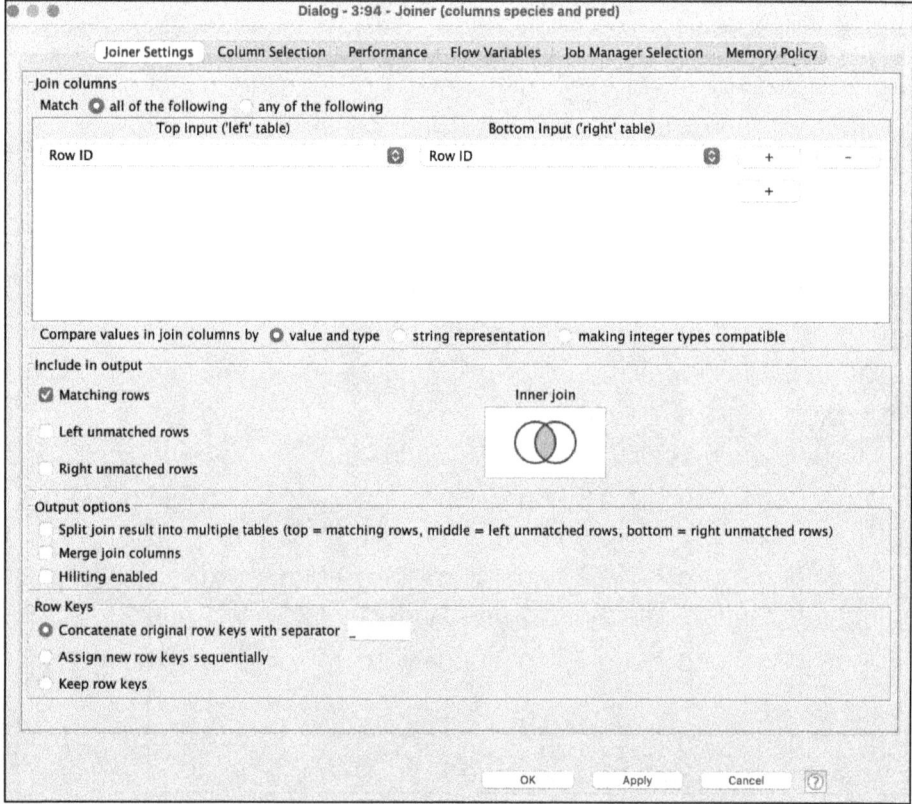

Figure 11.15 Settings Window of the Joiner Module

11.1 Simple ANNs

Each row in a table has a unique RowID. As both columns originally come from a single table, the rows of the matching columns have the same RowID. This information from both tables is used as the basis for joining them. The intersection (*inner join*) of the two columns is also formed. Only RowIDs that are contained in both tables are used for the new table. If one column has more entries than the other or if a column has no counterpart, it would be discarded. In the **Column Selection** tab of the **Joiner** module, we select **pred** from one table and **species** from the other.

```
Console    Node Monitor  X

Node:    Joiner (3:94)
State:   EXECUTED

Port Output   Port 0              Load data     Rows: 30, Columns: 2

ID                pred  species
Row1_Row1          0    0
Row6_Row6          0    0
Row30_Row30        0    0
Row32_Row32        0    0
```

Figure 11.16 Output of the Node Monitor for the Joiner Module

Now, we can go through this new results table row by row and see where a wrong prediction may have been generated. However, this is very cumbersome and it's not the point. The last module to be used is therefore the **Scorer**. Here, too, **pred** and **species** must be selected. Select the module, and read the description for it in the right-hand window (**node description**) to find out which variable is to be connected to which input. After executing, right-click and select **View: Confusion Matrix** to see the screen shown in Figure 11.17.

```
●  ●  ●    Confusion Matrix - 3:92 - Scorer (comparison)
File   Hilite
species \ p...  0      2      1
0               7      0      0
2               0      8      2
1               0      2     11

Correct classified: 26          Wrong classified: 4
Accuracy: 86.667%               Error: 13.333%
```

Figure 11.17 Confusion Matrix

With this first draft, a correct classification rate of nearly 87% was achieved.

11.1.2 Classification Using Python Node

You can also solve the task from the previous section using Python modules (see Figure 11.18).

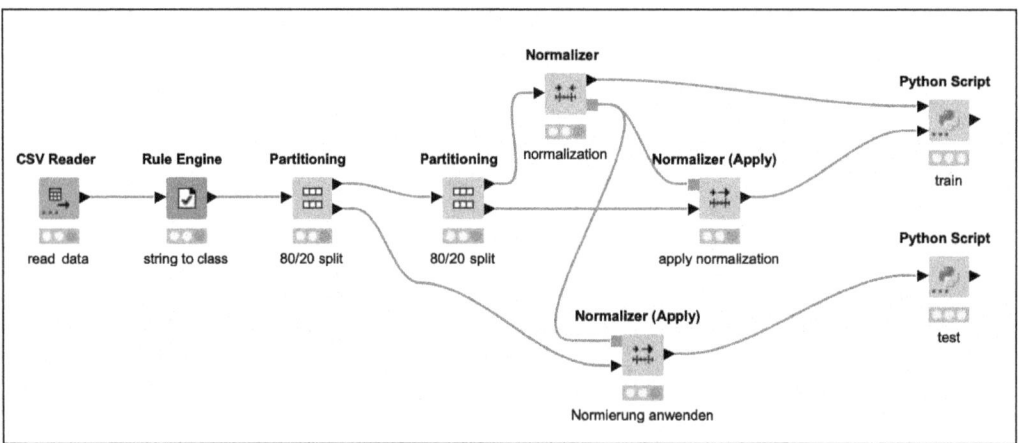

Figure 11.18 The 01-iris-py.knwf Program

The structure of the program remains similar with only the Keras modules missing. Two **Python Script** modules are used for training and testing.

```
Import knime.scripting.io as knio
import tensorflow as tf
import pandas as pd

# Training data
train_col = knio.input_tables[0][4].to_pandas().astype('category')
train_data = knio.input_tables[0][0:4].to_pandas()

# Test data
test_col = knio.input_tables[1][4].to_pandas().astype('category')
test_data = knio.input_tables[1][0:4].to_pandas()

# Build model
model = tf.keras.Sequential([
    tf.keras.layers.Dense(32, activation=tf.nn.relu, input_dim=4),
    tf.keras.layers.Dense(64, activation=tf.nn.relu),
    tf.keras.layers.Dense(10, activation=tf.nn.softmax)
])

# Configuration, training, and testing
model.compile(optimizer='adam', loss='sparse_categorical_crossentropy',
 metrics=['accuracy'])
```

```
cb_early = tf.keras.callbacks.EarlyStopping(monitor='val_loss', patience=3)
history = model.fit(train_data, train_col, epochs=100,
 validation_data=(test_data, test_col),callbacks=[cb_early])

# Save model
model.save("model1.keras")

# Output of the result
out = [history.history['val_accuracy'][-1]]
df = pd.DataFrame(out, columns=['acc'])
knio.output_tables[0] = knio.Table.from_pandas(df)
```

Listing 11.3 Python Script for Training the Model

The source code should be easy to interpret; only the reading and output of the data may appear unusual. The data types of this data must be transformed in each case so that the data flow to other blocks is guaranteed. Don't spend too much time with this transformation, as you'll see more examples that you can copy and use for your future programs.

> **Processing and Preparing Input and Output Data**
>
> If you're interested, you can find detailed documentation on processing and preparing input and output data at *http://r-wrk.de/9763-input-output*.

> **Last Element of a List**
>
> You can use `my_list[-1]` to read the last element of a list. Only the last value of the evaluation is output in this program.

```
Import knime.scripting.io as knio
import tensorflow as tf
import pandas as pd

# Load validation data
val_col = knio.input_tables[0][4].to_pandas().astype('category')
val_data = knio.input_tables[0][0:4].to_pandas()

# Load model
loaded_model = tf.keras.saving.load_model("model1.keras")

# Evaluation
val_loss, val_acc = loaded_model.evaluate(val_data, val_col)
```

11 Visual Programming Using KNIME

```
# Output of the result
df = pd.DataFrame([val_acc], columns=['acc'])
knio.output_tables[0] = knio.Table.from_pandas(df)
```

Listing 11.4 Python Script for the Test

The saved model is loaded and applied to the test data. The value of the correct classification is output so that you can see it in the **Node Monitor** screen.

11.1.3 Regression

In the next example, we want to determine the length of the sepal again. When predicting a continuous number using an ANN, the standardization of the data also often leads to better results, as is the case with classification. We'll have the ANN calculate the mean absolute error (MAE). This gives us an indication of how far the prediction is from the actual value.

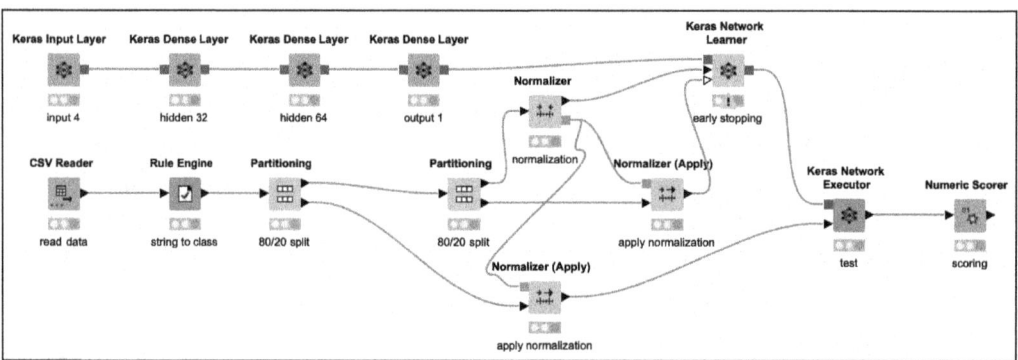

Figure 11.19 The 02-iris.knwf Program

To avoid having to create the entire workflow from scratch, you can copy and rename the previous one because much remains the same or is similar. Let's take a look at the differences next.

We don't need to aggregate the target column as we don't need a softmax function. Our aim is not to classify, but to determine the MAE.

The **Normalizer** module standardizes the training data; the column for the length of the sepal (target column) is excluded from the standardization. Otherwise, the determined MAE values of the ANN would also be standardized and would have to be scaled back to interpret the results (see Figure 11.20).

The output layer of the ANN now also has only one node with a linear activation function; the input and output values of this activation function are identical. You could also say that no activation function is used at all. We can't deactivate the activation function for the module, but by selecting this function, we can do just that, because setting a linear function implies that there is no special activation function.

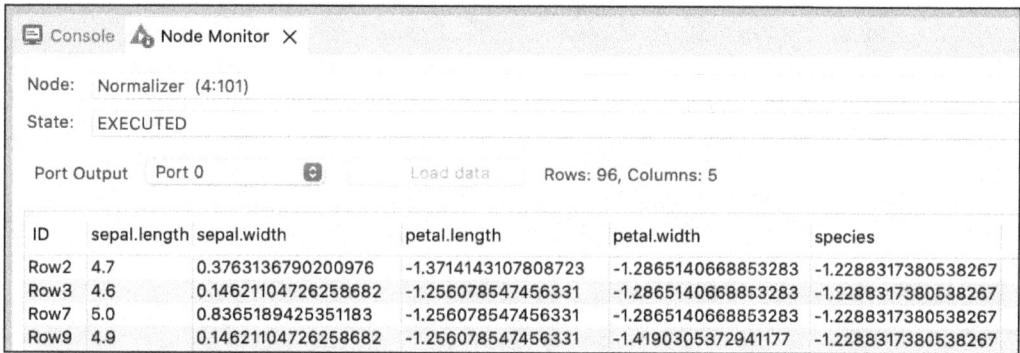

Figure 11.20 Output of the Node Monitor for the Normalizer Module

In the **Keras Network Learner** module screen, some settings must be made on the following tabs:

- **Input Data**
  From Number (double) must be set for **Conversion**. The length of the sepal is excluded from the dataset.

- **Target Data**
  For **Conversion**, **From Number (double)** must be set again. Only the length of the sepal is used from the dataset. **Mean absolute error** is set as the loss function.

- **Options**
  The number of epochs is set to **100** and the **Batch Size** to **10**. The data should be mixed so that it can be reproduced. The probability of overfitting is high with this number of epochs. But the problem is solved in the next point.

- **Advanced Options**
  Early Stopping is selected using **Training loss (total)**. The training process is therefore terminated if there is no further improvement.

The corresponding input data must also be selected for the **Keras Network Executor** module. The output of the last output layer is selected in the **Output** section. In addition, select **Keep input columns in output table**. The actual and set values are then still available at the output of the module. This means that you don't need a joiner to merge these columns from other tables for the subsequent calculation.

Among other things, the Numeric Scorer calculates the MAE, for which you only need to select the two columns for actual and target values. After the program has run, you can right-click and open **View: Statistics** to view the MAE (0.45), among other things.

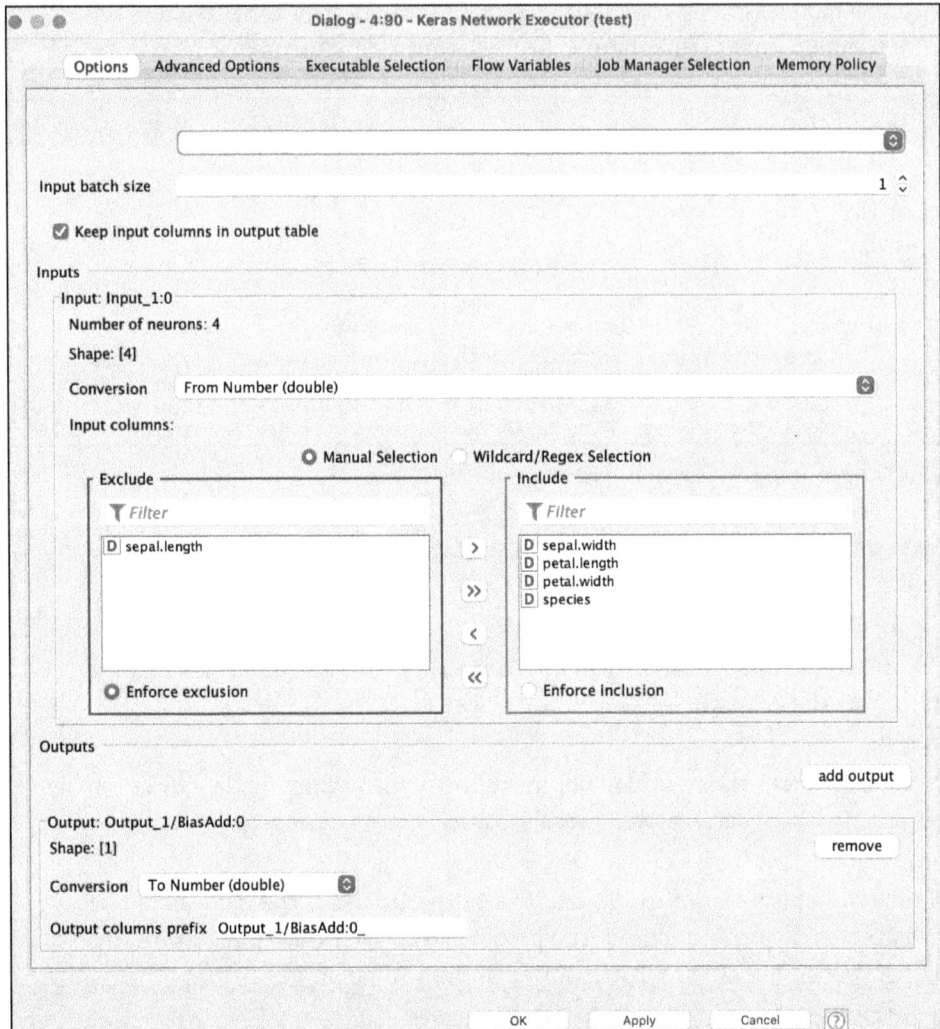

Figure 11.21 Settings Window of Keras Network Executor

During the process or afterwards, you can analyze the training process graphically in the **Learning Monitor** screen by right-clicking on **Keras Network Learner** and selecting **Learning Monitor** (see Figure 11.22).

You can smooth the curve and select certain areas using the mouse to enlarge them. If the training process takes too long, you can end the training via the **Stop Learning** button. This doesn't result in an error, so the rest of the program can continue to be executed.

11.1 Simple ANNs

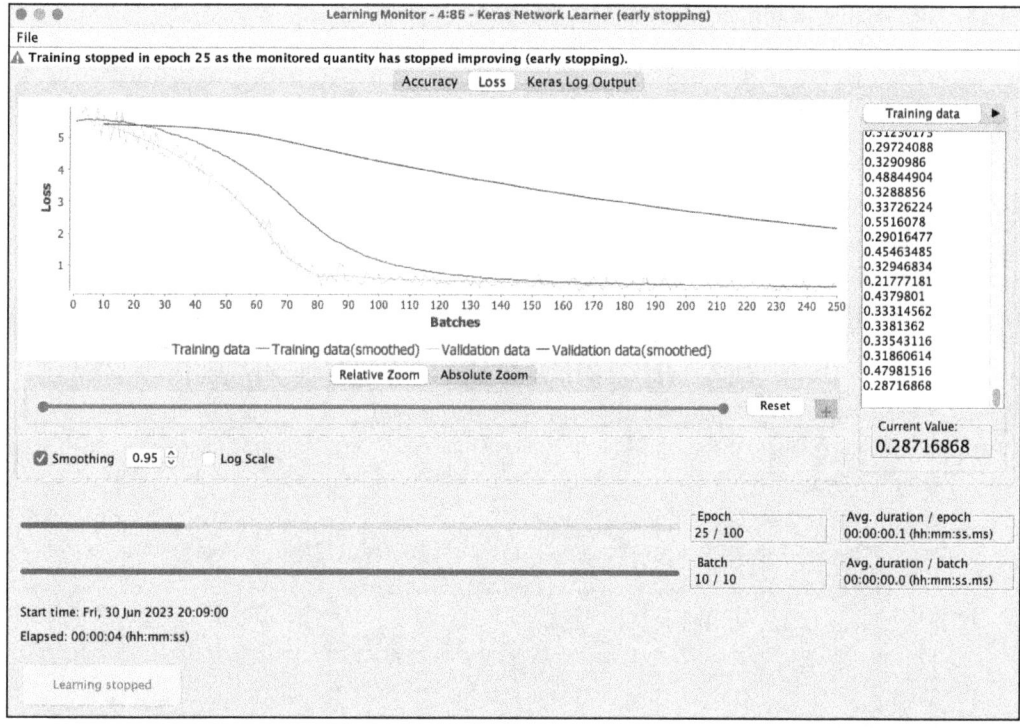

Figure 11.22 Learning Monitor

11.1.4 Regression Using Python Node

If you want to implement the program using Python modules, you only need to program the ANN yourself (see Figure 11.23).

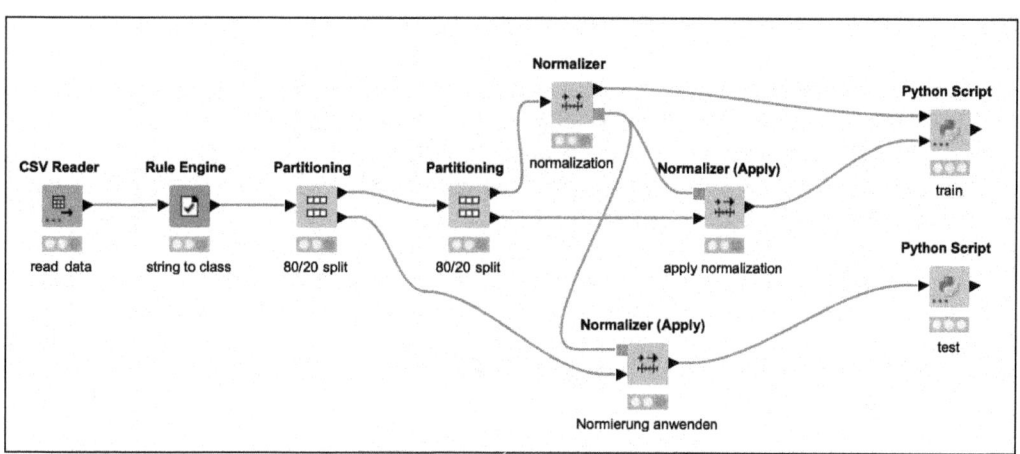

Figure 11.23 The 02-iris-py.knwf Program

221

Let's first look at the code for the training.

```python
import knime.scripting.io as knio
import tensorflow as tf
import pandas as pd

# Training data
train_col = knio.input_tables[0][0].to_pandas().astype('category')
train_data = knio.input_tables[0][1:5].to_pandas()

# Test data
test_col = knio.input_tables[1][0].to_pandas().astype('category')
test_data = knio.input_tables[1][1:5].to_pandas()

# Build model
model = tf.keras.Sequential([
    tf.keras.layers.Dense(32, activation=tf.nn.relu, input_dim=4),
    tf.keras.layers.Dense(64, activation=tf.nn.relu),
    tf.keras.layers.Dense(1)
])

# Configuration, training, and testing
model.compile(optimizer='adam', loss='mae', metrics=['mae'])
cb_early = tf.keras.callbacks.EarlyStopping(monitor='val_loss', patience=3)
history = model.fit(train_data, train_col, epochs=100,
 validation_data=(test_data, test_col),callbacks=[cb_early])

# Save model
model.save("model2.keras")

# Output of the result
out = [history.history['val_mae'][-1]]
df = pd.DataFrame(out, columns=['mae'])
knio.output_tables[0] = knio.Table.from_pandas(df)
```

Listing 11.5 Python Script for Training the Model

This model is also trained and saved. Feel free to change the hyperparameters.

```python
import knime.scripting.io as knio
import tensorflow as tf
import pandas as pd

# Load validation data
val_col = knio.input_tables[0][0].to_pandas().astype('category')
```

```
val_data = knio.input_tables[0][1:5].to_pandas()

# Load model
loaded_model = tf.keras.saving.load_model("model2.keras")

# Evaluation
val_loss, val_mae = loaded_model.evaluate(val_data, val_col)

# Output of the result
df = pd.DataFrame([val_mae], columns=['mae'])
knio.output_tables[0] = knio.Table.from_pandas(df)
```

Listing 11.6 Python Script for the Test

The test data is evaluated and output. The result can be viewed in the Node Monitor.

11.2 XGBoost

In this section, we'll implement regression and classification using XGBoost. You may remember that AI models based on XGBoost are clearer, and the data doesn't need to be prepared in a time-consuming process. You can also obtain very good results with these models in a simple way.

This library must first be installed. You can generally install libraries via **Help • Install New Software**. Search for **XGBoost**, and install **KNIME XGBoost Integration**.

11.2.1 Classification

We use the *03-iris* program again to classify irises. Compare the structure with the *01-iris* program. You can see how simple and clear the program is when you use XGBoost (see Figure 11.24).

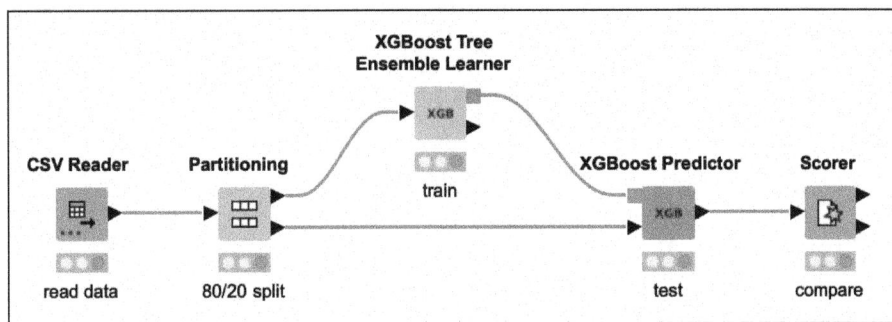

Figure 11.24 The 03-iris.knwf Program

Central settings must be made for the **XGBoost Tree Ensemble Learner** module (see Figure 11.25).

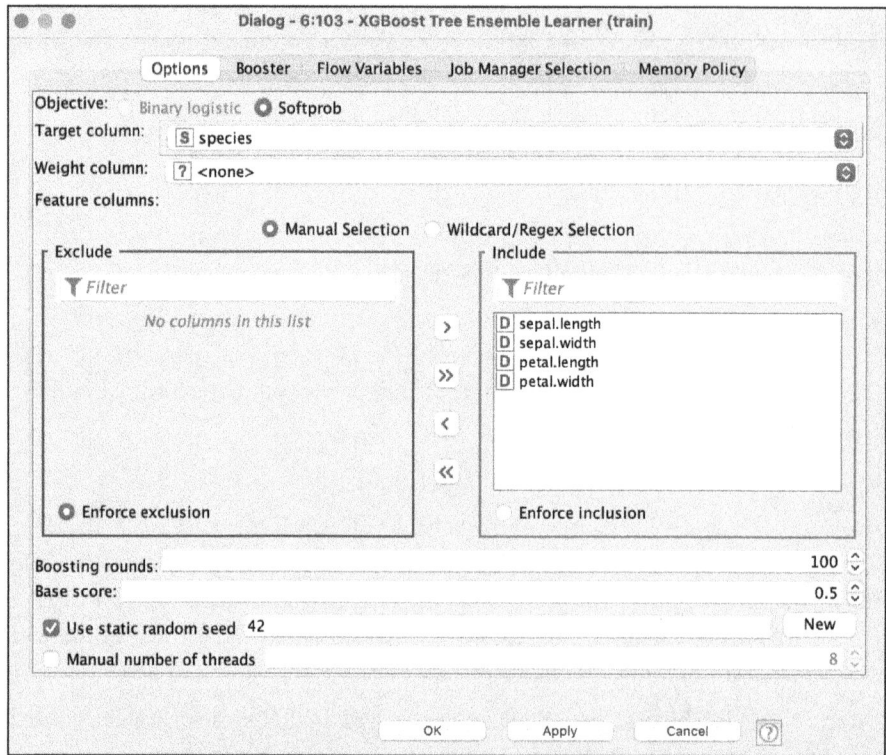

Figure 11.25 Settings Window of the XGBoost Tree Ensemble Learner Module

The most important settings are the selection of input and target columns. Using **Use static random seed**, the trees are built in such a way that they can be reproduced to make sure that a repeated call delivers the same results.

In the **XGBoost Predictor** module, we set the predictions to be saved in the new **pred** column. The output of this module contains all the data from the original table, supplemented by the prediction (see Figure 11.26).

Figure 11.26 Output of the Node Monitor for the XGBoost Predictor Module

The Scorer module is used to compare the **pred** and **species** columns. This program achieves a correct classification rate of over 93%.

11.2.2 Deployment

In this section, we'll not only train and test an AI (classification of irises) but also save the model (using **Model Writer**) and the validation data (using **Table Writer**). Another program, which you can see in Figure 11.27, loads this data and creates predictions.

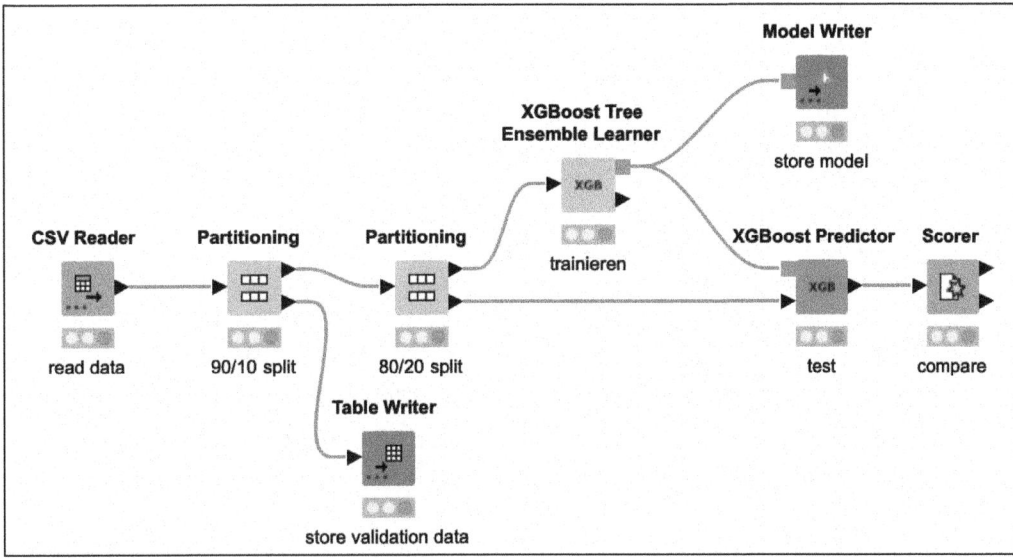

Figure 11.27 The 04a-iris.knwf Program

Of the total data, 10% is saved as validation data in binary format, as is the model. You must set the path and saving options for these saving modules. Make sure that the **Write options: overwrite** setting is set. Otherwise, running the program again will generate an error message stating that the file already exists, as shown in Figure 11.28.

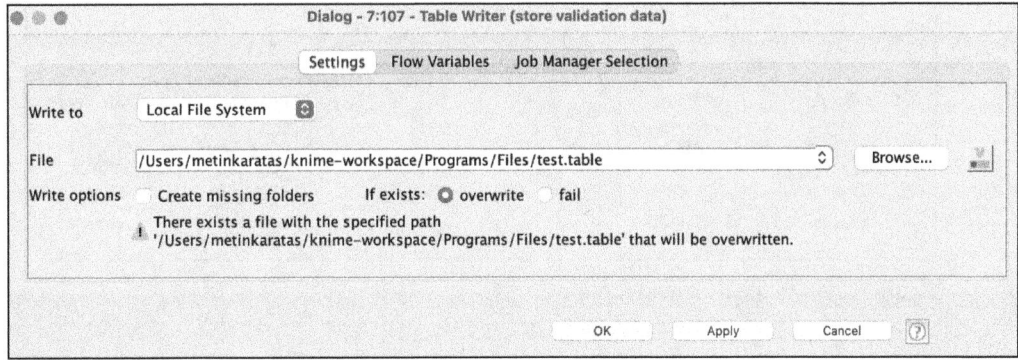

Figure 11.28 Settings Window of the Table Writer Module

What is the big difference between validation data and real data in production use? The validation data contains the target column (i.e., the information on what is to be predicted). The AI is actually there to determine this information. But the validation data enables us to evaluate the model by comparing the prediction with the actual values (see Figure 11.29).

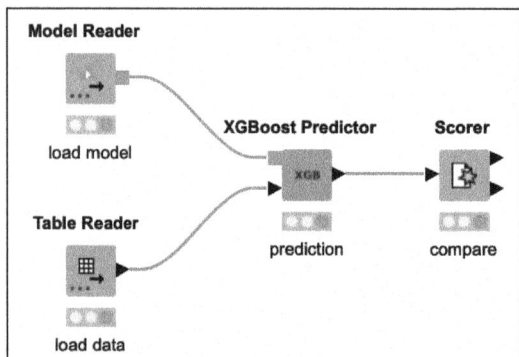

Figure 11.29 The 04b-iris.knwf Program

In this program, the module also creates a **pred** column with the predictions so that Scorer module can determine the correct classification rate.

In the case of a prediction for real data that doesn't contain the subspecies information, the Scorer module would be useless, and you'd have to trust the AI's predictions.

11.2.3 Regression

The prediction of continuous numerical values shouldn't be too difficult to implement with your current knowledge. We want to determine the width of the petal (see Figure 11.30).

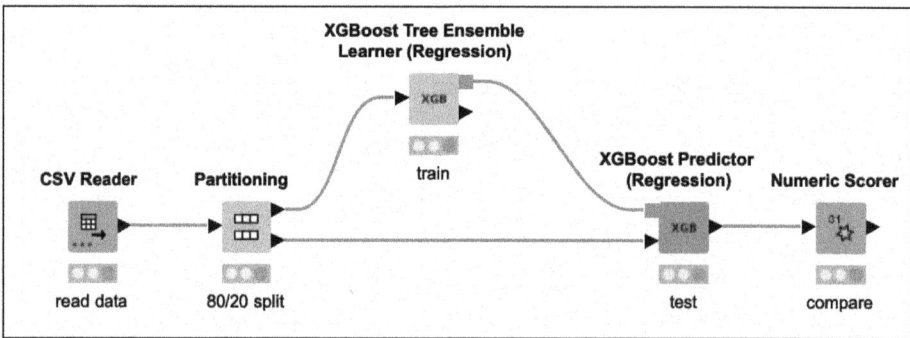

Figure 11.30 The 05-iris.knwf Program

It's important that you always have an overview of the data. Which columns are the input data, and which column is the target column? You should think about this before programming. In this program, the width of the petal is predicted with an accuracy of 0.17 cm or 0.07" (MAE).

11.3 Image Classification Using a Pretrained Model

In the next example, we'll use a pretrained model to classify an object in an image (teddy bear). There will be one program each with Keras and Python modules.

11.3.1 Image Classification Using Keras Node

This time, we're not using VGG19, but InceptionV3 (*https://arxiv.org/abs/1512.00567*) from Google. This model is faster in execution than VGG19 (at least with KNIME).

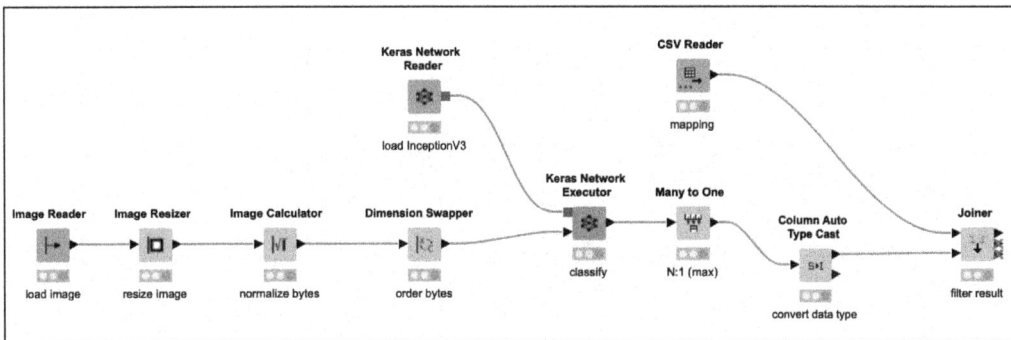

Figure 11.31 The 06-image-classification.knwf Program

First, we load the image of the teddy bear and scale it using the **Image Resizer** module. As shown in Figure 11.32, select **Linear Interpolation** and **Absolute Image Size** in the settings, and set the values **X = 299**, **Y = 299**, **Z = 0.0**, and **Channel = 3.0**.

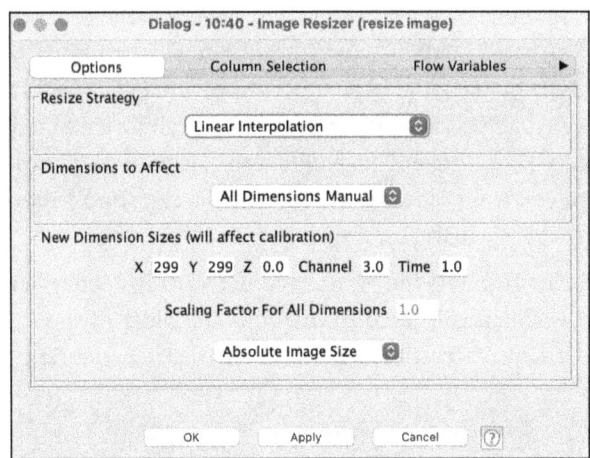

Figure 11.32 Settings Window of the Image Resizer Module

The model later expects the images in the size of 299 × 299 pixels and 3 bytes for RGB. The next two modules also process the image data so that the model can work with it. Under **Column Selection**, you can set that the existing image should be overwritten (and no new table gets created).

The **Image Calculator** module standardizes the pixel values for InceptionV3 between -1 and +1:

`(($Image$/255)-0.5)*2`

The existing image is also replaced, as the original image is no longer required in the rest of the process.

The **Dimension Swapper** module reorganizes the image information. The X, Y, Channel sequence results in Channel, X, Y. This is also an adaptation to the model used (see Figure 11.33).

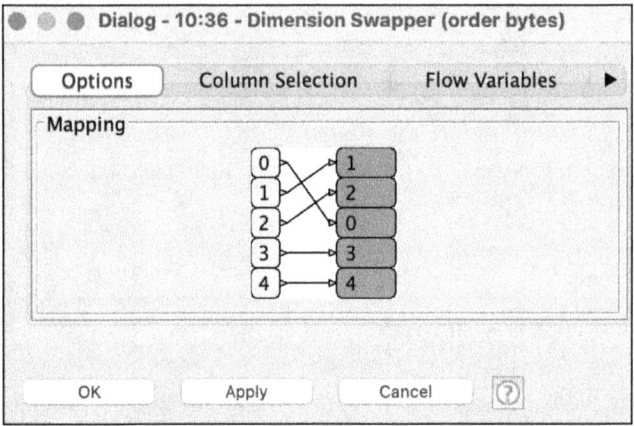

Figure 11.33 Settings Window of the Dimension Swapper Module

How do we know that the image data must be scaled as just described and the order of the image information adjusted? You could read the scientific publications or study the Keras source code, of course, but fortunately, other developers have already done this and made their KNIME programs available to the public. You can search for and view programs at the KNIME Community Hub (*https://hub.knime.com/*). If the description of a project sounds interesting and you want to test the program, you can simply drag and drop it from the browser into the KNIME workspace.

The **Keras Network Reader** module loads the H5 file of the model. You only need to set the path to the *inception_v3.h5* file, which is located in the *data* subfolder of the *06 image classification* program. You can also access the subfolder via the File Explorer of your operating system.

The image data can now be fed to the **Keras Network Executor** module (see Figure 11.34). Under **Option • Outputs • Conversion**, select **To Number (double)**, and leave **Output columns prefix** empty so that only numerical values are output at the output.

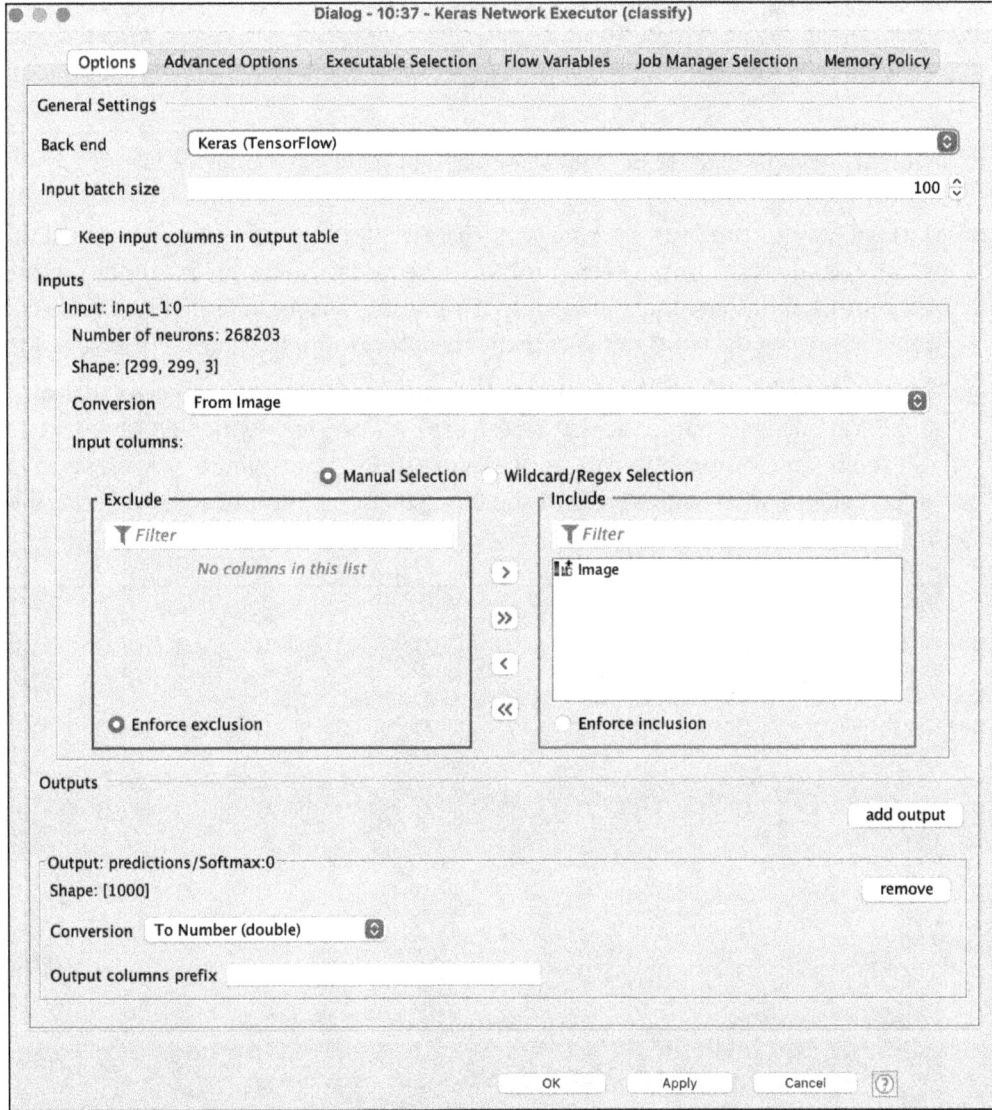

Figure 11.34 Settings Window of Keras Network Executor

When you execute the program, the **Keras Network Executor** module outputs 1,000 values containing the corresponding probabilities for the respective classes (formed using the `softmax` function). Select the module, and check the result in the **Node Monitor** screen, shown in Figure 11.35.

11 Visual Programming Using KNIME

Console	Node Monitor ×					
Node:	Keras Network Executor (10:37)					
State:	EXECUTED					
Port Output	Port 0		Load data	Rows: 1, Columns: 1000		
ID	0	1	2	3	4	5
teddy.png	2.9898808861617E-5	2.4395983200520277E-4	4.882254506810568E-5	4.861741399508901E-5	1.8909230129793286E-4	5.76337188249

Figure 11.35 Output of the Node Monitor for the Keras Network Executor Module

Now we have to continue working with these values at the output of the module. Which class does the image belong to? We have to determine the maximum of the table entries at the output of the module. If we were to use the **Rule Engine** module, we would have to set the 1,000 variables in the module to form the value we're looking for.

The setting we've selected in the **Keras Network Executor** module means that only numerical values are output, so that we can use the **Many to One** module. This module also determines the maximum value (see Figure 11.36). The maximum value is saved in the **col** column. When you execute the module, you'll see the number **850** at its output (in the **Node Monitor**).

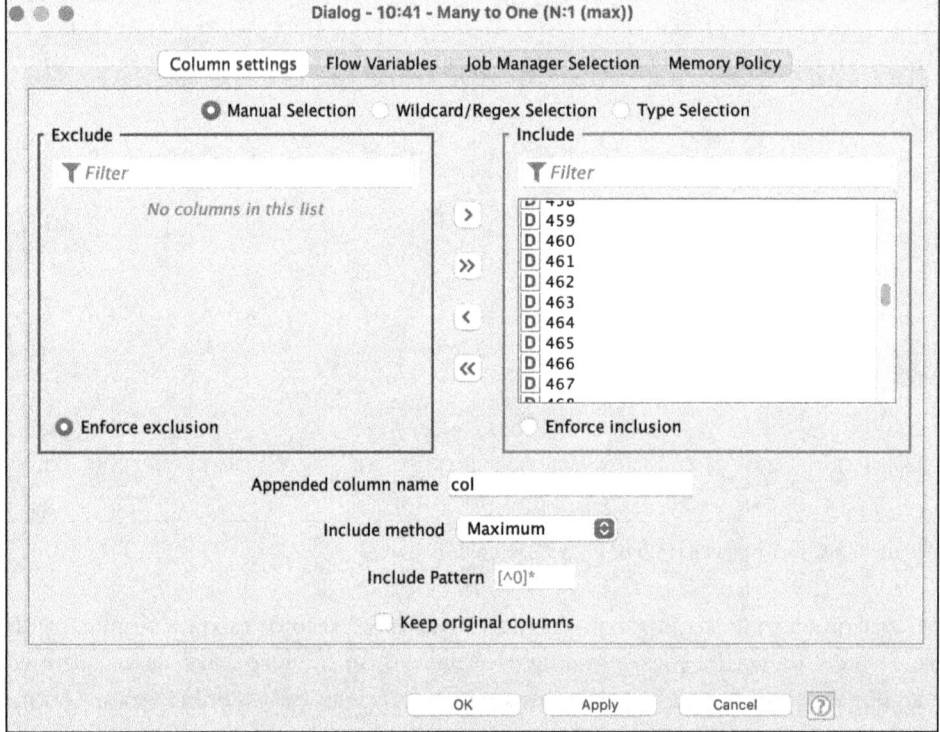

Figure 11.36 Settings Window of the Many to One Module

11.3 Image Classification Using a Pretrained Model

Although a number is displayed, the data type is string, so it must be transformed using **Column Auto Type Cast**. With this module, only the column needs to be selected because the remaining settings are determined automatically. The number 850 is output at the output of this module, but this time with the correct data type. We don't know which class is behind this number, but we can continue using it.

The allocation table is loaded using **CSV Reader**. Note that this file doesn't contain any column headings. This table contains the assignment of numerical values to class descriptions. The **Joiner** module forms the intersection of the table with the "col" column, a row (the value 850), and the assignment table ("Column0" column). In the **Node Monitor** of this last module, you'll finally see the important information as to which class the object in the image belongs to: "teddy, teddy bear".

11.3.2 Image Classification Using Python Node

The structure of the program is very clear. The image is loaded and resized, and the data is converted so that it can be processed in the **Python Script**. You can view the loaded image using **Image Viewer** (see Figure 11.37).

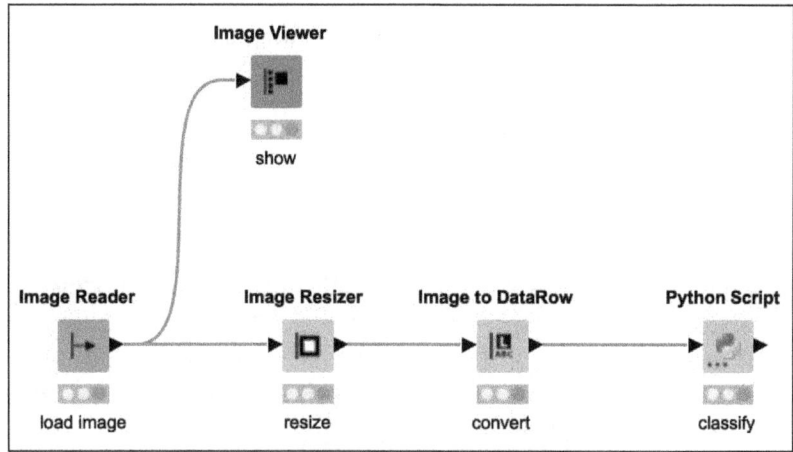

Figure 11.37 The 06-image-classification-py.knwf Program

Let's take a look at the source code in the Python module.

```
import knime.scripting.io as knio
import tensorflow as tf
import pandas as pd
import numpy as np
from tensorflow.keras.applications import vgg19

# Read data
pic = knio.input_tables[0].to_pandas()
```

11 Visual Programming Using KNIME

```
# Instantiate model
model = vgg19.VGG19()

# Adapt data to the model
img = np.array(pic['Image'][0])
img = img.reshape(1, 224, 224, 3)
img = tf.keras.applications.vgg19.preprocess_input(img)

# top ten
pred = model.predict(img)
top_ten = vgg19.decode_predictions(pred, top=10)

# Output of the result
df = pd.DataFrame(top_ten[0])
df.drop([0], axis = 1, inplace=True)

knio.output_tables[0] = knio.Table.from_pandas(df)
```

Listing 11.7 Python Script for Image Classification

The image is evaluated and the result—a list of the top 10 with the probabilities for the correct classification—is output.

11.4 Transfer Learning

Transfer learning can also be implemented using KNIME. Again, I'll present two options: implementation using Keras and implementation using Python modules. In the variant that uses Keras modules, the pretrained InceptionV3 model is used again, as the training is much more performant than VGG19 (nevertheless, the training procedure takes a very long time without the support of a graphics card). The variant that uses Python modules, on the other hand, uses VGG19.

11.4.1 Transfer Learning Using Keras Node

The goal is to modify and adapt this model in such a way that images of hand gestures for rock, paper, and scissors are classified. These three classes aren't originally included in the 1,000 classes of the pretrained model. You could also proceed in the same way for future image-classification projects. You collect images in different folders, one folder per class. The folder structure assigns the images to the classes. Then, you use a proven, pretrained model that has already been trained with over a million images. Although this model is trained for other objects, you can reuse the detection part (used to recognize features). You can rebuild the identification section (for classification) exactly as you need it. The finished model is then trained with the new images,

11.4 Transfer Learning

whereby only the weights of the new identification part are updated. The weights of the detection part are "frozen" and aren't changed during the training process.

You can see a special module in the program in Figure 11.38: **Metanode**. This module allows sections to be combined and programs to be structured. The creation wizard can be launched via **Menu Bar • Node • Add Metanode Wizard**. There you can set how many inputs and outputs you want the module to have (the number can also be changed later). You can open the module by double-clicking.

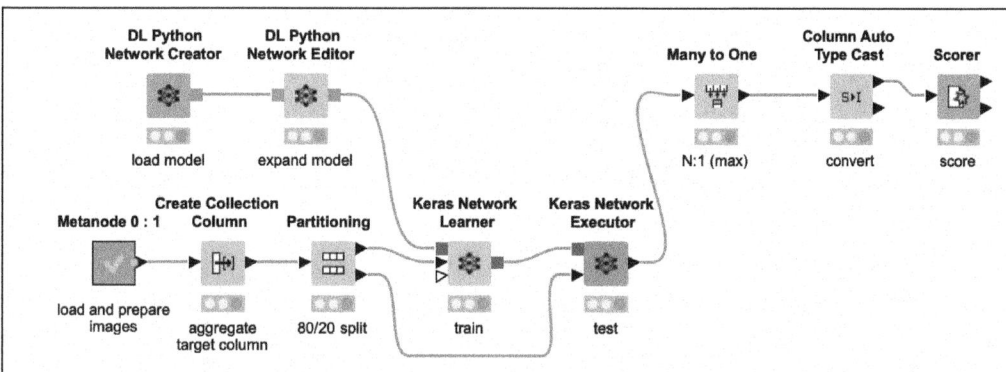

Figure 11.38 The 07-Transfer.knwf Program

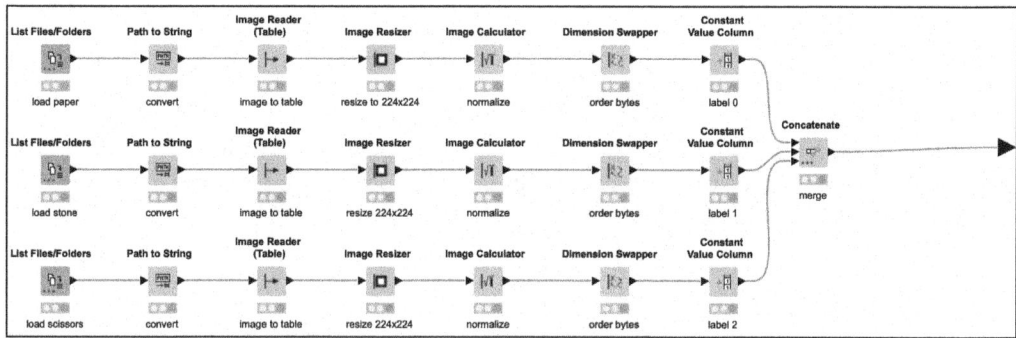

Figure 11.39 Metanode for Loading and Processing Images

The images are stored in directories. The paths to the folders are set, the Path data type is transformed to String, and then the images are saved in tables. The images must then be scaled and adapted for the InceptionV3 model. The pixels are standardized, and the order of the image information is adjusted. Each table is given a column with a number as the target column for the ANN. A large table is created from the three tables, and the result is linked to the output of the module.

In the main program, we need to transform the target column into a list with only one element. As you already know, this procedure is necessary if the `softmax` function will be used for the output of the ANN.

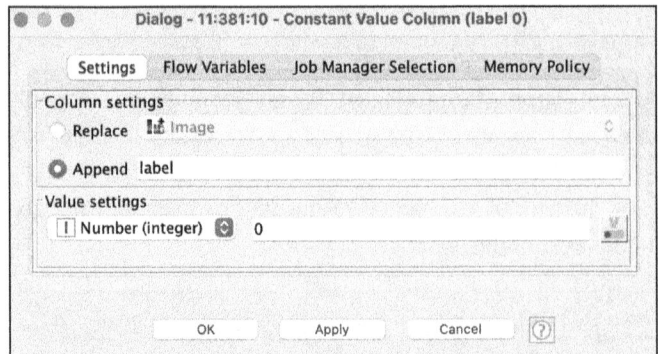

Figure 11.40 Settings Window of the Constant Value Column Module

The module is imported using the **DL Python Network Creator** module, in which you can enter Python commands.

```
# variable name of the output network:  output_network
from keras.applications.inception_v3 import InceptionV3

output_network = InceptionV3(include_top=False,input_shape=[224,224,3])
output_network.trainable = False
```

Listing 11.8 Loading the Model and Freezing the Detection Section

The default output variable of the module is output_network. Our model is stored in this variable.

The DL Python Network Editor module is used to extend and adapt the model. Here, input_network is the input variable, and output_network is the output variable.

```
# variable name of the input network: input_network
# variable name of the output network:  output_network
from keras.models import Model
import keras

x = keras.layers.Dropout(0.3)(input_network.output)
x = keras.layers.Flatten()(x)
x = keras.layers.Dense(256,activation='relu')(x)
x = keras.layers.Dropout(0.3)(x)
x = keras.layers.Dense(3,activation='softmax')(x)

output_network = Model(inputs=input_network.input, outputs=x)
```

Listing 11.9 Building the Final Model with the New Identification Part

11.4 Transfer Learning

You should be able to interpret the rest of the main program by now. You only need to train the ANN for five epochs (batch size = 32), but this will also take a lot of time. Note this important detail: Set the learning rate to 0.0001, or you'll only get a correct classification rate of approximately 33%. With the settings selected here, you'll achieve approximately 98%.

11.4.2 Transfer Learning Using Python Node

In this variant, the main program remains similar, but the Keras modules are also missing here (see Figure 11.41).

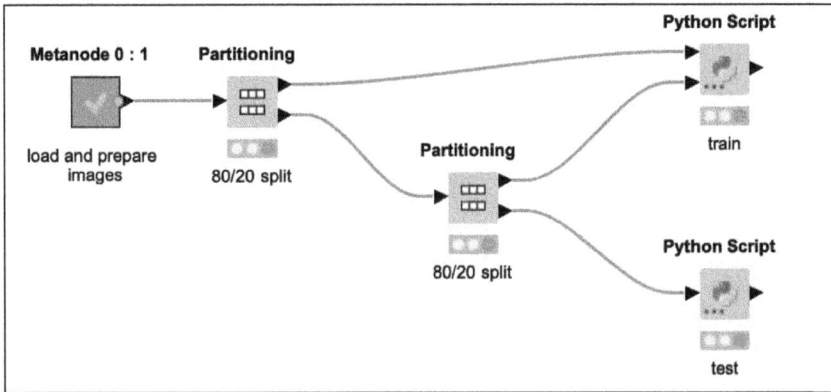

Figure 11.41 The 07-Transfer-py.knwf Program

When preparing the data, the Image Calculator and Dimension Swapper modules must be omitted, as the VGG19 model is used here. For this purpose, the data must be transformed using **Image to DataRow** for use in **Python Script** (see Figure 11.42).

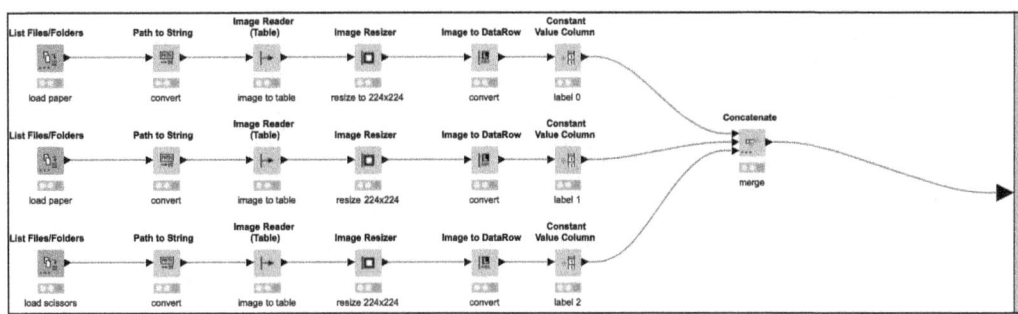

Figure 11.42 Metanode for Loading and Processing Images for VGG19

Let's first look at the Python Script module for training. Apart from the data type transformation, the program should look familiar to you.

235

```python
from keras.models import Model
from tensorflow.keras.applications import vgg19

# Read data
data_0 = knio.input_tables[0].to_pandas()
data_1 = knio.input_tables[1].to_pandas()

# Prepare data
train_data = data_0['Image'].to_list()
train_col = data_0['label'].astype('category')

test_data = data_1['Image'].to_list()
test_col = data_1['label'].astype('category')

train_data = np.array(train_data).reshape(-1, 224, 224, 3)
test_data = np.array(test_data).reshape(-1, 224, 224, 3)

# Instantiate model
base_model = vgg19.VGG19(include_top=False,input_shape=[224,224,3])
base_model.trainable = False

x = tf.keras.layers.Dropout(0.3)(base_model.output)
x = tf.keras.layers.Flatten()(x)
x = tf.keras.layers.Dense(256,activation='relu')(x)
x = tf.keras.layers.Dropout(0.3)(x)
x = tf.keras.layers.Dense(3,activation='softmax')(x)

model = Model(inputs=base_model.input, outputs=x)

# Configuration and training
cb_early = tf.keras.callbacks.EarlyStopping(monitor='val_loss', patience=3)
# model.compile(loss='sparse_categorical_crossentropy',
# optimizer=tf.keras.optimizers.Adam(learning_rate=0.0001), metrics=[
'accuracy'])
# Apple M1/M2
model.compile(loss='sparse_categorical_crossentropy',
 optimizer=tf.keras.optimizers.legacy.Adam(learning_rate=0.0001), metrics=[
'accuracy'])
history = model.fit(train_data, train_col, epochs=10,
 validation_data=(test_data, test_col),callbacks=[cb_early])

# Save model
model.save("model7.keras")
```

```python
# Output of the result
out = [history.history['val_accuracy'][-1]]
df = pd.DataFrame(out, columns=['acc'])
knio.output_tables[0] = knio.Table.from_pandas(df)
```

Listing 11.10 Python Script for Training the Model

The reshape(-1, 224, 224, 3) method call is interesting here as it forms data packets with the dimensions 224, 244, and 3 (for x, y, and RGB). We don't want to determine or calculate ourselves how many data packets can be generated. The parameter -1 specifies that all data should be converted into such data packets and as many as possible. Calling the model.compile method generates a warning on M1/M2 chips from Apple, so the following is a variant with an adapted optimizer call.

```python
import knime.scripting.io as knio
import tensorflow as tf
import pandas as pd
import numpy as np

# Load data
data = knio.input_tables[0].to_pandas()

# Prepare data
val_data = data['Image'].to_list()
val_col = data['label'].astype('category')
val_data = np.array(val_data).reshape(-1, 224, 224, 3)# Load model
loaded_model = tf.keras.saving.load_model("model7.keras")

# Evaluation
val_loss, val_acc = loaded_model.evaluate(val_data, val_col)

# Output of the result
df = pd.DataFrame([val_acc], columns=['acc'])
knio.output_tables[0] = knio.Table.from_pandas(df)
```

Listing 11.11 Python Script for Training the Model

The program for testing the model does nothing other than load the trained model and evaluate the test data.

11.5 Autoencoder

The advantages of visual programming are particularly evident in AI projects such as autoencoders. The data flow is clearly recognizable, and the structure is easier to understand. It's important not to get lost in the details. Take a look at the program, and try to understand the process and how it works. The aim here is to recognize conspicuous credit card transactions. Because the ratio of normal to abnormal transactions is very unequal, an AI based on an autoencoder can be a way of detecting these special cases.

11.5.1 Autoencoder with Keras Node

In this chapter, too, we first look at the program with Keras modules, as shown in Figure 11.43.

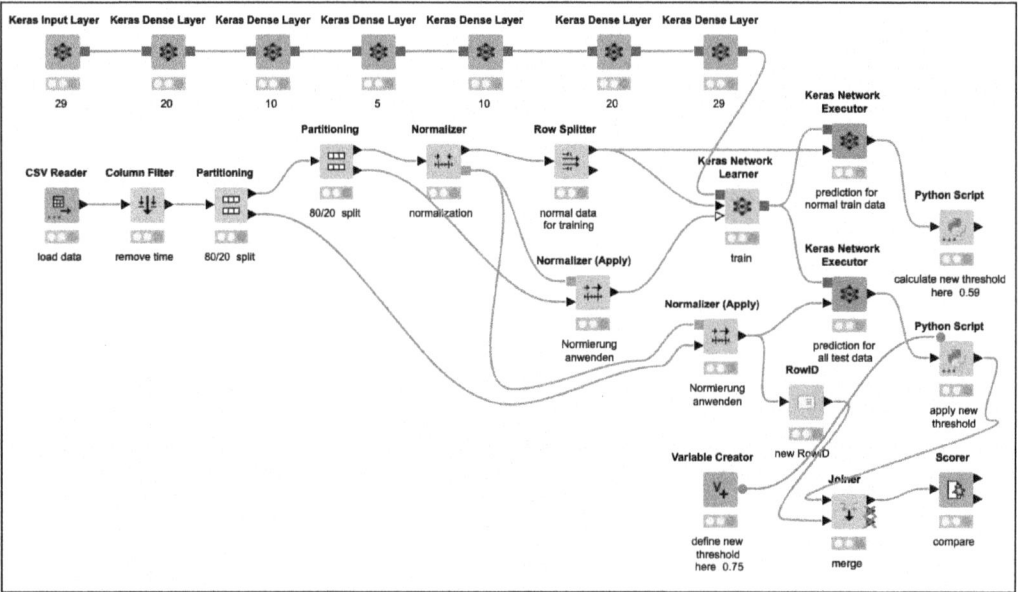

Figure 11.43 The 08-Autoencoder.knwf Program

The Time column is removed from the table, and then this table is split into training and test data. The training data is split up again: during the training process, the ANN is only trained with normal data, but tested with normal and abnormal data.

The **Row Splitter** module (see Figure 11.44) is configured in such a way that the data without special cases is output at the upper output, and the special cases are output at the lower output. As we only need the data without special cases for the training, the upper output is used. The special cases are thus filtered out.

This division means that the **Keras Network Learner** module can also be trained with early stopping. The calculation of the required values (mean value and standard deviation) for the standardization of the Amount column as well as the application of the standardization takes place with the training data; for the remaining data, the standardization is only applied with the values already determined.

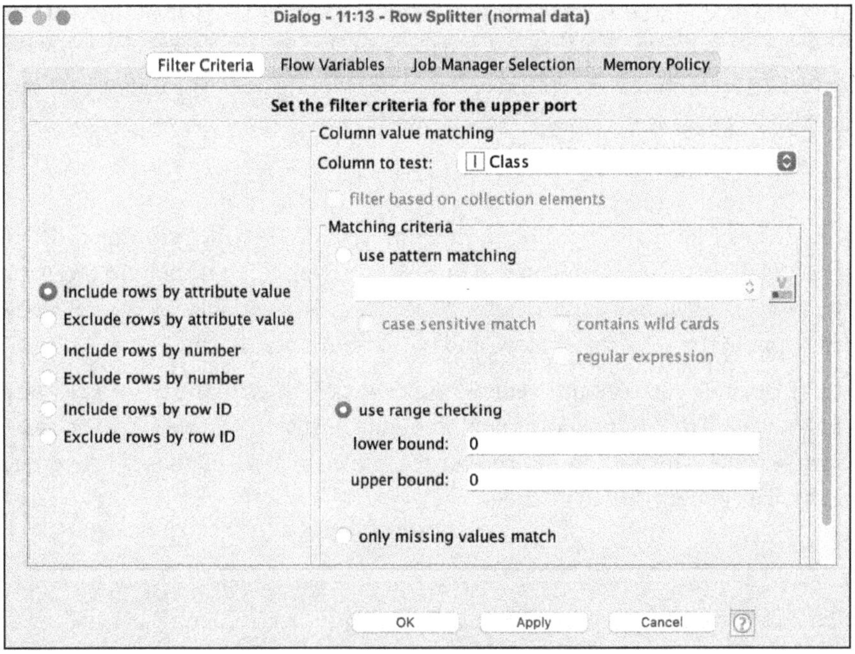

Figure 11.44 Settings Window of the Row Splitter Module

The ANN consists of various layers that compress the data (encoders) and then decompress it again (decoders). The aim of the training is to ensure that the same values come out at the output as go in at the input. The structure with a bottleneck ensures that the ANN is optimized for this data and that other data leads to a large scattering at the output. If the values at the output exceed a certain threshold value, different data is available than the data with which the ANN was trained.

To calculate the first draft for the threshold value (Threshold), we need the predictions for the training data (without anomalies). We use the normal training data again for this. The filtered training data is used to calculate the threshold value. To do this, we use Python commands in the Python Script module.

```
import knime.scripting.io as knio
import pandas as pd
from sklearn.metrics import mean_absolute_error
```

Listing 11.12 Importing the Modules

The required modules must also be imported here. If the import results in error messages, you should verify whether the respective module is installed via Anaconda.

```
train_data = knio.input_tables[0][:29].to_pandas()
train_pred = knio.input_tables[0][30:].to_pandas()
```

Listing 11.13 Lists with Training and Test Data

The input data of the module is saved in a list. More specifically, the first element of this list contains another list in which the data is stored. The first 29 elements with the index 0 to 28 (i.e., up to the exclusive 29) contain the training data. The following notation would also be possible here:

```
train_data = knio.input_tables[0][0:29].to_pandas()
```

The first element of `knio.input_tables` (i.e., `knio.input_tables[0]`) contains a list in which we access the elements 0 (inclusive) to 29 (exclusive), convert them to the DataFrame data type (pandas), and assign them to the `train_data` variable. Look at the variables at the input in the settings window, and try to reproduce this line.

All elements from index 30 contain predictions, so we save them in the `train_pred` variable. We don't want to determine how many elements the list has, so we leave the space after the colon blank. The Python interpreter recognizes that all elements (up to and including the last element) are to be used.

```
mae = mean_absolute_error(train_pred, train_data)

df = pd.DataFrame([mae], columns = ['mae'])

knio.output_tables[0] = knio.Table.from_pandas(df)
```

Listing 11.14 Calculating the Result

The MAE gets calculated. The output of the module is a table, so we create a variable of type DataFrame with the mae column and save this value in the only row. The output table is then generated from the variable. Once the ANN has been trained and the predictions have been generated, you'll see the threshold value (0.59) at the output of the Python Script module.

Based on this, we calculate the predictions for new test data (including anomalies) and compare whether the MAE is greater than the threshold calculated previously. If so, there is an anomaly. It's always necessary to check how much data has been classified correctly. If you're not content with the result, you should increase the threshold and reevaluate. It's very likely that the new threshold value used to detect actual anomalies must be higher than the previously calculated one.

To avoid having to open the module and change the variable in the Python source code every time, we save the new threshold value in the Variable Creator module. Here, we create the `newThreshold` variable of type double (float) with a value above 0.59. Remember that for the first draft of the new threshold value, we take the previously calculated MAE and increase it until we're happy with the result. I hope I'm not robbing you of the excitement by telling you that 0.75 provides good results.

You can configure modules with flow variables. To do this, right-click on the second **Python Script** module, and click on **Show Flow Variable Ports**. Then, you can link the variable to the module and access it in the module.

```
import knime.scripting.io as knio
import tensorflow as tf
import pandas as pd

new_threshold = knio.flow_variables['newThreshold']
```
Listing 11.15 Saving the New Threshold Value in a Variable

Here, we access the newThreshold flow variable and save the value in a new variable.

```
Test_data = knio.input_tables[0][:29].to_pandas()
test_pred = knio.input_tables[0][30:].to_pandas()
```
Listing 11.16 Training and Test Data

The test data and predictions are again saved in two variables.

```
Maes = tf.keras.losses.mae(test_data, test_pred)
pred_col_bool = tf.math.greater(maes, new_threshold)
pred_col = tf.cast(pred_col_bool, dtype=tf.int32)
```
Listing 11.17 Determining the Predictions

The system checks row by row whether the MAE is greater than the threshold value. If yes (in the case of fraud), the number 1 is saved; otherwise, 0 is saved. The CSV file also contains the number 1 for a fraud case.

If the CSV file contained the number 0 for a fraud case, we would have to reverse the logic and check whether the deviation is smaller than the threshold value. If yes (usually), the number 1 is saved; otherwise, 0 is saved.

```
Df = pd.DataFrame()
df['pred_col']=pd.Series(pred_col.numpy())

knio.output_tables[0] = knio.Table.from_pandas(df)
```
Listing 11.18 Outputting the Result at the Output of the Module

A new variable of type DataFrame is created here, and the predictions are saved in the pred_col column and then transferred to a table. After executing the module, you'll see this column with the predictions at the output of the module. This column must be compared with the actual values in the CSV file. We'll create a table with these two columns and create a confusion matrix.

The Partitioning module outputs the test data at the lower output. We use the RowID module to set a new, ascending numbering of the rows. This means that the corresponding rows of the test data and the predictions have the same row numbers and can

11 Visual Programming Using KNIME

be merged by using the Joiner module. The Scorer module can now calculate the confusion matrix (see Figure 11.45).

Confusion Matrix - 11:17 - Scorer (compare)		
File Hilite		
pred_col \ Class	0	1
0	48129	9
1	8725	99

Figure 11.45 Output of the Confusion Matrix

In this case, 99 cases of fraud are recognized correctly, and 9 aren't. Almost 92% of fraud cases can therefore be identified, and 15% of normal transactions (8725) are classified as fraud. The following principle applies here: if customers receive an email that a transaction has been completed, the result may be acceptable.

11.5.2 Autoencoder with Python Node

In this section, we'll only analyze the deviations from the previous program. In the area at the top right (see Figure 11.46), the model is trained, and the first draft for threshold is calculated; at the bottom right, this is modified and applied to the test data.

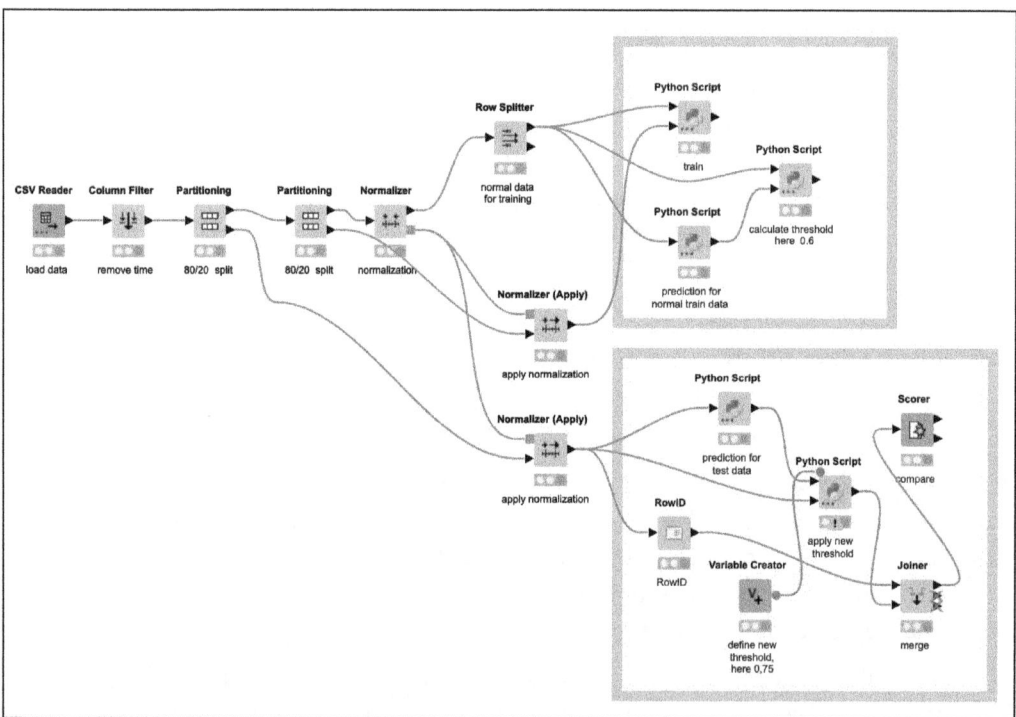

Figure 11.46 The 08-Autoencoder-py.knwf Program

```
import knime.scripting.io as knio
import tensorflow as tf
import pandas as pd

# Training data
train_data = knio.input_tables[0][0:30].to_pandas()

# Test data
test_data = knio.input_tables[1][0:30].to_pandas()
test_col = knio.input_tables[1]['Class'].to_pandas()

encoder = tf.keras.Sequential(name='encoder')
encoder.add(layer=tf.keras.layers.Dense(units=20, activation=tf.nn.sigmoid,
 input_shape=[30]))
encoder.add(layer=tf.keras.layers.Dense(units=10, activation=tf.nn.sigmoid))
encoder.add(layer=tf.keras.layers.Dense(units=5, activation=tf.nn.sigmoid))

decoder = tf.keras.Sequential(name='decoder')
decoder.add(layer=tf.keras.layers.Dense(units=10, activation=
tf.nn.sigmoid, input_shape=[5]))
decoder.add(layer=tf.keras.layers.Dense(units=20, activation=tf.nn.sigmoid))
decoder.add(layer=tf.keras.layers.Dense(units=30, activation=tf.nn.sigmoid))

autoencoder = tf.keras.Sequential([encoder, decoder], name='autoencoder')
autoencoder.compile(optimizer='adam', loss='mae', metrics=['mae'])

# Configuration, training, and testing
autoencoder.compile(optimizer='adam', loss='mae', metrics=['mae'])
cb_early = tf.keras.callbacks.EarlyStopping(monitor='val_loss', patience=1)
history = autoencoder.fit(train_data, train_data, epochs=100,
 validation_data=(test_data, test_col),callbacks=[cb_early])

# Save model
autoencoder.save("model8.keras")

# Output of the result
out = [history.history['val_mae'][-1]]
df = pd.DataFrame(out, columns=['mae'])
knio.output_tables[0] = knio.Table.from_pandas(df)
```

Listing 11.19 Python Script for Training the Model

The model is trained with early stopping. Change some hyperparameters to improve the model and compare the results.

11 Visual Programming Using KNIME

```python
import knime.scripting.io as knio
import tensorflow as tf
import pandas as pd

# Training data
data = knio.input_tables[0][0:30].to_pandas()

print(data)
# Load model
loaded_model = tf.keras.saving.load_model("model8.keras")

test_pred = loaded_model.predict(data)

print(test_pred.shape)
# Output of the result
df = pd.DataFrame(test_pred)
knio.output_tables[0] = knio.Table.from_pandas(df)
```

Listing 11.20 Python Script for Generating Predictions with Training Data

The trained model is used to generate predictions for the training data. The predictions are compared with the actual data in the next module to determine the first draft for the threshold.

```python
import knime.scripting.io as knio
import numpy as np
import pandas as pd
from sklearn.metrics import mean_absolute_error

train_data = knio.input_tables[0][:30].to_pandas()
train_pred = knio.input_tables[1][:].to_pandas()

mae = mean_absolute_error(train_pred, train_data)

df = pd.DataFrame([mae], columns = ['mae'])

knio.output_tables[0] = knio.Table.from_pandas(df)
```

Listing 11.21 Python Script for Calculating the Threshold

Using the training data and the predictions, the threshold is calculated based on the MAE. We'll adjust and apply this value to the test data to detect anomalies. The code of the module for predicting the test data corresponds to the code of the module for predicting the training data, only with different input data (namely the test data).

```python
import knime.scripting.io as knio
import tensorflow as tf
import pandas as pd
import numpy as np

new_threshold = knio.flow_variables['newThreshold']

test_data = knio.input_tables[0][:30].to_pandas()
test_pred = knio.input_tables[1][:30].to_pandas()

maes = tf.keras.losses.mae(test_data, test_pred)
pred_col_bool = tf.math.greater(maes, new_threshold)
pred_col = tf.cast(pred_col_bool, dtype=tf.int32)


df = pd.DataFrame()
df['pred_col']=pd.Series(pred_col.numpy())

knio.output_tables[0] = knio.Table.from_pandas(df)
```

Listing 11.22 Python Script for Using the Threshold

The deviation between the predictions and the actual values of the training data is determined and checked to see whether the newly set threshold value is exceeded. Accordingly, the output column is filled with a 0 or 1. This output is compared again with the test data, and a confusion matrix is created and output.

11.6 Text Classification

In Python, you need good programming skills to edit texts or prepare them for AIs. In KNIME you can do everything using modules or use Python modules. The following AI is intended to classify movie ratings again. Here, too, you can compare the program with the Python code from Chapter 8.

11.6.1 Text Classification with Keras Node

Here, the text classification is completely realized with visual modules. The CSV Reader module has problems loading the large *IMDB-Dataset.csv* file, so the **File Reader** module was used (far left in Figure 11.47). Then 5,000 characters are read from each row, and the rest is truncated if necessary. For text processing, the data type must be converted to Document.

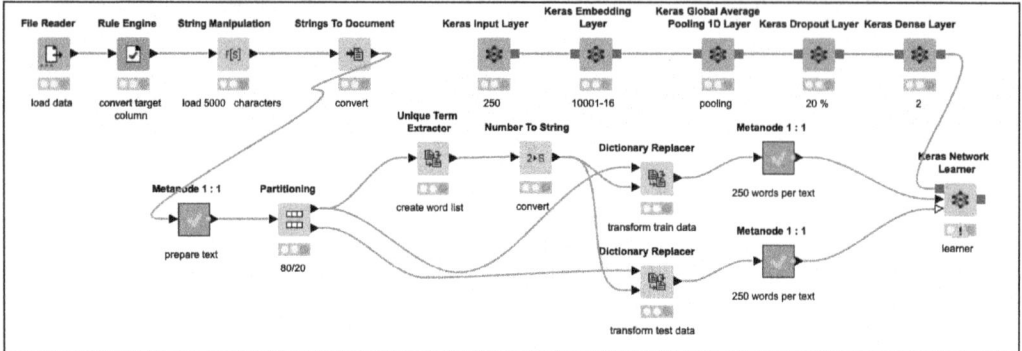

Figure 11.47 The 09-Textclassification.knwf Program

The text processing modules are organized in a **Metanode**, which keeps the main program clear (see Figure 11.48). The aim is to define an assignment of words to integers. If a word is sometimes capitalized (e.g., at the beginning of a sentence) and sometimes written in lower case, there would unnecessarily be two entries in the list. When training the AI model, this would be two different pieces of data. The assignment list should therefore not contain any redundant entries.

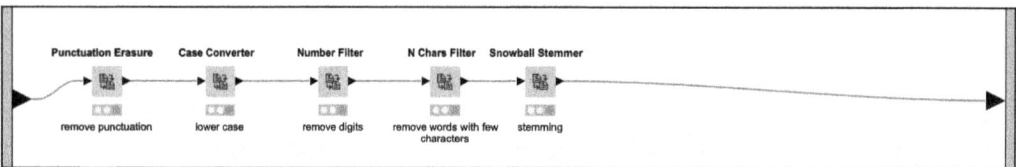

Figure 11.48 Prepare Text Metanode

Any punctuation is removed and all characters are written in lower case. In addition, numbers and words with fewer than three characters (e.g., "a" and "or") are removed. Stemming removes some letters from the word if necessary. For example, the algorithm replaces the word "watching" with "watch". Sometimes, no meaningful words come out of it. But that's not a bad thing at all because the result is intended for the AI, not for human eyes. Before the text processing module, there was a sentence that read as follows:

One of the other reviewers has mentioned that after watching just 1 Oz episode you'll be hooked.

After processing, we get the following result:

one the other review has mention that after watch just episod you hook

The data is divided into training and test data, and the dictionary (the assignment list) is created with the training data. This test data simulates "real" data from real life, and you would also have to transform this using the existing dictionary. The test data is therefore not used to create the dictionary; instead, the existing dictionary is used for transformation. The **Unique Term Extractor** module creates the dictionary for those 10,000 words that occur most frequently in the ratings. The newly created **Index** column with the numerical values is converted to the String data type. Finally, the words are replaced by the numerical values (as a string) using the **Dictionary Replacer** module. We replace words that aren't contained in the dictionary with "10000" (see Figure 11.49).

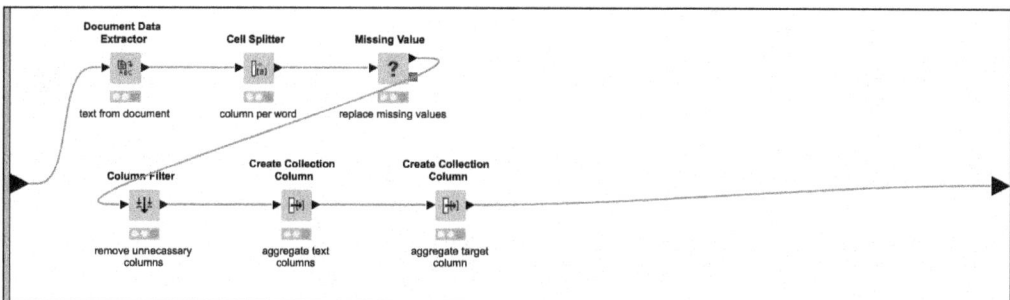

Figure 11.49 The "250 Words per text" Metanode

Now the texts must be prepared so that they each contain 250 words. If these are longer, the rest will be cut off. Shorter texts must be filled with "10000". Why's that? This is the word with the fewest occurrences, so we use it as a placeholder for everything "unnecessary". This processing takes place in the next metanode. There you can also see another advantage of metanodes: when you're done with the module for the training data, you can copy it and use it for the test data.

First, the texts must be extracted from the data type. We then create a large table in which each word is contained in a column. The number of columns (starting with index 0) corresponds to the words in the longest text. Now missing values can be replaced by our placeholder 10000, then there are no more empty cells. All superfluous columns (including columns >= 250) are then removed, and the necessary columns are merged again. The last block aggregates the target column again so that we can build the ANN using the `softmax` function.

The ANN has 250 input nodes. The embedding layer has 10,001 input nodes, as this module requires a list with elements from 0 to n, where n = 10,000 in our program. That is a total of 10,001 elements. You know the remaining layers from the corresponding Python program. You can look at it again and compare it with this visual program. The program achieves a correct classification rate of approximately 89%.

11.6.2 Text Classification with Python Node

Here, too, the preparation of the data is completely done with KNIME modules. There is only one significant difference to the variant with Keras nodes: the target column doesn't need to be aggregated (see Figure 11.50).

11 Visual Programming Using KNIME

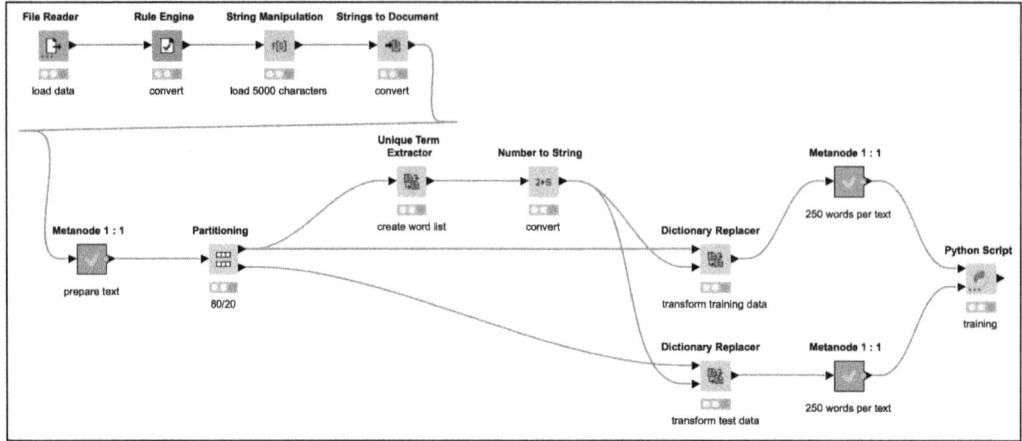

Figure 11.50 The 09-text-classification-py-iris.knwf Program

There is only a single Python module in which the model is built, trained, and tested.

```
import knime.scripting.io as knio
import tensorflow as tf
import pandas as pd
import numpy as np

# Read data
data_0 = knio.input_tables[0].to_pandas()
data_1 = knio.input_tables[1].to_pandas()

# Prepare data
train_data = data_0['AggregatedValues'].to_list()
train_col = data_0['rating'].astype('category')

test_data = data_1['AggregatedValues'].to_list()
test_col = data_1['rating'].astype('category')

train_data = np.array(train_data).reshape(-1, 250)
test_data = np.array(test_data).reshape(-1, 250)

#print(train_data)

model = tf.keras.Sequential([
  tf.keras.layers.Embedding(10001, 16),
  tf.keras.layers.GlobalAveragePooling1D(),
  tf.keras.layers.Dropout(0.2),
  tf.keras.layers.Dense(2, activation=tf.nn.softmax)
])
```

248

```
model.compile(optimizer='adam', loss='sparse_categorical_crossentropy',
 metrics=['accuracy'])

cb_early = tf.keras.callbacks.EarlyStopping(monitor='val_loss', patience=3)

history = model.fit(train_data, train_col, validation_data=(test_data, test_
col),
 epochs=100, callbacks=[cb_early])

# Output of the result
out = [history.history['val_accuracy'][-1]]
df = pd.DataFrame(out, columns=['acc'])
knio.output_tables[0] = knio.Table.from_pandas(df)
```

Listing 11.23 Python Script for Text Classification

This model with the Python Script module also achieves a correct classification rate of 89%. Try to improve this value through hyperparameter optimization.

11.7 AutoML

KNIME also provides a way to create AI models for you with the AutoML module. This module standardizes data and fills any empty cells with mean values from this column. For categorical data, empty cells are replaced by the most frequent class that occurs in the data. The data is also divided into test and training data.

AutoML is used to apply and evaluate various AI models and approaches. You can then view the results and choose one of these models. The AI is then tested and evaluated with validation data.

11.7.1 Installation

Go to the KNIME Community Hub website (*https://hub.knime.com*), search for "AutoML", and filter the results by **Components**. Select the **AutoML** module from KNIME, as shown in Figure 11.51.

Components are another option (as an alternative to metanodes) for structuring programs. If you want to use locally encapsulated flow variables within your own structure modules, for example, you should convert the metanode into a component block (right-click on the block, and select **Convert to Component**). To open a component module, right-click, and select the corresponding entry. Double-clicking on the block doesn't work; instead, a configuration window appears (if one is set). You can find a detailed comparison of these structuring options at *www.knime.com/blog/metanode-or-component*.

11   Visual Programming Using KNIME

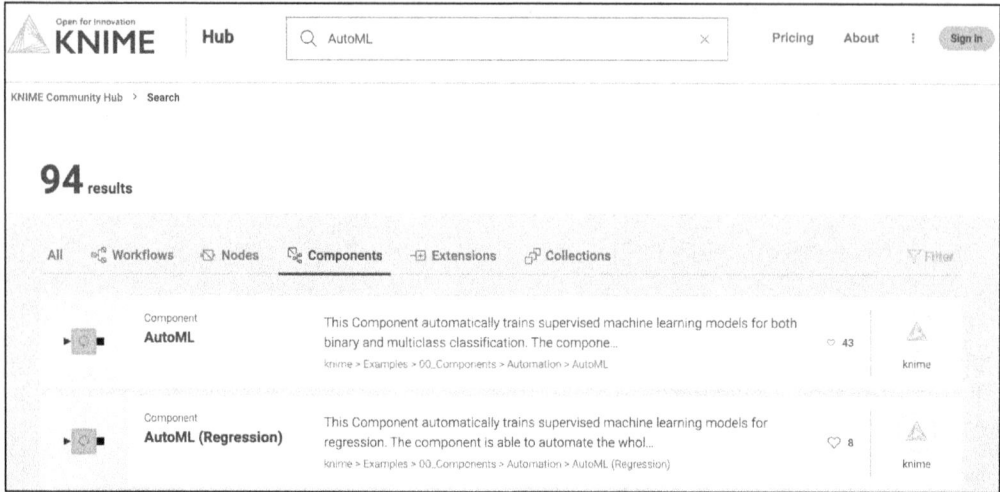

Figure 11.51 Search for AutoML

If you select the **AutoML** entry in the results window, you'll be taken to the description of the module (see Figure 11.52).

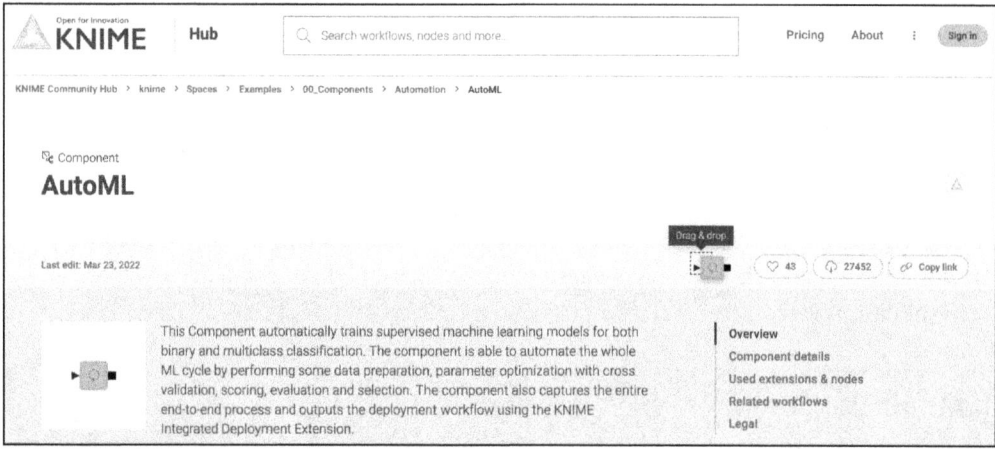

Figure 11.52 Description of the AutoML Module

From the information page, you can drag and drop the module into your workspace (workflow). KNIME suggests that you install other required packages, which you can confirm.

11.7.2 Classification

You can now use the module. Connect the **AutoML** module to the data source after you've split the data as usual. When we classify the irises again, we can compare and evaluate the results with other models (see Figure 11.53).

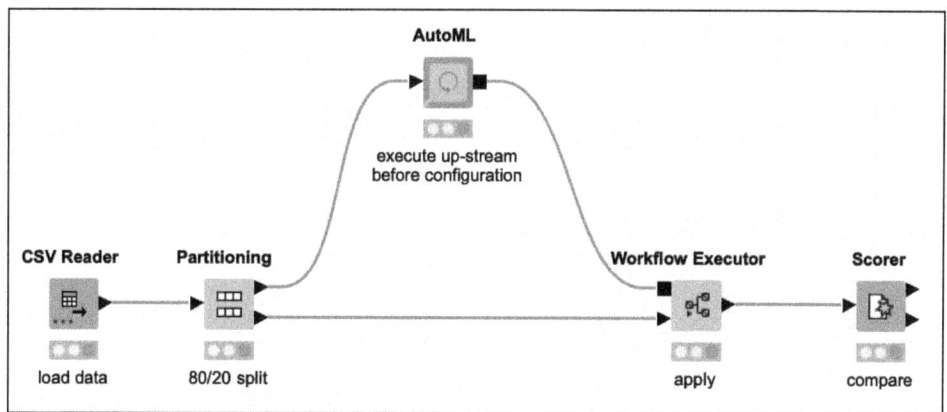

Figure 11.53 The 10-AutoML-iris.knwf Program

Once you've connected the new module to **Partitioning**, you can configure it by right-clicking. Specify the models to be tested, the target column, and the metric.

You've already become familiar with most of the suggested models. Some are new and haven't been covered in this beginner's book:

- **Naive Bayes**
  During training, the probabilities of a class are calculated for all attributes. During application, the probabilities of the attributes (for the classes) are multiplied with each other. The class with the highest product of probabilities is the class you're looking for.
- **Logistic Regression**
  You try to find a function that calculates the classes depending on attributes.
- **Generalized Linear Model (H2O)/H2O AutoML - StackedEnsemble**
  These are based on the H2O open-source library from the H2O.ia company.

Once you've made the required settings, you can run the module. Then, right-click, and select **Interactive View: Select AutoML**. The result is shown in Figure 11.54.

You can move your mouse over the bars to display more details about the training and test data. Finally, select a model of your choice on the right, and click **Apply** at the bottom right. In the message window that opens, confirm that this is only a temporary selection, and then close the window. If you don't make an explicit selection, the best model will be used for you.

You can now link the validation data and AutoML to the **Workflow Executor** module. In the configuration window, select **Auto-adjust ports (carried out on apply)**, as shown in Figure 11.55. Then, you can feed the validation data to the function block and connect the output to the scorer.

11   Visual Programming Using KNIME

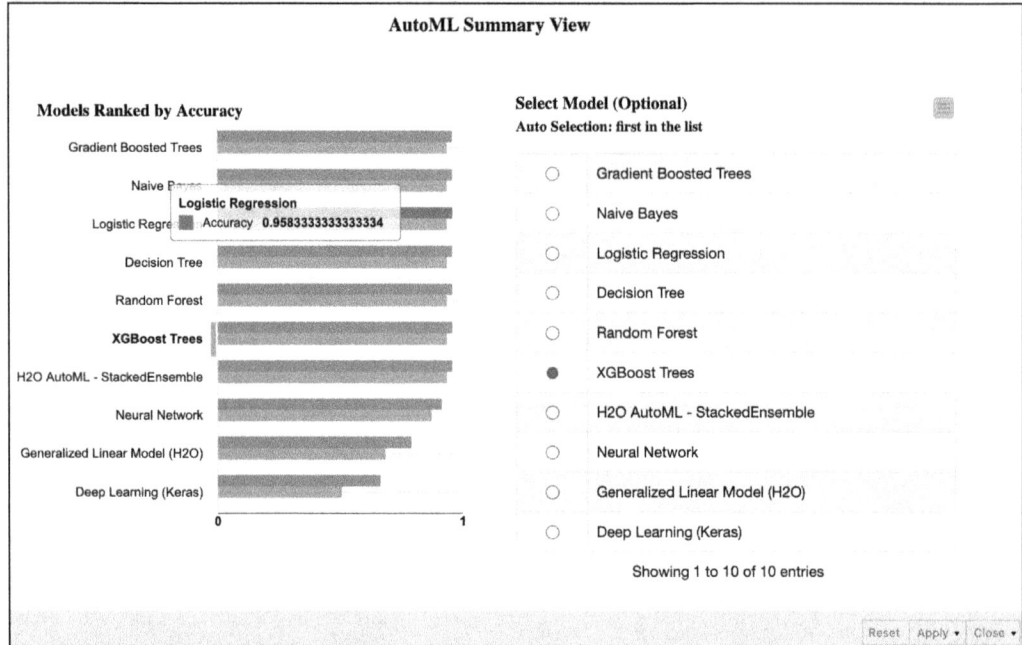

Figure 11.54 Interactive View

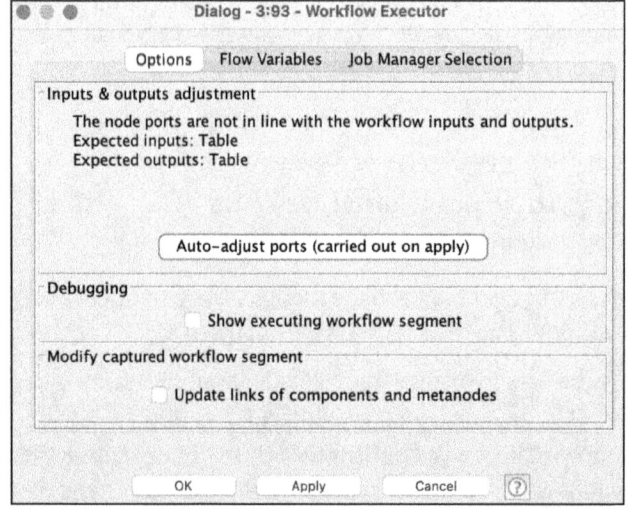

Figure 11.55 Settings Window of the Workflow Executor Module

The AutoML module is therefore used to prepare the data, create different AI models, and compare them with each other. This module saves you a lot of work during development and also delivers good results.

11.8 Cluster Analysis

Using cluster analysis, we can search for similarities in datasets and group them together. In this section, the iris species are grouped together. We don't remove the species column with the information about the subspecies so that we can compare the results. Remember, with cluster analysis, you can't expect the grouping to be as you would like it to be. In this example, the subspecies of iris would ideally be recognized and the data grouped accordingly. You'll see that this doesn't work very well.

The data is grouped based on the existing attributes or columns. This method can be used to search for patterns in an unknown, unlabeled dataset and form groups accordingly. We also use the iris dataset here, as we can better classify the results thanks to the experience we've gained with this dataset. You can experiment with other datasets during the exercises.

11.8.1 Manual Cluster Setting

In the first example, the number of clusters is set manually (i.e., three). The goal is actually to have the algorithm determine the number and size of groups. This workflow should first familiarize you with the required modules (see Figure 11.56).

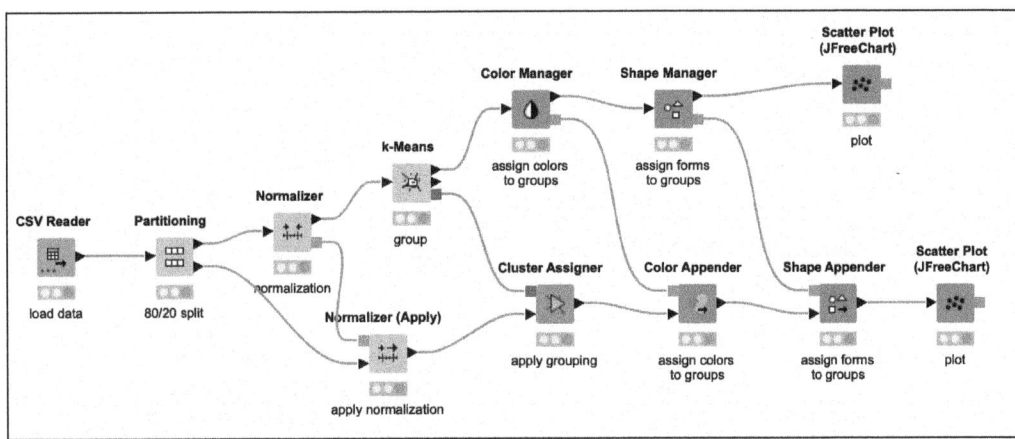

Figure 11.56 The 11-cluster-iris.knwf Program

You want to set the four columns with the size specifications in the **k-Means** module; the number of clusters is 3.

The **Color Manager** module assigns a color to each cluster, while the **Shape Manager** module assigns a shape. Once these colors and shapes have been set, the data points can be plotted using the **Scatter Plot (JFreeChart)** module, as shown in Figure 11.57. Plot the sepal and petal information one after the other, and analyze the output of the **k-Means** and **Cluster Assigner** modules in the Node Monitor. You'll see that a new Cluster column has been created, and the group information (e.g., "cluster_0") is saved here.

11 Visual Programming Using KNIME

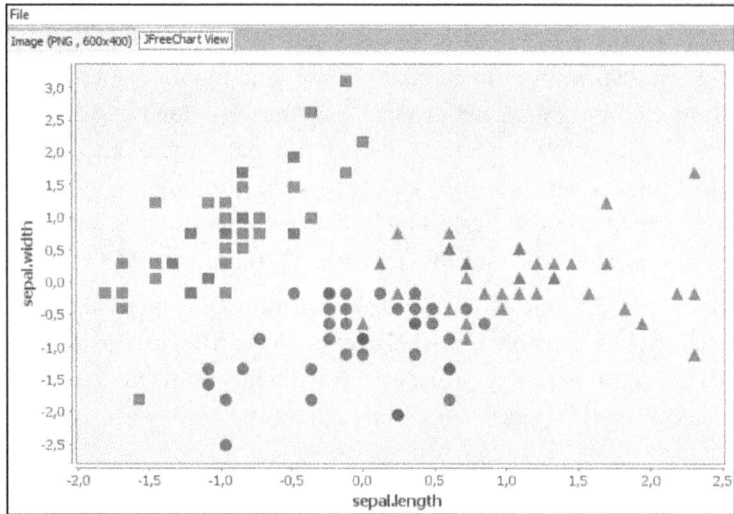

Figure 11.57 Plot of the Training Data

The training data and the **k-Means** module are used to iteratively position the three centers so that all data points that are closest to a center are assigned to the corresponding group. The **Cluster Assigner** module is used to group the test data, whereby the position of the centers is no longer changed.

The process was analyzed in detail in Chapter 9. If you have any unanswered questions, you should take another look at that chapter.

11.8.2 Cluster Setting with a Loop

We'll program the first loop in this workflow. In the loop, **k-Means** is executed with the $k = 2$ variable. The result is evaluated, and k is increased by one. The procedure is then repeated up to and including $k = 11$. After the loop, the evaluation results are plotted so that we can analyze the elbow and determine the optimal number of clusters. The first point at which the curve flattens out (here, the value for k) is the optimum value for k.

The **Empty Table Creator** module creates a table with only RowIDs and 10 rows. **Counter Generation** creates a new Counter column and fills the cells from top to bottom with the values 2 to 9. **Column Rename** renames the Counter column to k. After running this module, you should see the table with the column k and the values 2 to 11 in the Node Monitor. The counting variable k is thus created as a column.

In the **Table Row to Variable Loop Start** module, we select the k column as the data column. For the **k-Means** module, we activate **Flow Variable Ports** with a right-click and connect it to the aforementioned module. Using **k-Means**, you also select the four columns of the Iris dataset, but leave the number of clusters empty. In the **Flow Variables** tab, the number of clusters can be linked to the flow variable k (see Figure 11.59). This means that k is automatically increased with each iteration.

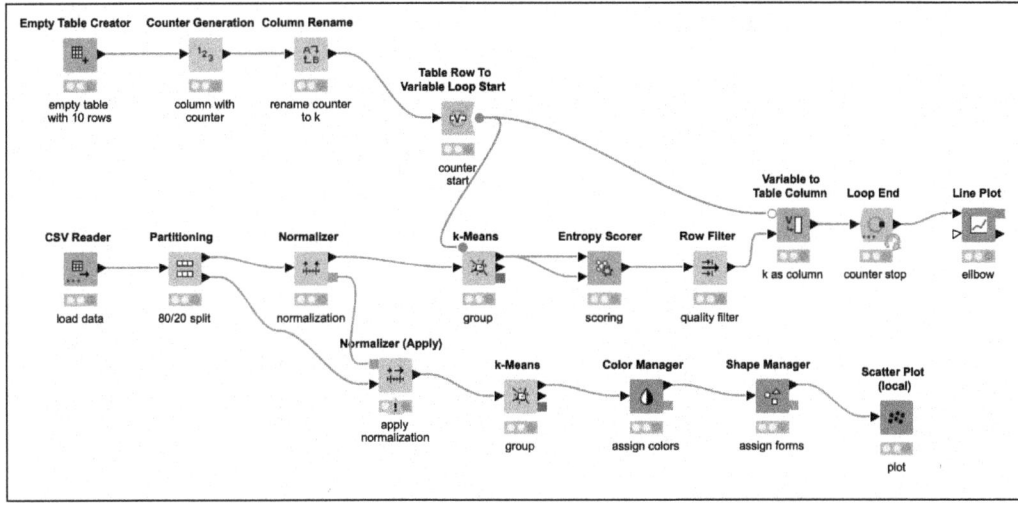

Figure 11.58 The 12-cluster-iris.knwf Program

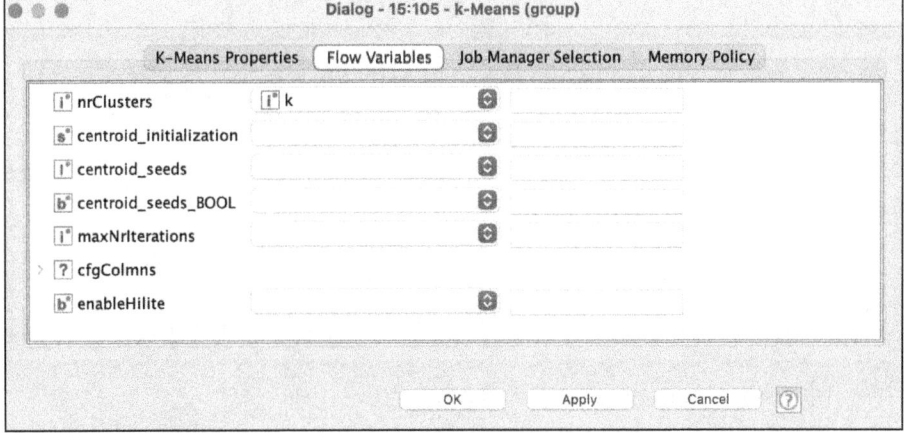

Figure 11.59 Settings Window of the k-Means Module

Entropy Scorer evaluates the Cluster and species columns; quality values less than 0 must then be filtered out or the plot is unusable. Finally, the loop value is saved in a column in the loop. After running the **Loop End** module, the **Entropy** column can be plotted (k for the x-axis, Entropy for the y-axis), as shown in Figure 11.60.

You can see from the plot that the entropy curve flattens out at $k = 3$. But let's be honest: if we didn't know the dataset and only interpreted the curve, a cluster number of 5 would also be possible with this data.

The test data is grouped with this determined number 3 for k, and the result is plotted again (see Figure 11.61).

11 Visual Programming Using KNIME

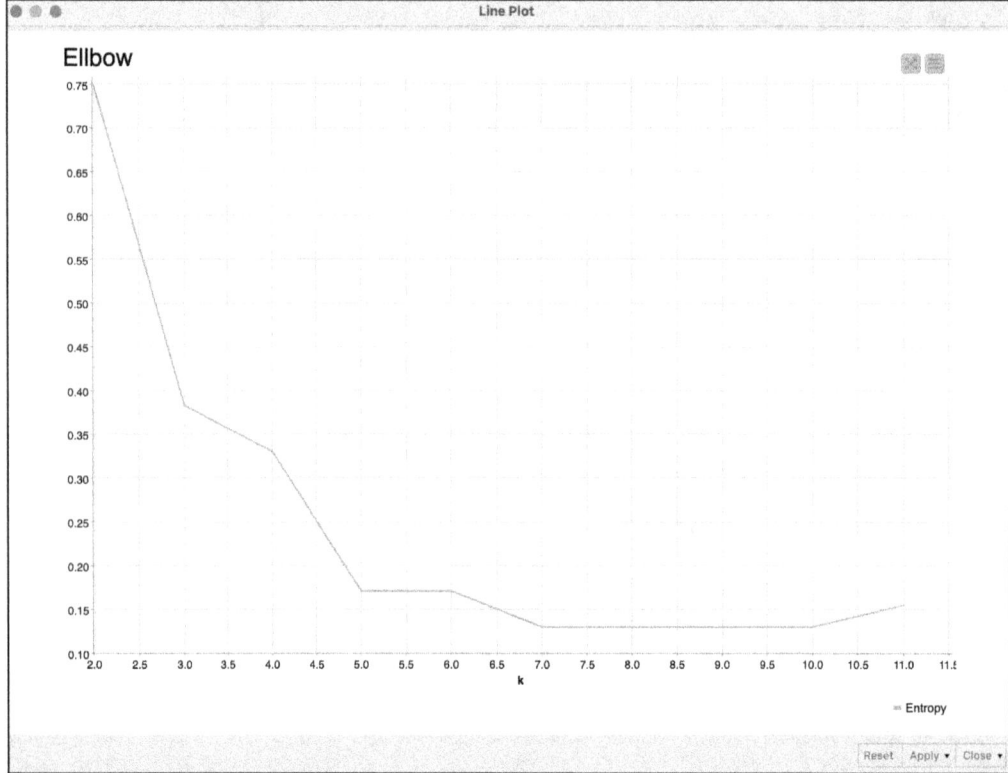

Figure 11.60 Elbow Line

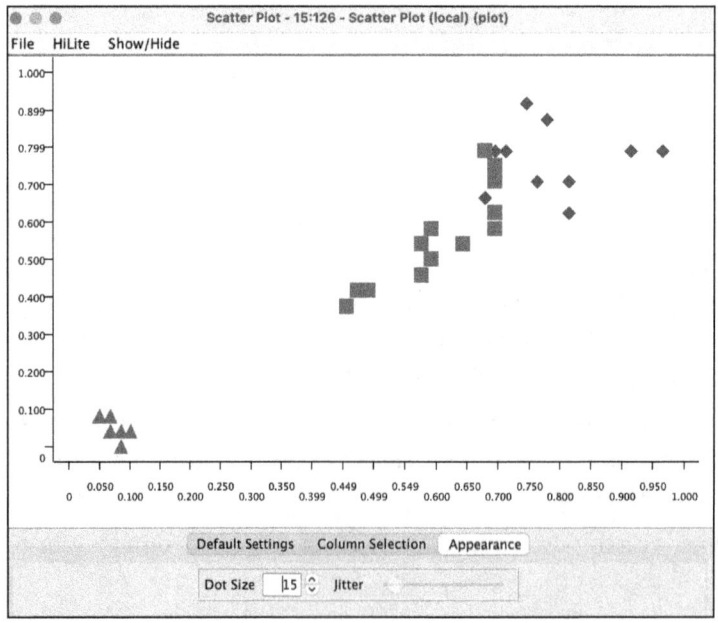

Figure 11.61 Plot of the Grouping: Petals (x = length, y = width)

11.9 Time Series Analysis

As you know, the black box for ANN in the field of supervised machine learning includes the areas of feed forward, backpropagation, and gradient methods. The data is processed from input to output. The input data is multiplied by weights, the values are added up, and then they are fed to the activation function. We can say that there is a function f that calculates the weights and the input, namely f(W, i). You also learned that a bias *b* is added to each node, so the function would be f(W, i, b).

The results form the input data for the next layer in the ANN. At the output, the results are compared with setpoint values, and the control difference or the results of the loss function are distributed proportionally to the weights in the ANN. The old and new weights are then offset against each other using a gradient method so that the deviation at the output becomes smaller and smaller over time. You can imagine the process as a data wave that moves from the input to the output and then back to the input. The entire process is repeated several times (number of epochs).

The input data in the programs were mostly table data, whereby one column was declared as the target column to be predicted. A row of the table without a target column was fed to the ANN, the entry in the target column was compared with the output of the ANN, the deviation was calculated, the weights were updated, and so on. It was similar with image data: the pixels were filtered and lined up one after the other so that a line (similar to the rows in a table) with pixel data was available at the input of the ANN. This means that up until now, the output has always been directly dependent on the input. The system had no memory for old results.

Let's consider the following problem: You record the electricity consumption of the past few months in a table. Now you want to develop an AI that is trained with this data and then predicts what electricity consumption can be expected in the near future. AI needs a memory for this. It should also be obvious that this training data must not be mixed, as the sequence is important. Our model in the black box must therefore be expanded.

11.9.1 Recurrent Neural Networks

For problems that require a "memory function," the recurrent neural network model (*RNN*) was developed. This model is often visualized as follows: The input and output layers are each displayed as individual nodes. The hidden layers are also combined into a single node with a connection to itself (see Figure 11.62). Specifically, the values from the previous call are weighted and included in the current call.

The iteration level the network is at is now relevant too, as the previous values are also taken into account. To illustrate the relevance of iteration stages, a series connection of cells is shown using the compressed model. Each link in the chain symbolizes a call of the layer at time t. The variable x stands for the input values, y for the output values, and h for the weights in the hidden layers or for the status information.

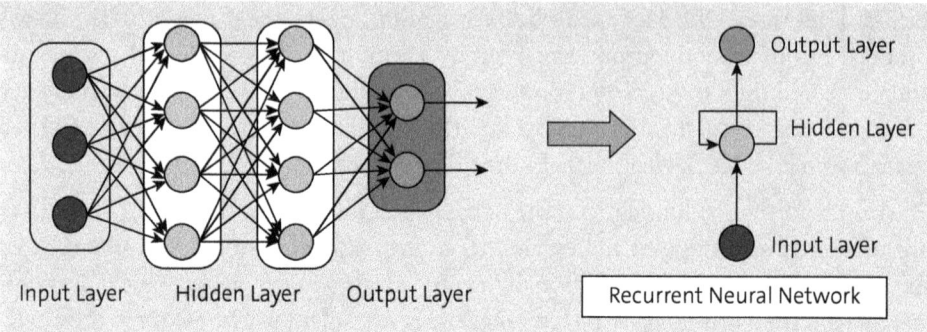

Figure 11.62 Model for RNN

How are the values offset against each other, and how are the results obtained? The status information from the previous cell, the actual input values, and the current calculated status information are multiplied by the respective weights, followed by adding a bias. The block diagram in Figure 11.63 shows the connections.

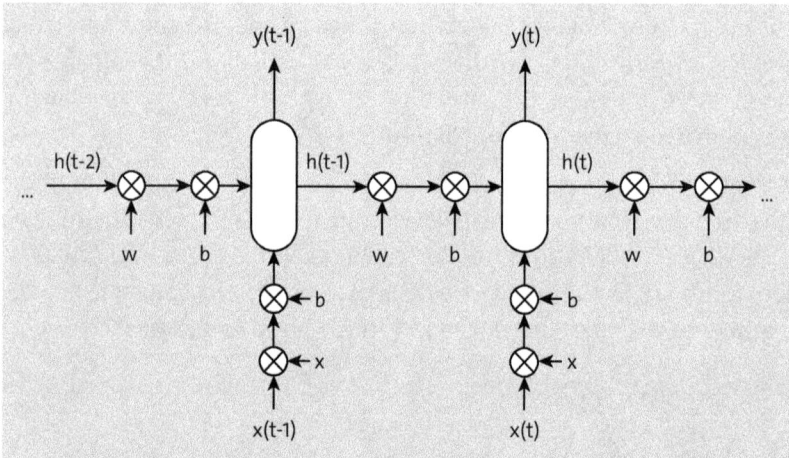

Figure 11.63 Series Connection of RNN Layers

Mathematical Procedures

If you're interested in the exact mathematical procedures, you can study them, for example, on the Stanford University website (*http://r-wrk.de/9763-stanford-rnn*).

A major disadvantage of simple RNNs is that they forget the empirical values over a longer period of time. You could also say that the model doesn't have a very good memory.

11.9.2 Long Short-Term Memory

The Long Short-Term Memory (LSTM) model (presented by Hochreiter and Schmidhuber in 1997 in their paper by the same name) has a better memory than simple RNNs. The upper path in Figure 11.64 represents the memory trace, the lower one the state. The memory trace and the status information are passed on to the next cell, whereby the status information is also the output of the cell, there is no additional *y* output.

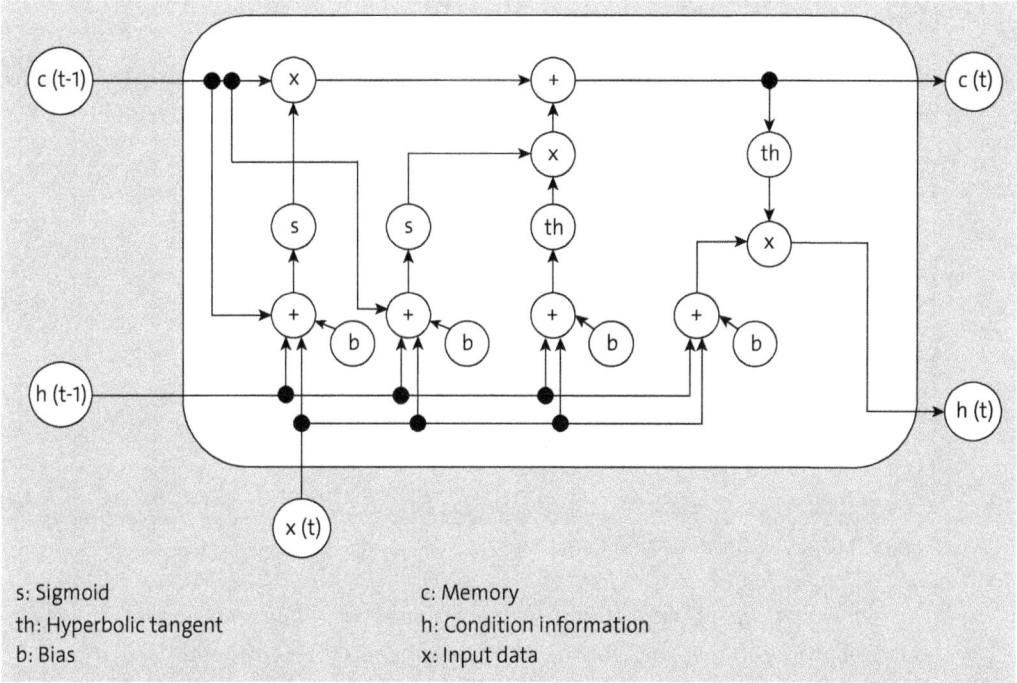

s: Sigmoid
th: Hyperbolic tangent
b: Bias

c: Memory
h: Condition information
x: Input data

Figure 11.64 LTSM Cell

The sigmoid functions in the figure should be our reference points in the block diagram:

- **First sigmoid function**
  The state information, the input values, and the bias are added up and fed to the sigmoid function. The memory values are multiplied by the result. As the sigmoid function only outputs values between 0 and 1, the memory values are scaled down by this multiplication. This is where you set how much experience should be forgotten.

- **Second sigmoid function**
  This sigmoid function has the same input variables (with their own bias) as the first one. The output is multiplied by the output of the hyperbolic tangent function. This hyperbolic tangent function has similar input variables, except for the experience

values. The product is added to the memory values. Previously forgotten, new experience values are then added to the memory trace in Figure 11.64 (upper path). However, if the summand is negative, the memory values become even smaller.

- **Third sigmoid function**
  Experience is no longer taken into account here. However, the output of the function is again multiplied by the output of a hyperbolic tangent function, which receives the new experience values as input variables. The product forms the new status information based on new empirical values.

The function curve of the hyperbolic tangent function, which you can see in Figure 11.65, is similar to the sigmoid function, but the value range is between -1 and 1.

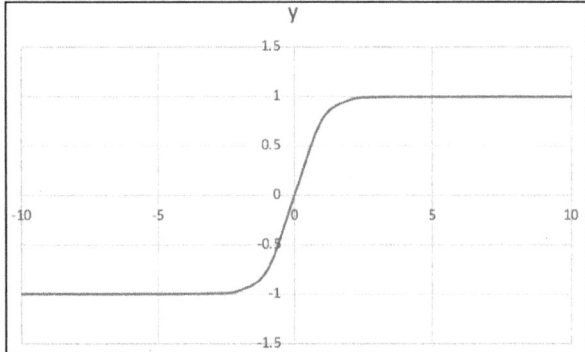

Figure 11.65 Hyperbolic Tangent Function

The following principle applies here: you don't need to be able to recalculate the mathematical procedures. A "feel" for the interrelationships is all you need.

Now we can program ANNs with memories that can learn from past experiences for the future.

11.9.3 Prediction of Energy Consumption (Next Hour) Using Keras Node

The next AI should predict the next hour's energy consumption. The data was collected in Ireland in 2009 for various areas (*http://r-wrk.de/9763-bigdata-energy*). The AI is supposed to predict the energy consumption of the stores (Cluster 26 column of the dataset) for the next hour (see Figure 11.66).

> **Energy Consumption**
> Of course, energy can't really be "consumed". But we don't want to get bogged down in electrotechnical details.

The data must be loaded and prepared. This is again done in a metanode so that the main program remains clear.

11.9 Time Series Analysis

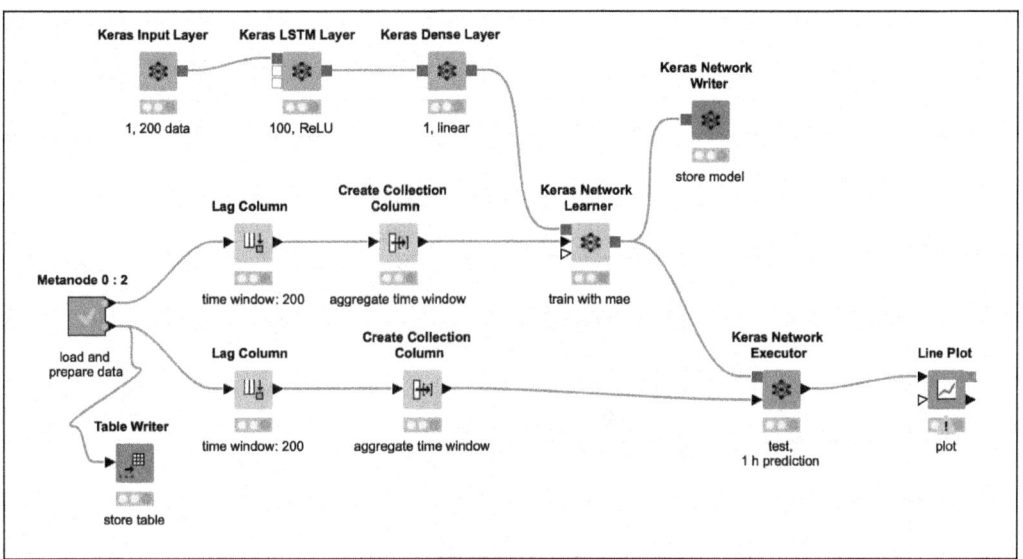

Figure 11.66 The 13a-time-series.knwf Program

Always think about how you can structure parts of the program, in this case, loading and preparing the data. Then, think about how many inputs and outputs a metanode should have. Here, I opted for two outputs. In addition to the training and test data, the evaluation data is also output and saved in the main program. However, the evaluation data could also be saved in the metanode.

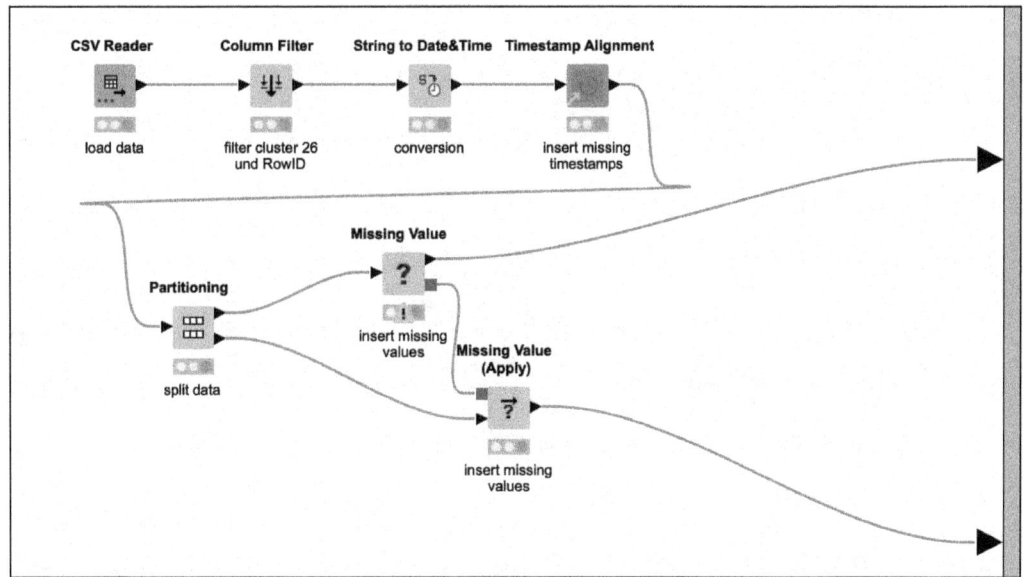

Figure 11.67 Metanode for Loading and Preparing the Data

The CSV file is loaded, and the rowID (date and time) and cluster_26 columns are filtered. Now, analyze the output of the module in the Node Monitor for **CSV Reader** and get an overview of the data. The data type of the **row ID** column must be transformed to **Date&time** (see Figure 11.68).

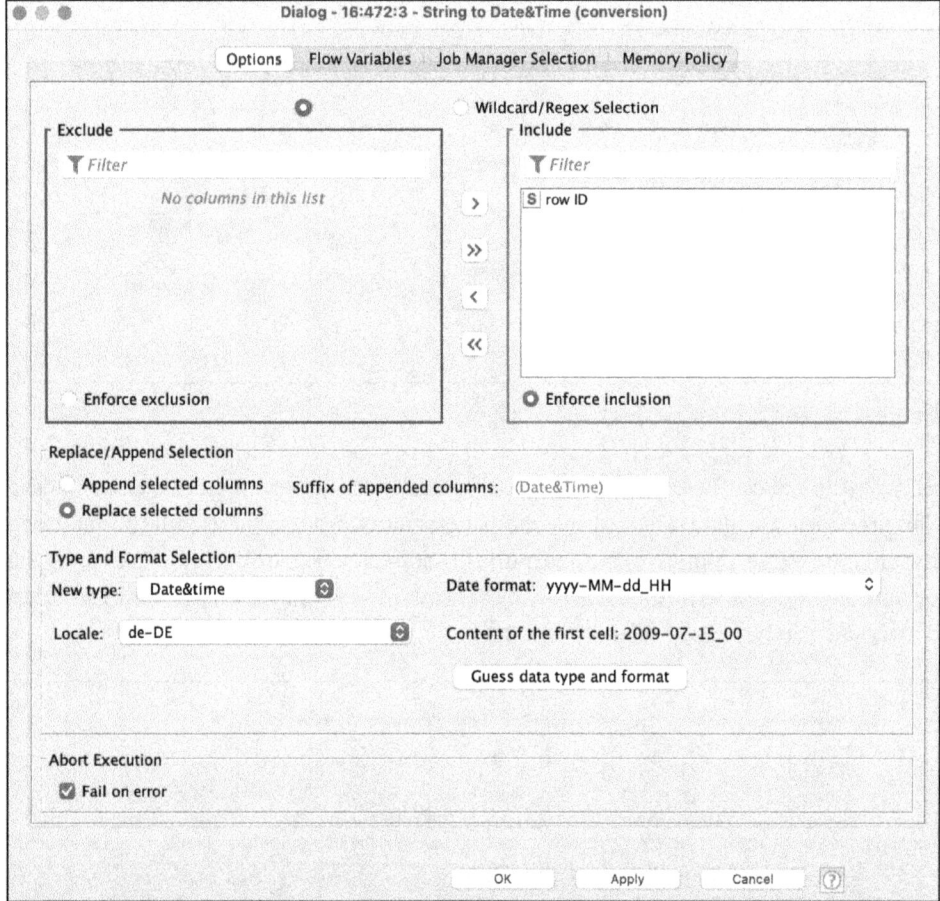

Figure 11.68 Settings Window of the String to Date&Time Module

Use the **String to Date&Time** module to select the relevant column, and click on **Guess data type and format**. The correct date format is then recognized, and the data type is transformed. For other files, you may have to change the strings in the column so that the date format can be recognized correctly.

The **Timestamp Alignment** module recognizes missing timestamps (in our case, missing hourly lines) and inserts new ones if necessary. After executing this module, you'll have a table with one line per hour. You can set the time period and the column. If you can't find the block in the node repository, you should download it from the KNIME website (via drag and drop). The table is then split into training data and test data (but not mixed).

11.9 Time Series Analysis

If new rows have now been generated in the table (by **Timestamp Alignment**), these have no energy data, or the new cells are empty. We can use the **Missing Value** module to set the data with which these empty cells are to be filled. If we select **Linear Interpolation** here, the data is generated from the information in the previous and subsequent cells. We also apply this setting to the test data.

The **Lag Column** module is used to transform the rows in the cluster_26 column into columns, resulting in time slots. You must set how large **Lag** is (how many elements the time slot has) and which nth element (**Lag interval**) from **Lag** should be used. Our time slot has 200 elements, one element per hour, as shown in Figure 11.69.

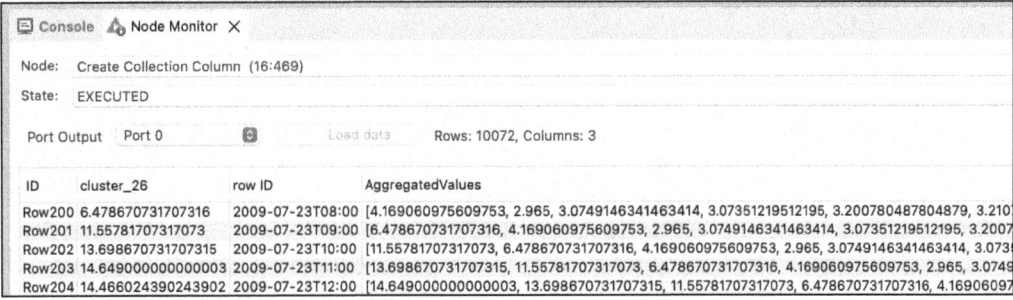

Figure 11.69 Settings Window of the Lag Column Module

The columns of the time slots are then combined into one column. Pay attention to the aggregated values in the **Node Monitor** screen (see Figure 11.70). Here, you can see how the time slots are structured.

Figure 11.70 Node Monitor for the Create Collection Column Module

Note the first element under **AggregatedValues: [4.169....** This value is the second element in the next line, then the third, and so on. The data packets are therefore structured in such a way that the last element is removed in the next line, and a new element is added in its place. The direction of the data packets (chronological order) therefore

11 Visual Programming Using KNIME

runs from right to left. The elements are the energy consumption data for the respective hours. The ANN must predict one value for each line with 200 elements, namely, the consumption for the next hour.

Next, we set up the ANN. The input layer has a shape of 1,200. This describes the structure of the data packets because a line also has 200 elements. This is followed by an LSTM layer (**ReLu** as **Activation** function, and **Sigmoid** as **Recurrent Activation** function), and finally a dense layer with a node and a linear activation function. For the LSTM layer, we must select **Go backwards** (see Figure 11.71), as the time sequence of the data is in reverse order.

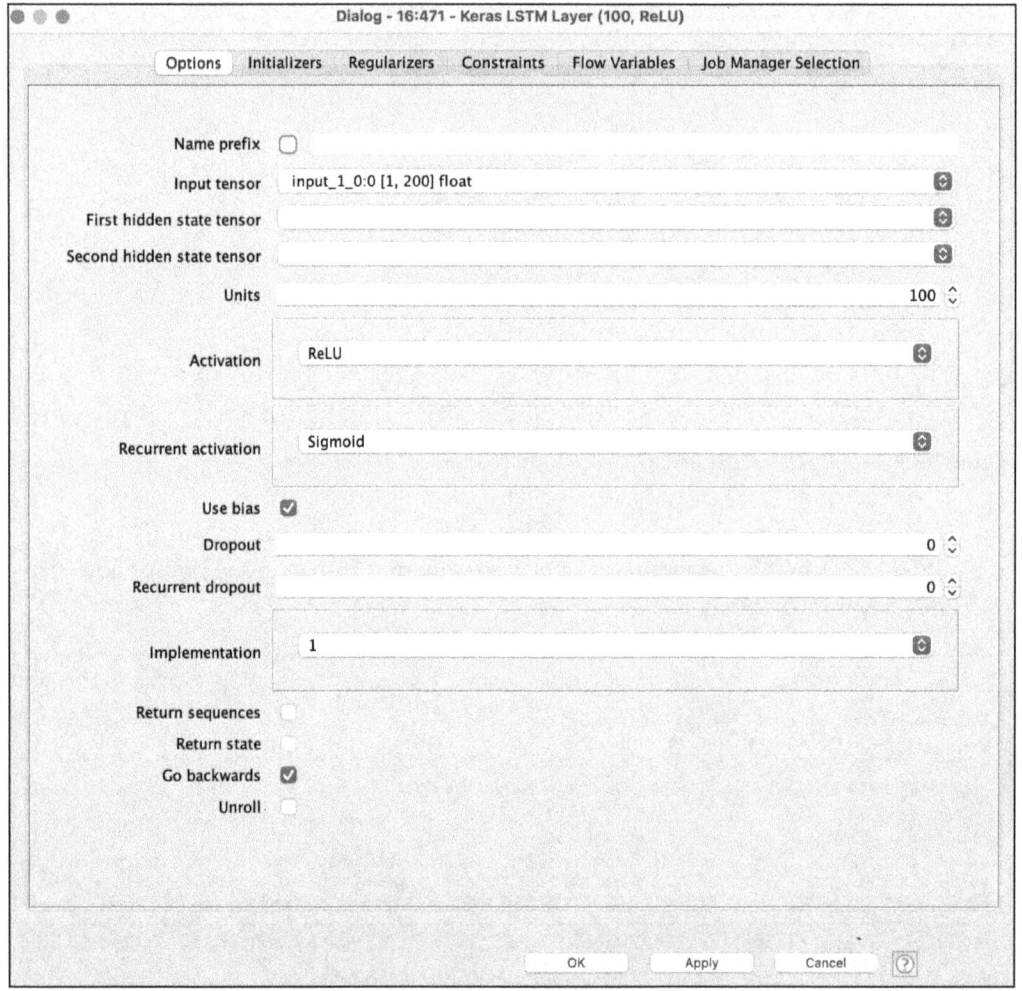

Figure 11.71 The Settings Window of the Keras LSTM Layer Module

An epoch number of 50, MAE as a loss function, and a training batch size of 128 should be set in the **Keras Network Learner** module. In addition, make sure that the row is the input as a data packet and the *cluster_26* column is the output.

The ANN can now be trained and tested. We save the model and the test data for the next program.

The Line Plot module, which is shown in Figure 11.72, is used to visualize the result. For the x-axis, you want to select the *rowID* column (Date&Time data type), while the *cluster_26* (as a reference from the original table) and *dense_1_0:0_0* (the prediction) columns are supposed to be displayed.

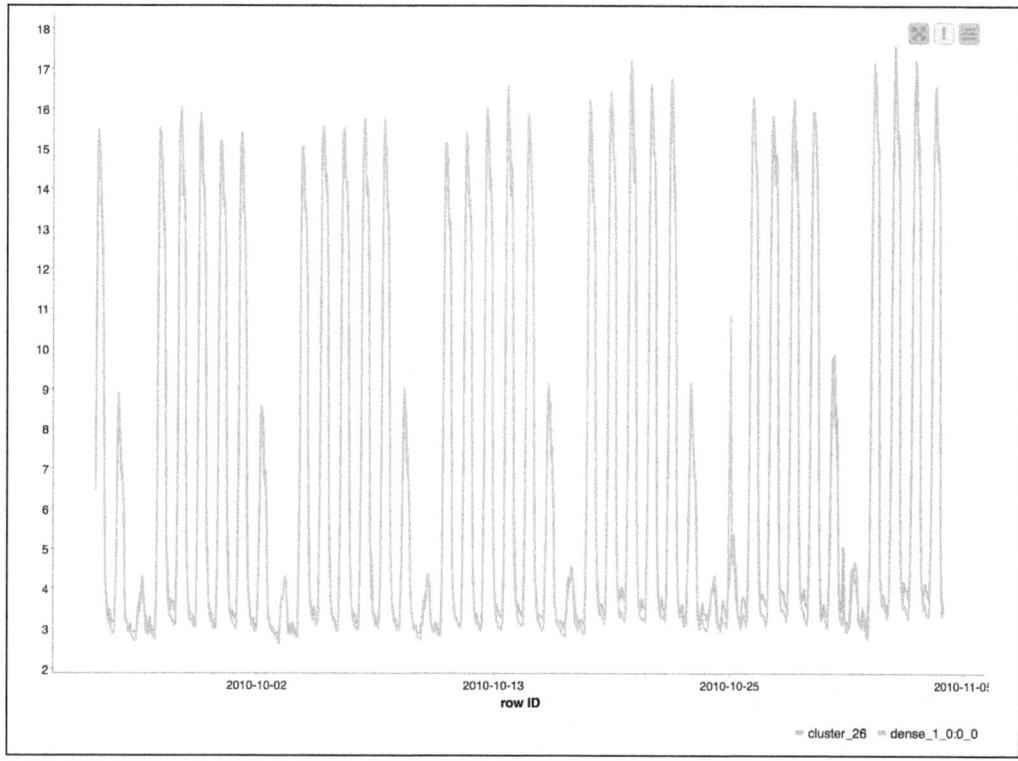

Figure 11.72 Line Plot of Target and Actual Values

The result is quite good: target and actual values match acceptably. The next forecast (energy consumption for the next hour) was generated for each data packet or time slot with 200 elements.

11.9.4 Prediction of Energy Consumption (Next Hour) Using Python Node

Only minor changes are necessary for the variant with Python Node, as shown in Figure 11.73:

- The Keras blocks are replaced by the **Python Script** block.
- The **RowIDs** of the predictions and test data must be renewed so that they can be merged into a new table.

11 Visual Programming Using KNIME

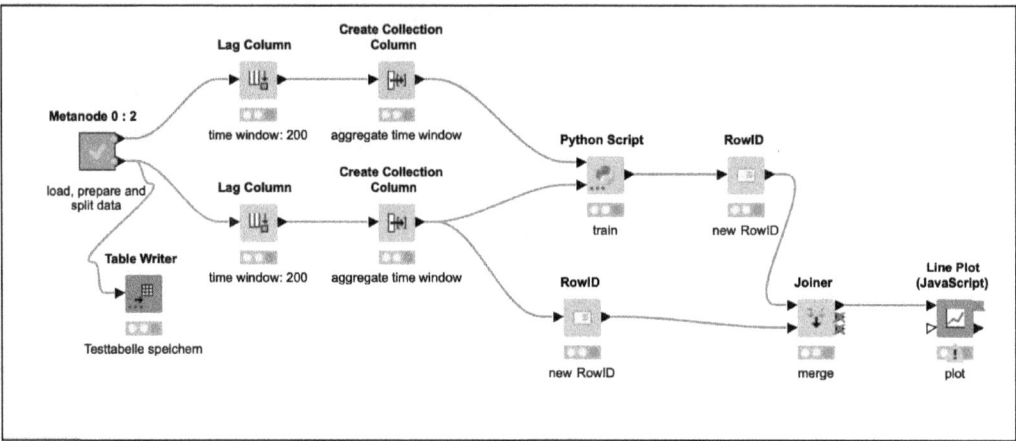

Figure 11.73 The 13b-time-series-py.knwf Program

The Python code is structured as shown in the following listing.

```
import knime.scripting.io as knio
import tensorflow as tf
import pandas as pd
import numpy as np

# Read data
data_0 = knio.input_tables[0].to_pandas()
data_1 = knio.input_tables[1].to_pandas()

# Prepare data
train_data = data_0['AggregatedValues'].to_list()
train_col = data_0['cluster_26']

test_data = data_1['AggregatedValues'].to_list()
test_col = data_1['cluster_26']

train_data = np.array(train_data).reshape(-1, 1, 200)
test_data = np.array(test_data).reshape(-1, 1, 200)


model = tf.keras.Sequential()
model.add(tf.keras.layers.LSTM(8,input_shape=(1,200),go_backwards=
True, activation="relu"))
model.add(tf.keras.layers.Dense(1))
```

```
# Configuration, training, and testing
model.compile(optimizer='adam', loss='mae', metrics=['mae'])
cb_early = tf.keras.callbacks.EarlyStopping(monitor='val_loss', patience=5)
history = model.fit(train_data, train_col, epochs=100,
 validation_data=(test_data, test_col),callbacks=[cb_early])

model.save("../model13.keras")

test_pred = model.predict(test_data)

# Output of the result
df = pd.DataFrame(test_pred, columns=['pred'])
knio.output_tables[0] = knio.Table.from_pandas(df)
```

Listing 11.24 Python Script for LSTM

You've already seen how you can import, process, and output data in the previous examples, but what's new here is the Python code for the LSTM model. Basically, it's not much different from the design of other models. If you want to implement this example completely in Python (without KNIME), you must also take care of organizing the data (as a replacement for **Lag Column**).

11.9.5 Prediction of Energy Consumption (Next 500 Hours) Using Keras Node

The next program will predict the energy consumption for the next 500 hours. We start with a data packet or time slot with 200 elements, predicting the energy consumption for the next hour. The last element in the list is then removed, and the prediction is inserted. This procedure is repeated 500 times in a loop. The time slot is shifted by one element at a time. We only need a time slot of test data; the rest is generated by our AI.

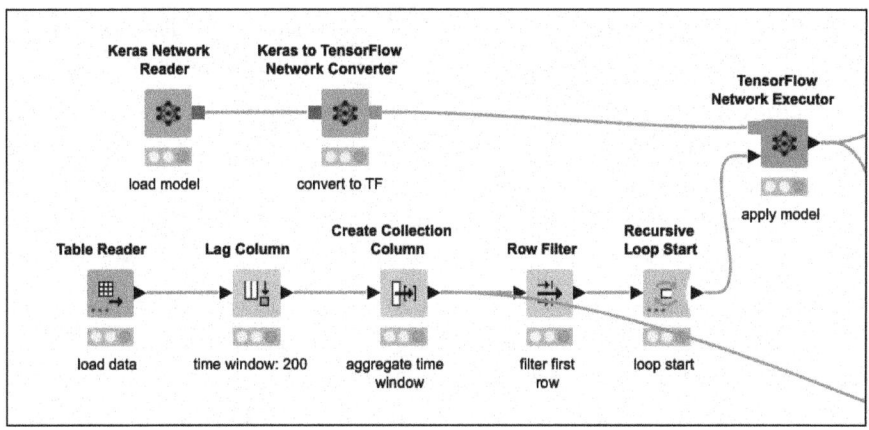

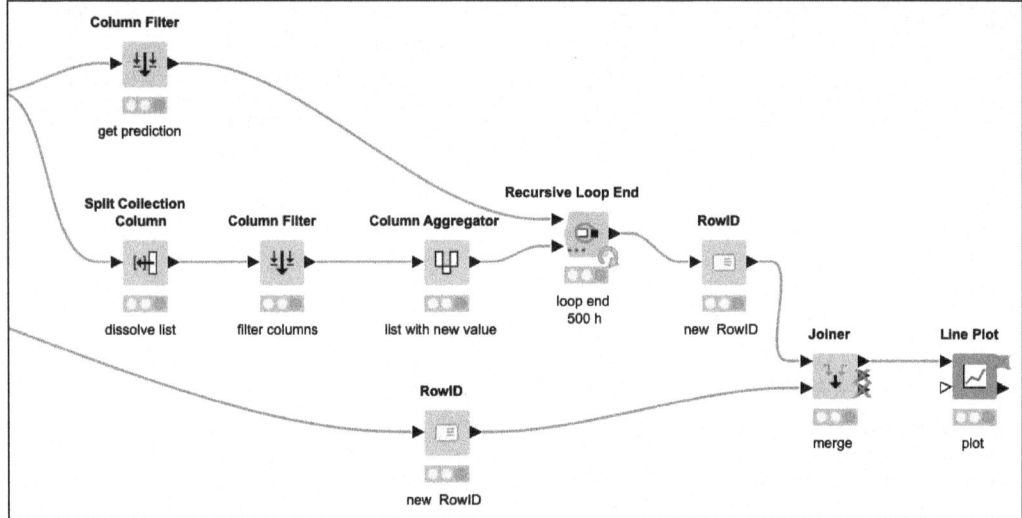

Figure 11.74 The 13b-time-series.knwf Program

The model and the test data are loaded. Rename the aggregated list in the **Create Collection Column** module to "List", as the loop uses this list. In this example, we use the **TensorFlow Network Executor** module because the execution in the loop is highly performant.

The loop starts with just one line of test data. A prediction is then generated and fed to the **Recursive Loop End** in the upper path. This module saves the upper values in a column (please also read the description of the module). In the lower path, the list is exploded, the last column is removed, the new value is added, and a new list is generated. This shifts the time slot by one element. The procedure is repeated via the loop.

Now, it's important that you execute the modules individually up to **Recursive Loop End**. You must execute the **Recursive Loop End** module step-by-step (right-click, and select **Step Loop Execution**), and then the loop execution starts. Right-click again to execute an iteration step. You must then check which data is selected for all modules in the loop. The values you've set should be updated by the loop with each iteration.

The values in the list are therefore replaced by new values (e.g., split value 1). After this iteration step and the correctly set values, you can execute the program completely. It's necessary to carry out this one step and adjust the data in the modules. You'll certainly need a few attempts at the beginning to get this right. In the next example with Python nodes, you can also see how you can filter columns by name as an alternative approach.

The data in the *cluster_26* column and the prediction generated by the loop are merged to visualize the comparison. However, new RowIDs must first be assigned so that an intersection can be formed. With just one data packet of 200 values, we've generated 500 values. The result is acceptable, as you can see from the plot in Figure 11.75.

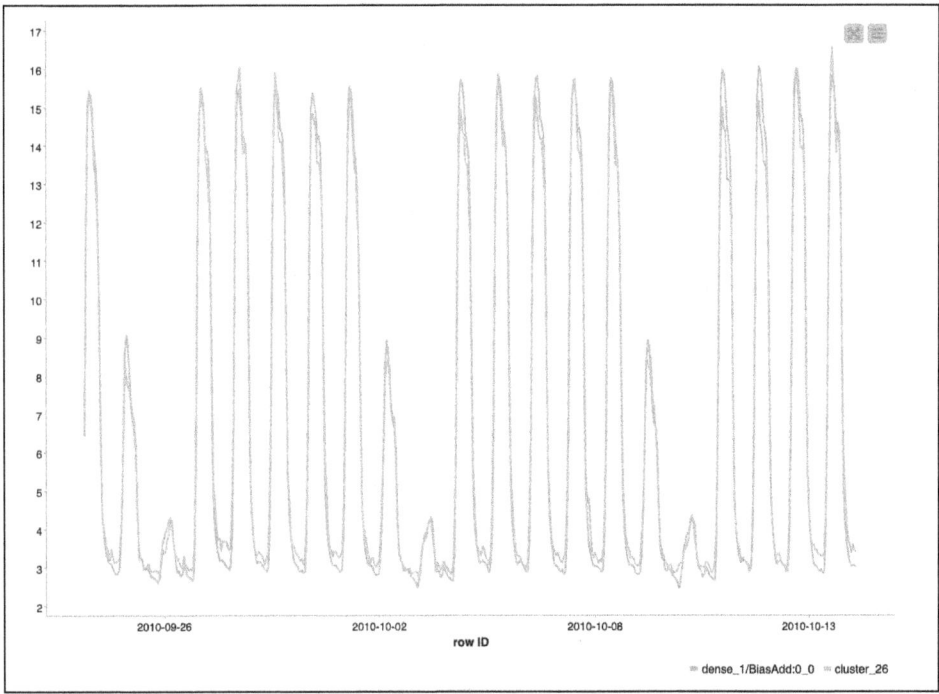

Figure 11.75 Line Plot of Target and Actual Values

11.9.6 Prediction of Energy Consumption (Next 500 Hours) Using Python Node

The following program has a similar structure to the previous one, with a few differences, as shown in Figure 11.76:

- The Keras modules have been replaced by a **Python Script** module.
- The target column doesn't need to be aggregated as the processing is done in Python.
- The superfluous column is filtered out when the packets are created.

Let's start with the source code in the **Python Script** module.

```
import knime.scripting.io as knio
import tensorflow as tf
import pandas as pd
import numpy as np

# Read data
data_0 = knio.input_tables[0].to_pandas()

# Prepare data
train_data = data_0['List'].to_list()
```

```
train_data = np.array(train_data).reshape(-1, 1, 200)

# Load model
loaded_model = tf.keras.saving.load_model("../model13.keras")

pred = loaded_model.predict(train_data)

# Output of the result
df = pd.DataFrame(pred, columns=['pred'])
knio.output_tables[0] = knio.Table.from_pandas(df)
```

Listing 11.25 Python Script for the Prediction

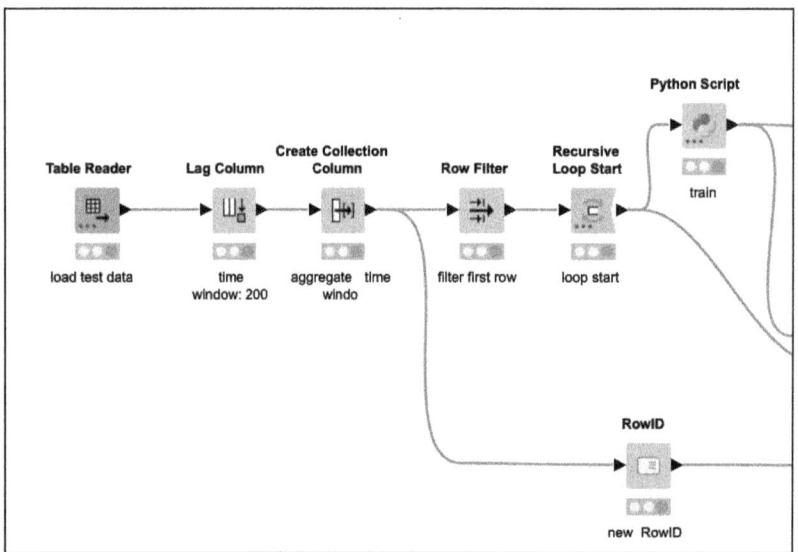

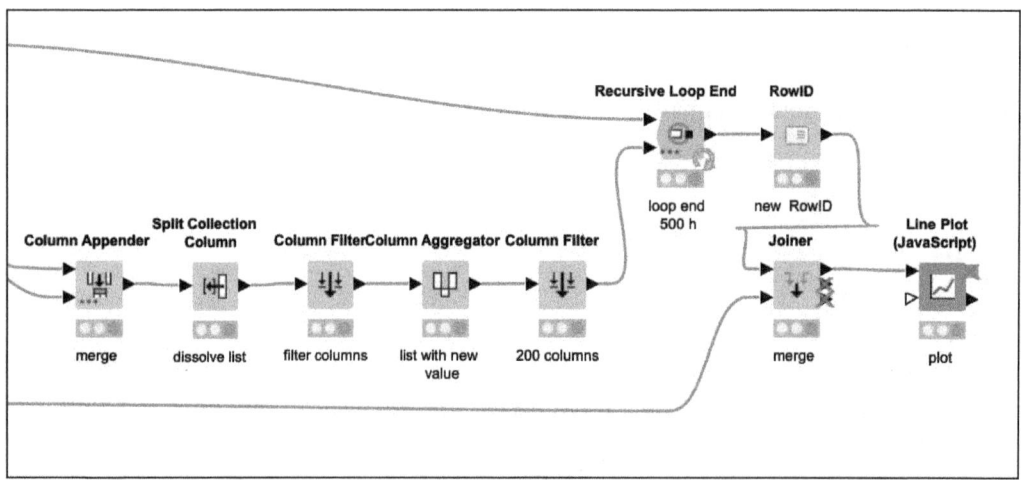

Figure 11.76 The 13b-time-series-py.knwf Program

The trained model is loaded, a prediction is generated for the input data or the data packet, and the result is output. In the repeat loop, this result is used again to create the new data packet for the next iteration. In comparison to the previous example, a value is filtered out by name in the module (see Figure 11.77).

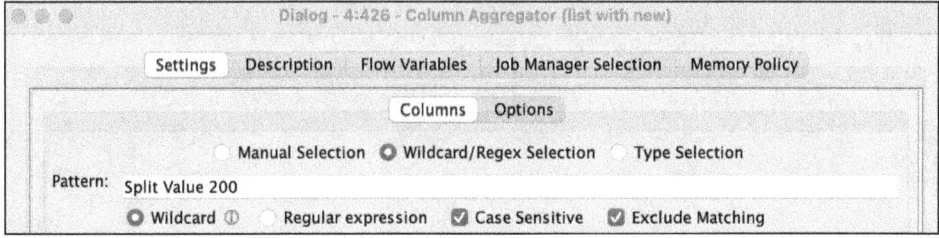

Figure 11.77 Settings Window of the Column Aggregator Module

All columns whose names begin with "Split Value 200" (the last element of the previous data packet) are excluded from the new data packet. Then **Columns** filter is used, and these columns are also excluded from the results table.

11.10 Text Generation

In this section, we'll program an AI that is trained with the text of Goethe's *Faust* (or an excerpt from it). The AI should then generate text in this style itself.

Before we start programming, it's advisable to think about the structure of the program. We must first load and prepare the text. On the Gutenberg Project website (*www.gutenberg.org*), you'll find many literary texts for free download.

In the text classification program, we transformed words into numbers before feeding them to the AI. A similar approach could be taken here, but the dictionary would be very long. A number would have to be generated for each word or word variation. For this reason, in this program, we'll work on the level of letters and punctuation marks: each letter and each punctuation mark are represented by a number. You'll see that our mapping table (character to number) will have just over 70 entries.

Let's now take the program for predicting energy consumption as a guide. An ANN with an LSTM layer is trained with these numbers and should then generate numbers again. When handling characters, the assignment table is reused, but this time in the opposite direction. The numbers are transformed back into characters and joined together.

In addition, note that we're working with integers, so it's advisable to encode the characters using OHE so that the characters aren't weighted unintentionally. Just because a character has a higher value in the assignment table doesn't mean that it's more valuable or more meaningful.

11 Visual Programming Using KNIME

After these preliminary considerations, we'll analyze the program in the *14a-Textgeneration.knwf* file, which you can see in Figure 11.78.

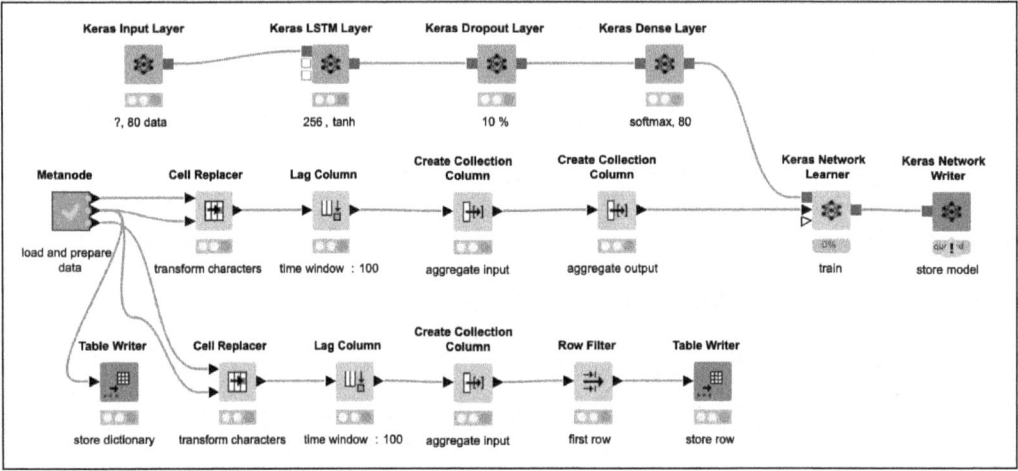

Figure 11.78 The 14a-Textgeneration.knwf Program

11.10.1 Data Preparation

The data is loaded in a metanode, the characters are extracted, and the mapping table is created.

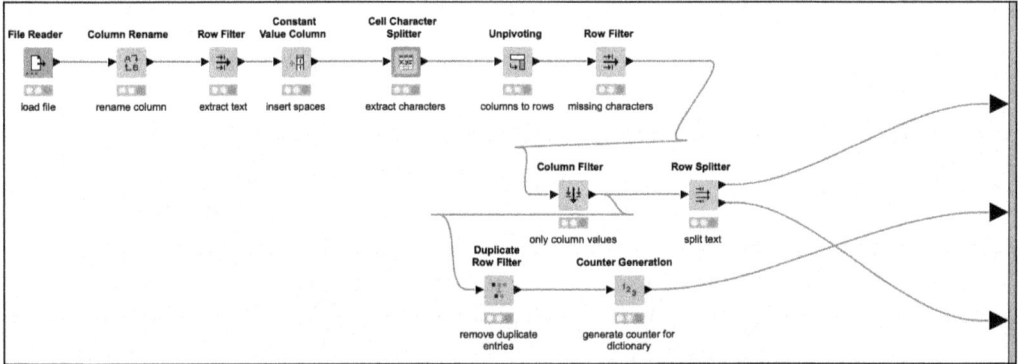

Figure 11.79 Metanode for Loading and Preparing the Data

First, the *Faust.txt* text file is loaded, the Column0 column (text content) is converted to TXT, and a space is inserted at the end of each line. The meta information (source, etc.) is then removed from the text, and only part of the entire text is filtered. This is done using the Row Filter module, but only rows 24 to 3,000 are filtered, while the rest are removed.

11.10 Text Generation

Figure 11.80 shows the **Cell Character Splitter** module, which splits the words in the TXT column into individual characters, whereby each character is mapped in a column. If this module isn't available in the node repository, download it from the website.

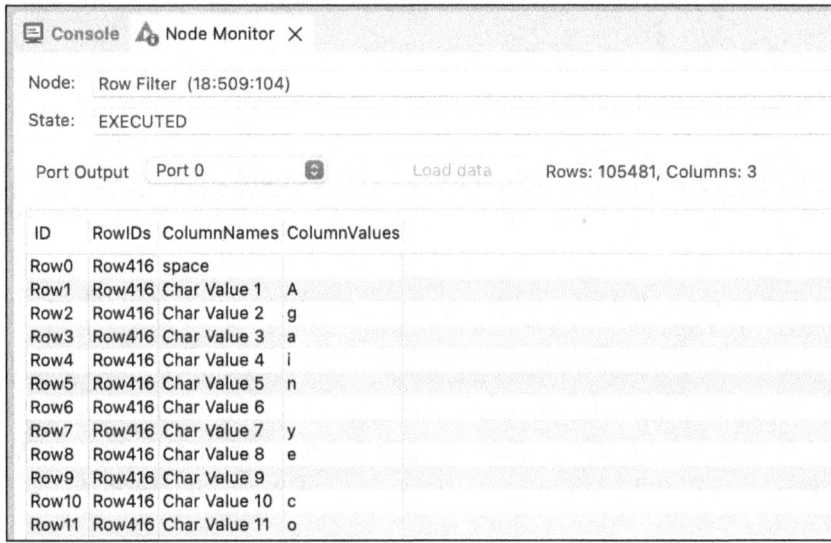

Figure 11.80 Node Monitor for the Cell Caracter Splitter Module

These individual columns are then merged into one column so that the text gets displayed in one column from top to bottom. This is done using the Unpivoting module.

The empty cells are represented by a question mark. These empty cells are filtered out using the **Row Filter** module and the **Exclude rows by attribute value** and **Only missing values match** in its settings. The result is located in the **ColumnValues** column, which displays the text letter by letter, as shown in Figure 11.81.

Figure 11.81 Node Monitor for the Row Filter Module

The Column Filter module discards all columns except **ColumnValues**. This result is split into two parts using Row Splitter; the first 60,000 characters are put out of the node at the top, and the rest are put out at the bottom.

273

Duplicate Row Filter creates a column with all characters that appear in the text (73 characters in total). The next module—Counter Generation—creates an additional column with the numbers 0 to 72. This also completes the assignment table, which we lead out of the node with the middle output.

11.10.2 Trainings

The ANN has an input layer, which should encode the 73 characters via OHE. The data packets are 100 characters long, but with a question mark, we can keep the layer variable, so "Shape=?,73". The LSTM layer has 256 nodes, Tanh as the activation function, and Sigmoid as the recurrent activation function. This is followed by a dropout layer (10%) and a dense layer (73 nodes, Softmax as the activation function). Keras Network Learner is trained for 75 epochs, and the batch size is 256.

Before the data is fed into the ANN, these characters must be transformed into numbers. To do this, we use the **Cell Replacer** module. **Lag Column** generates 100 columns, each with one character, and **Create Collection Column** merges these characters into a list with 100 elements (refer to Section 11.9). Here, too, the chronological order of the data packets is reversed, that is, from right to left. The **Go backwards** option must therefore be activated for the LSTM layer. Because we use OHE, the target column must also be transformed into a list with one element each.

The model, the assignment table, and one line of the data packet are saved for generation.

11.10.3 Generation

The program for generating text is similar to the program for predicting energy consumption; the only difference is that the numerical values have to be transformed back into characters. The model is loaded and converted to a TensorFlow model to speed up execution (see Figure 11.82). A line of the data packet is also loaded so that the AI can generate a text based on it.

Now let's take a look at the loop. First, the data packet is used as input. The lower branch takes the following number with the highest probability, resolves the list, converts the data type of the prediction, removes the last element, and inserts the prediction. The new data packet is then shifted by one element and serves as input for the next iteration.

In the upper branch, the number with the highest probability is filtered out, and the data type is converted and transformed back to a character. The data at the upper input of the **Recursive Loop End** module is saved in a column one below the other. This procedure is repeated 1,000 times; that is, 1,000 characters are generated.

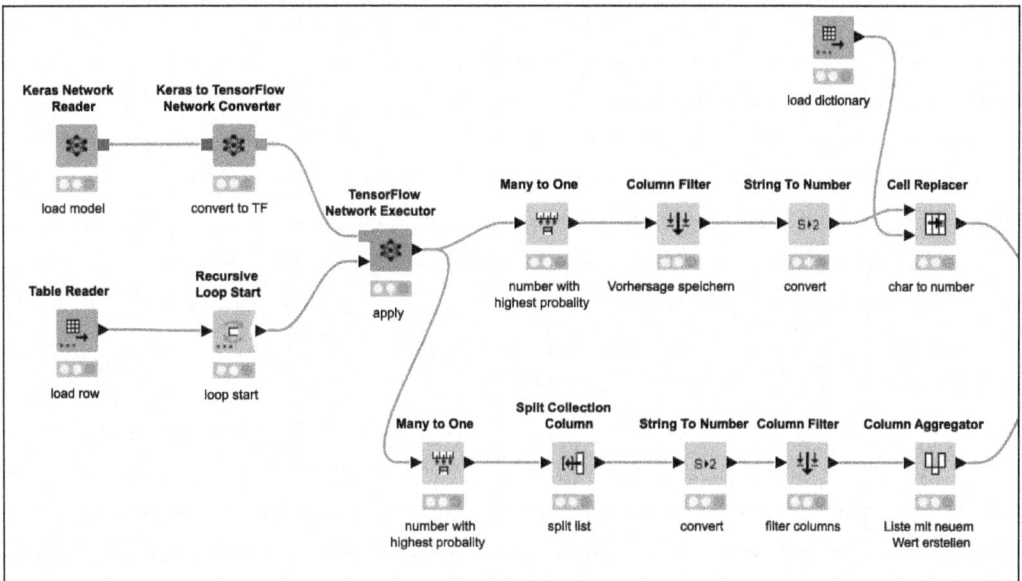

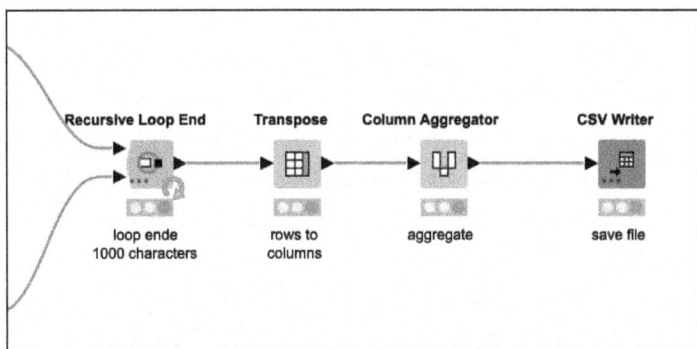

Figure 11.82 The 14b-Textgeneration.knwf Program

The rows with the predictions are then transformed into columns and merged into one column. The **CSV Writer** module is used to save the result in a file. Let's take a look at the result:

> he welemerthOnly impelsed,--with creature uses itsela!And the Ged auth meason vill bristly their paysient,From degibs me dows to heart like thee may be.MEPHIS-TOPHELESI must confess that forth I may not wander,My steps by one slight obstacle controlled,--The wizard's-foot, that on your threshold made is.FAUSTThe pentagram prohibits thee?Why, tell me now, thou Son of Hades,If that prevents, how cam'st thou in to me?Could such a spirit be so cheated?MEPHISTOPHELESInspect the thing: the drawing's not completed.The outer angle, you may see,Is open leds he seeving hare and dain,Who dring thou thus, whon the posers of Hell!With all that I have undertaken,Not been by ac lives he pleasure.FAUSTThou canst not compass the aigne,The soars, ever arain, and sel, sthelf me.MEPHISTOPHELESThe

mocest truth for me it dicks to the van.To soon ey the cheerful heart, itso many.Who stand the feeth, dowI the selven sersably pance?When these open where bedow themblass,--The wittle t

Let's try this with other parameters: Let's read only the first 2000 lines, use 40000 characters for training, the dictionary has 80 characters, the LSTM layer has 512 nodes:

shosting theme a cile toredWith rever tands the Spirit of the Ages plante,Whinct illy be ore to papiousMasper, works arashingFrom the less now thull beales?Coundoun, though by that all belling,Eapthe deer the glass of heavenly light and feeling,Ander wards and mindsest walls its stall,My fature's plays, with all its pleasures,And still a frumune, sheply ground your be.That's gish me frood to tence lust bread,Around weresh daze is fines and peeringBrom that's desire, chimprofed all inself lent,That not is forgs platen, Senter and from tly dely conornather.I forlot, that all the fines from the less show,And hame to seeving but the mines a teeply?Your I, that's go plaine, as its coners;The dest, and lively liss and reveloug!POOTHGELChTull st, the subsing goth to them!Sang, be drein and the bus on panceWhich bense, and lovely ladies my nature.SOLOND tAGERChisums wo! the Hines himClianing, of dur ta mostorThough solled, grows greed, and undermory.This might of the

And one more attempt: Let's read only the first 500 lines, use 15000 characters for training, the dictionary has 67 characters, the LSTM layer has 128 nodes, no dropout:

and the sures a pantesting,And then the grease of the world to me to watted,And ever the gate the gacks of and and to me to meation,And in the surgs a corest a poration the lealen.If laks to the cores the mong to seave to mearing,And the seares a there ong and and the be the gast.THE LORDThough seall the bosth the beat ond the be ongThe sores a peet and greave ingFrow hat on the lare the rowe to me.Watt not the world and that the greasting,And with the surgh be deare and there!If bost for the bead of the world meating,And the seares theres bestile foy then,And all seaven the sure to mentirged,And and the still and greavent aspres,And the serengant and with a pang,And thes the great and thengers with the manging,And then soment and the be the reast.THE LORDTh ughe stall not the browe to me tay.THE LORDThough still thend int and the berellessing his to me!What seall the world and thengers is then in and.THE LORDThouth still you past the browd the greasting,And

One last attempt: Let's read only the first 3000 lines, use 60000 characters for training, the dictionary has 80 characters, the LSTM layer has 512 nodes, 10% dropout:

in her thy skillusion,Wo lett the digh on the Master near,--The sige the stirits there fort in feacherWAGNERWhy, for a poodle who has lost his master,And scents about, his track to find.FAUSTSeest thou the spiral circles, narrowing faster,Which he, approaching, round us seems to wind?A streaming trail of fire, if I see rightly, Follows his path of mystery.WAGNERIt may be that your eyes deceive you slightly;Naught but a plain black poodle do I see.FAUSTIt seems to me that with enchanted cunningHe snares our feet, some future chain to bind.WAGNERI see him timidly, in doubt, around us running,Since, in his master's stead, two strangers doth he find.FAUSTThe circle narrows: he is near!WAGNERA dog thou seest, and not a phantom, here!Behold him stop--upon his belly crawl--Histail set wagging: canine habits, all!FAUSTCome, follow us! Come here, at least!WAGNER'Tis the absurdest, drollest beast.Stand still, and you will see him wait;Address him, and he gambols str

Well, you see, Goethe can't be replaced that easily by an AI. You can change various hyperparameters and test whether the text becomes more realistic. However, this is very time-consuming. But the important thing is that you now have an idea of how chatbots work. Of course, the structure of such AIs is more complex, and they are trained with much more data. A lot of time is also spent on training and optimization. But the basic principle remains similar: the next character is determined for a character string, this new string is used again to determine the next character, and so on. This is how texts are generated by AI models.

If you want to learn more about this topic, you can find many examples of text generation on the Keras and KNIME websites. However, you'll need a powerful computer and a good knowledge of Python, as well as time for a lot of fun experimenting. Since the introduction of ChatGPT, you can generate texts very quickly and conveniently at a completely different level without having to train a—let's admit it, modest—AI yourself at great expense. This can spoil your mood when experimenting.

> **Text Generation Using Python Node**
> I've deliberately omitted text generation using Python nodes here, as the execution of the program would take far too long. You can see from the quality of the generated text that this time investment isn't really justified. If you want to generate texts using Python, I suggest that you don't use KNIME. For this purpose, however, you should deepen your knowledge of Python. You can find very good examples of this on the Keras website at *https://keras.io/examples/generative/text_generation_gpt*.

11.11 Further Information on KNIME

You've now become familiar with various AI programs with KNIME. My students really enjoy visual programming. If you want to delve deeper into this software, you should consider configuring GPU support on your computer. The setup can be time-consuming (if your graphics card is supported at all), but saves a lot of time when you run the programs. However, you can also rent KNIME fully setup on a server for a fee. Details can be found on the KNIME homepage.

It's also possible to run your KNIME workflows on a server so that users can access them via a web screen. Further details can also be found on the KNIME website at *http://r-wrk.de/9763-knime-intro*.

11.12 Exercises

11.12.1 Exercise 1: XGBoost for Classification, Mushrooms

Program an AI using XGBoost to classify mushrooms into edible and poisonous. Use the *mushrooms.csv* file. Information on the attributes can be found at *www.kaggle.com/datasets/uciml/mushroom-classification*. You can also search for information on the data on other pages on the internet.

11.12.2 Exercise 2: XGBoost for Regression, Diamonds

Program an AI with ANN to determine the price of diamonds. Use the *diamonds.csv* file. Information on the attributes can be found at *www.kaggle.com/datasets/shivam2503/diamonds*, for example.

11.12.3 Exercise 3: Image Classification Using InceptionV3

Create an image or download it from the internet, and have it classified by the pretrained model.

11.12.4 Exercise 4: Transfer Learning, Horses or Humans

Download the Horses or Humans data, and have it classified using transfer learning. You can find the data at *www.kaggle.com/datasets/sanikamal/horses-or-humans-dataset*.

11.12.5 Exercise 5: Anomaly Detection Using an Autoencoder, ECG

Use the *ecg.csv* file to detect anomalies in the data (heart rhythms). The value of the last column is to be predicted. Warning: In this data, the number 0 classifies the anomaly. After loading the columns, I was automatically assigned the data type Double (floating point number). You can use the Double to Int module to transform the data type of the column to Integer. The Scorer module requires two integer columns to create the confusion matrix.

11.12.6 Exercise 6: Text Classification

Program an AI that reads the *spam.csv* file and classifies the texts as ham or spam. Note: You can use the SMOTE module for upscaling. Research information and possible use cases for this module.

11.12.7 Exercise 7: AutoML for Regression

Program a workflow to determine the length of the sepal using the AutoML (regression) module. You must first install the module for this purpose.

11.12.8 Exercise 8: Cluster Analysis

Try to cluster the class column of the *mushrooms.csv* file using the elbow method. Categorical data can be converted into numerical values by using Category to Number. The counter variable *k* should start at 1. The Column Resorter module can re-sort columns so that clusters and classes are next to each other and can be analyzed. It becomes clear once again: the result of the grouping won't match what you expect, namely, the values in the class column. The procedure only groups data according to a specific method. You can analyze such generated groups and possibly identify correlations. In addition, try removing other columns and grouping the remaining data. Compare the results of the groupings.

> **Mushrooms**
> Please don't eat mushrooms just because your AI says they are nontoxic.

11.12.9 Exercise 9: Time Series Analysis

Program an AI that predicts the amounts that will be traded on the stock market the next day. Use the *yohoo_stock.csv* file with the target column as Volume. Change hyperparameters to improve the prediction.

11.12.10 Exercise 10: Text Generation

Change the hyperparameters of the *14a-Textgeneration.knwf* program, and evaluate the generated text. There is no sample solution for this task.

Chapter 12
Reinforcement Learning

This type of AI learns by way of trial and error to independently find a way to maximize rewards and minimize punishments.

> **What This Chapter Is About**
> - The consequences of possible moves for each condition in the training phase
> - The optimum move carried out for each state in the test phase or application

The approaches to AI development presented so far have always required a dataset to train an AI model. Reinforcement learning does without this. Let's start from a specific example again so that the explanations don't remain too abstract. In a computer game, the character is guided through a labyrinth. The aim is to collect treasures and avoid deadly dangers.

If you're the player controlling the character, you'll probably not be very successful at the beginning. But over time, you'll become more and more practiced at avoiding the dangers and amassing treasures. The character is supposed to independently learn the optimal moves.

How can a program be developed that independently and successfully controls the game character? One possibility is to analyze the entire game and establish rules, such as the following:

- Go forward two spaces.
- Go three spaces to the right.
- Take the treasure with you.
- Turn around.

The results of such a program would be impressive for outsiders: the character would master the game completely independently. However, the developer must analyze the entire game beforehand (every step) to define an exact path. If something changes in the program, the analysis and programming must be carried out again. This is a lot of work and also unnecessary if you familiarize yourself with *reinforcement learning*. Let the character learn to win the game with the help of an AI.

The specific task now is the following: A character is to be guided through a corridor. Depending on where it is and in which direction it's moving, there are deadly dangers or the exit is the character is looking for. In this simple game, you can only move to the right or left. Using AI, the character is supposed to learn to master the game by avoiding the dangers and finding the exit.

Figure 12.1 Deadly Dangers in One Direction and the Exit You're Looking for in the Other

There must be a training phase in which the character learns the moves. This is followed by a test phase in which what has been learned is applied.

12.1 Q-Learning

Q-learning is a possible algorithm for the implementation of reinforcement learning. "Q" stands for quality, specifically for the quality of an action in terms of the reward. However, the quality of an action isn't fixed from the outset, but is determined over time during the training phase by way of trial and error. After the training phase, the qualities of all possible actions in all states are known.

In the subsequent test or application phase, a trained agent observes the state of the environment and performs the action that returns the greatest reward. In other words, what has been learned is now simply applied. In our example, the character is the agent, and the game is the environment. If the character finds the exit or the direction to the exit, it must be rewarded. If, on the other hand, the wrong direction is taken (which leads to death or a deduction of points), a penalty must be imposed.

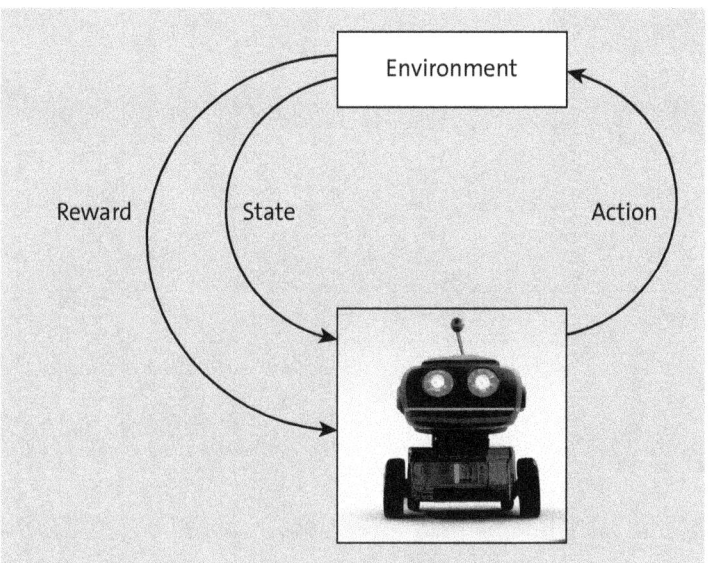

Figure 12.2 Reinforcement Learning Scenario

Let's look at the game from a more abstract level. For example, there are 10 fields (the corridor is divided into fields) that contain the exit, the deadly danger, or neither of these.

state	0	1	2	3	4	5	6	7	8	9
door	**exit**	nothing	nothing	**death**	nothing	nothing	nothing	nothing	nothing	**exit**

Figure 12.3 States of the Game

Let's assume that the character is in state 6, that is, on the sixth square. A movement to the left should reduce the reward or trigger a punishment, and a movement to the right should increase the reward.

So, there has to be a training phase to learn these moves. But even during training, the character needs to know when a reward or punishment is due. The first step is to create a *rewards table* of the environment with the extreme values (+100 or -100 points). This table contains the information that matters in the end, such as when the game is won or lost. If the character is in state 1 and moves one door to the left, full points are awarded. This also applies to the movement to the right in state 8. In state 2, a movement to the right means the death of the character (the opening of the door is neglected in this model), as is a movement to the left in state 4. The character has to learn everything else by itself. The rewards table therefore records the cases in which you gain or lose certain points, whereby it always depends on the combination of condition and action. You could also say that the game's points system is stored in the rewards table depending on the status and action. This table doesn't get changed during the game.

State	left	right
0	0	0
1	100	0
2	0	-100
3	0	0
4	-100	0
5	0	0
6	0	0
7	0	0
8	0	100
9	0	0

Table 12.1 Rewards Table

We still need a dynamically created table that lists points for each state in the training phase depending on the direction of movement, the *Q-table*. This table contains not only extreme values, but all points depending on the state and direction of movement. Points are then listed for each state and each action, which provide information on which direction of movement is the most promising. The path that leads to the goal is "paved" with higher scores than others. The table should be created automatically by the program during training, the values are updated iteratively after each game.

During the first run, the fields in the Q-table next to the extreme values in the rewards table are updated. This means that all fields that are only one step away from the extreme values are updated. All other fields remain unchanged. In the next run, the fields that are one step away from the last updated fields and therefore two steps away from the extreme values are also updated, and so on. After each run, there are more and more fields that indicate the path to the exit or death.

This means that we don't need to fill the table with the correct numerical values ourselves. But how is this supposed to work and how are these figures determined?

The initial Q-table only has cells that contain the value 0. The values of the cells are updated during training.

State	left	right
0	0	0
1	0	0

Table 12.2 Initial Q-Table

State	left	right
2	0	0
3	0	0
4	0	0
5	0	0
6	0	0
7	0	0
8	0	0
9	0	0

Table 12.2 Initial Q-Table (Cont.)

We need a simple strategy to determine or update the values of the Q-table. Depending on the state, we'll randomly select an action (move right or left), execute it, and update the Q-table and the current state. The procedure will be repeated several times.

Here is a formula for calculating the updated values for the Q-table:

$$Q_{new}(s,a) = Q_{alt}(s,a) + \alpha \times [R(s,a) + \gamma * \max(Q(s',a')) - Q_{alt}(s,a)]$$

Following are the meanings of the formula items:

- s is the current status.
- a is the current action.
- s' is the following status.
- a' is the following action.
- $Q_{new}(s,a)$ is the new Q-value, depending on the status and the action.
- $Q_{old}(s,a)$ is the old (previous) Q-value, depending on the status and the action.
- α is the learning rate.
- R is the reward.
- γ is the reduction rate.
- $\max(Q(s',a'))$ is the maximum Q-value that can be achieved in the subsequent state with a subsequent action.

Let's analyze the formula in detail to understand how the calculation is actually made. This is the previous value from the Q-table for the step and the action:

$$Q_{old}(s,a)$$

Here again, the Q-table is used to determine which action contains the highest value for the next status. If an action can be carried out in the next state that leads to the goal, the value in the state/action combination is increased. A repeated execution ensures that the path to the goal is "paved" with corresponding scores:

$$\max(Q(s', a'))$$

Take another look at the expression in the square brackets:

$$[R(s, a) + \gamma * \max(Q(s', a')) - Q_{old}(s, a)]$$

The highest possible Q-value that can be achieved with an action from the next state is added to a reward R (taken from the rewards table with the extreme values depending on the state and action). This maximum value of the next state is scaled with the reduction rate γ. If the value of γ were 0, the focus would only be on the present. With a value of 1, the values in all directions would be very large (close to the extreme values), so that a clear directional recommendation could not be derived, especially for games with multiple possible directions. You can test this later in the program and view the values of the squares. In many applications, the value of γ is around 0.9. The values are updated in relation to the current value, which is therefore subtracted.

The learning rate α can be used to set the weighting of the newly calculated value. If the value of α were 0, the previous value would simply be retained. With a value of 1, the jumps would be very large. A typical value for α is 0.1. With a small learning rate, you ensure that the value progression to exit or death is smooth. You can also experiment with this hyperparameter afterwards and analyze the values of the squares.

Let's get back to our game. If you play the game 1,000 times and update the values using the formula, you'll determine the final Q-table.

State	left	right
0	0	0
1	100	80.7
2	89.9	-100
3	0	0
4	-100	65.6
5	59.05	72.9
6	65.6	81
7	72.9	90
8	81	100
9	0	0

Table 12.3 Finished Q-Table

After creating the Q-table, you can say that the AI has been trained. Based on this table (let's think of it as a memory), a character can now successfully master the game. In the test phase or during use, this table no longer gets changed but is only used. The rewards table is no longer needed, as it was only relevant for training. This is a completely different approach for an AI application because you don't need collected training data with a target column as with the models you learned about in previous chapters. Let's therefore review the following points:

- A rewards table is required that only contains the extreme values (reward or punishment).
- The rewards table is used in the training phase. A score is determined for each state and each possible move and recorded in another table, the Q-table. Compared to the rewards table, the Q-table not only includes extreme values but also points for each state and action.
- In the test phase or in the application, only the Q-table is used to control the character. The quality of the action (score) can be taken from the Q-table for each possible move. The rewards table no longer plays a role here.

Analyze the Q-table, and compare it with the doors of the game. No matter where the character is, it's always guided to the exit with the help of the scoring system.

12.2 Python Knowledge Required for the Game

In this section, we'll discuss the basic knowledge required so that you can understand the source code of the game. You can find the examples discussed here in the *K12_python.ipynb* file.

12.2.1 Lists

As already discussed in previous sections, multiple related values can be saved in lists. The values can also have different data types. Two lists are created here. Then, the first list is output, followed by the number of elements from the second list (in this case, three).

```
list_1 = [0,1,2,3]
list_2 = ["hello", 1, 2.3]

print(list_1)
print(len(list_2))
```

Listing 12.1 Initialization and Output of Lists

Remember that indexing starts with 0. To access the first element, you must select the index 0 ("hello"). The last element has the index 2.

```
print(list_2[0])
```

Listing 12.2 "hello" Output

A list can in turn contain lists. `list_3` consists of two lists.

```
list_3 = [list_1, list_2]
print(list_3)
```

Listing 12.3 A List Containing Lists

```
print(list_3[1])
```

Listing 12.4 Output of the Second Element

Now it's getting interesting. What is output here? `list_3[1]` outputs the second element (i.e., `list_2`). `list_3[1][0]` accordingly outputs the first element of this list ("hello").

```
print(list_3[1][0])
```

Listing 12.5 "hello" Output

12.2.2 Branches

Very often, a part of the source code should only be executed if certain conditions are met. This is where branches come into play.

```
grade = 3

if grade < 5:
    print("passed")
print("end")
```

Listing 12.6 Simple Branch

The text "passed" is only displayed if the grade is less than 5. All indentations after an `if` query are only executed if the condition is fulfilled. The subsequent "end" output takes place in any case.

The condition can also be reformulated.

```
if grade <= 4:
    print("passed")
print("end")
```

Listing 12.7 Alternative Variant for the Condition

A "passed" is only output if the grade is less than or equal to 4. These program lines have exactly the same effect as the previous ones.

```python
if grade < 5:
    print("passed")
else:
    print("failed")
print("end")
```

Listing 12.8 Branch with "if" and "else" Paths

A "passed" is only awarded if the grade is less than 5; otherwise, a "fail" is awarded. The final output "end" takes place in any case.

```python
if grade < 5:
    print("passed")
elif note == 5:
    print("failed, grade 5")
elif note == 6:
    print("failed, grade 6")
else:
    print("invalid grading")
```

Listing 12.9 Linked Branches

Here, you can see how to check multiple conditions using concatenated queries. A "passed" is only output if the grade is less than 5. If this condition isn't met, the system checks whether the grade is 5. If yes, the corresponding output is made. However, if this condition isn't met either, a check is made to see whether the grade is 6. If yes, the corresponding output is made. In all other cases, an "invalid grade" is output.

```python
if 1 in list_2:
    print("Element included")
else:
    print("Element not included")
```

Listing 12.10 Checking whether a Specific Element Is Included in the List

It's checked whether the number 1 is an element of list_2 (with the keyword in). If yes, "Element included" is displayed. If not, the output is "Element not included". You can therefore use a query to determine whether the value is included in the list or not.

12.2.3 Loops

Sometimes, parts of the source code must be executed multiple times. This can be achieved by using loops.

12 Reinforcement Learning

```
i = 0

while i < 5:
    print(i)
    i = i + 1
```

Listing 12.11 "while" Loop

The value of the i variable is 0 at the start. With while, the condition is checked whether i is less than 5. The indented code is executed from top to bottom until the condition is no longer met. Here, you must not forget to increase the variable by one at each iteration stage; otherwise, you'll have implemented an infinite loop.

```
for i in range(5):
    print(i)
```

Listing 12.12 "for" Loop

This for loop does exactly the same as the previous lines of code. The i variable is automatically increased with each iteration step.

12.2.4 Random Choice

Later, during training in the final program, we need to select the direction at random, for example. Let's take a look at how this can be implemented.

The lines in the for loop are executed six times (for i = 0 to i = 5). Each time, a number (either 0 or 1) is selected at random, and this number is then output.

```
import random

for i in range(6):
    number = random.choice([0,1])
    print(number)
```

Listing 12.13 Random Selection of the Numbers 0 or 1

At each iteration step, an element of the list_1 list is randomly selected and output. At each iteration step, an integer between 0 and 2 is randomly selected and output.

```
for i in range(6):
    element = random.choice(liste_1)
    print(element)
```

Listing 12.14 Random Selection of List Elements

At each iteration step, an integer between 0 and 2 is randomly selected and output.

```
for i in range(6):
    number = random.choice(range(3))
    print(number)
```

Listing 12.15 Random Selection of the Numbers 0, 1, or 2

The `list_1` list contains four elements, so `len(list_1)` results in 4. At each iteration step, an integer between 0 and 3 is randomly selected and output.

```
for i in range(6):
    index = random.choice(range(len(list_1)))
    print(index)
```

Listing 12.16 Random Selection of the Numbers 0 to 3

12.2.5 Functions

You can use functions to swap out parts of the program and reuse them if necessary.

```
def sayHello():
    print("Hello")
```

Listing 12.17 Definition of a Function

A function is defined here that does nothing other than output "Hello". Every time you call this function, "Hello" is output.

```
sayHello()
```

Listing 12.18 Calling the Function

This function increases the number given as a parameter by 1 and returns the result to the caller.

```
def plus_one(number):
    return number + 1
```

Listing 12.19 Definition of a Function with a Parameter

The function is called with 5, and the result (the return value of the function) is saved in n1. If you output this variable, you'll see that n1 contains the number 6.

```
n1 = plus_one(5)
print(n1)
```

Listing 12.20 Calling the Function with a Parameter

12.3 Trainings

Let's analyze the *K12_game.ipynb* program. The game is kept very simple so that the focus is on the reinforcement learning algorithm.

```
import random

gamma = 0.9
learning_rate = 0.1

ACTION_LEFT = 0
ACTION_RIGHT = 1
TRAINING_ROUNDS = 1000
```

Listing 12.21 Definition of Required Variables or Constants

The `random` module is loaded (we'll need it later), and the `gamma` and `learning_rate` hyperparameters are set. We set constants for the movement to the left and right and the number of training rounds.

```
rewards = [[0, 0],[100, 0],[0, -100],[0, 0],[-100, 0],[0, 0],[0, 0],[0, 0],
 [0, 100],[0, 0]]

q_table = [[0, 0],[0, 0],[0, 0],[0, 0],[0, 0],[0, 0],[0, 0],[0, 0],[0, 0],[
0, 0]]

gameOver_states = [0,3,9]
```

Listing 12.22 Initialization of Required Lists

The rewards table is filled with reward or punishment values, the extreme values, according to the playing field. This list is required for the training phase. The initial Q-table only contains the value 0 and must be updated during training. These lists in turn contain lists. The first element in the respective list represents state 0, for example, which is again a list with two elements ([0, 0]). This inner list has two elements each: The first element represents "left", and the second one represents "right". Compare the rewards and q_table lists with the Q-table or the rewards table that we discussed earlier.

The squares that represent the end of the game are stored in the `gameOver_states` list. Compare the `rewards` and `gameOver_states` lists. If the character is on square 2 and moves to the right, the game is lost. This move is therefore awarded -100 points. On square 8, you receive 100 points for moving to the right, as the game is then won.

To keep the main program even clearer, we define a few helper functions.

```
def getNextAction(state):
    if state == 0:
        action = ACTION_RIGHT
    elif state == len(q_table)-1:
        action = ACTION_LEFT
    else:
        action = random.choice([ACTION_LEFT,ACTION_RIGHT])
    return action
```

Listing 12.23 Direction Selection

Possible directions of movement are determined depending on the current position. If the character is on the far left, it can only move to the right. However, if it's on the far right, it can only be moved to the left. For all other fields, both directions are basically possible, so the direction of movement is determined randomly.

It's important to note that chance plays a major role in determining the optimum. This is the only way to explore and evaluate new paths.

```
def getNextState(state, action):
    if action == ACTION_LEFT:
        return state - 1
    else:
        return state + 1
```

Listing 12.24 State Update

Depending on the current state and selected action, a subsequent state is assumed. It's quite simple here: the character is then one square to the left or right.

```
def isGameOver(state):
    return state in gameOver_states
```

Listing 12.25 Checking whether the Game Is Over

The game is over when the character is in certain predefined squares. It's not determined here whether you've won or lost. You'll recognize this later from the output of the position.

Now let's take a look at the training program. If you subtract the empty lines, it only consists of eight lines of source code.

```
for i in range(TRAINING_ROUNDS):
    # Take up starting position
    state = random.choice(range(len(q_table)))
    
    while not isGameOver(state):
```

12 Reinforcement Learning

```
        action = getNextAction(state)
        next_state = getNextState(state, action)

        q_table[state][action] = q_table[state][action] +
 learning_rate * (rewards[state][action] +
 gamma * max(q_table[next_state]) - q_table[state][action])

        state = next_state

print(q_table)
```

Listing 12.26 Training Phase

The game is played 1,000 times, and the values in the Q-table are updated. The starting position is determined at random. After the last game round, the final values of the Q-table are displayed. The output for the Q-table is as follows:

```
[[0, 0], [99.99858841836136, 80.92014975146536], [89.98454388096631,
 -99.99382963480826], [0, 0], [-99.9999999998435, 65.60999991019521],
 [59.04899982514381, 72.8999999990858], [65.60999999863937, 80.99999999948363],
 [72.89999999911996, 89.99999999970404], [
80.99999999703546, 99.99999999991682],
 [0, 0]]
```

This means that one character can now successfully complete the game.

12.4 Test

Based on the Q-table created, the AI is now supposed to play the game independently. For this purpose, we need a new function that determines an action depending on the current status.

```
def getNextActionPlay(state):
    if state == 0:
        action = ACTION_LEFT
    elif state == len(q_table)-1:
        action = ACTION_RIGHT
    else:
        action = max(q_table[state])
    return action
```

Listing 12.27 Direction Selection

Here, too, care is taken to ensure that the character doesn't move left or right over the edge of the playing field. Otherwise, the most promising direction from the Q-table is simply adopted. Chance no longer plays a role in the choice of direction. This is our simple but well-functioning AI.

```
state = random.choice(range(len(q_table)))
print("Startposition:", state)

while not isGameOver(state):
    action = getNextActionPlay(state)
    next_state = getNextState(state, action)
    state = next_state

    print("Position:", state)
```

Listing 12.28 Gameplay with Random Starting Position

You can execute this last code cell multiple times in succession. The output is, for example, is as follows:

```
Starting position: 5
Position: 6
Position: 7
Position: 8
Position: 9
```

The game is always won by the AI. However, if the randomly determined starting position corresponds to a state that means death, the game is over immediately.

12.5 Outlook

Let's assume that the game is expanded to a 2D field of the size 1,000 × 1,000. There will also be doors to rooms and treasures that give the character special abilities. A Q-table to store all this information would be very large. Artificial neural networks are therefore used in such applications to determine Q-values. This is referred to as *deep Q-learning*.

You can find examples of more complex games and applications with GUIs at *https://keras.io/examples/rl*.

If you want to use KNIME to implement a game with deep Q-learning, take a look at *www.knime.com/blog/an-introduction-to-reinforcement-learning* for a ready-made example in the form of the Tic-Tac-Toe game. It also explains how you can run the game on a server so that the game can be played via a browser.

12.6 Exercises

12.6.1 Exercise 1: Hyperparameters

Change the hyperparameters of the game and analyze the outputs, especially the values that lead to death or exit. There is no sample solution for this task.

12.6.2 Exercise 2: Expansion of the Game

Expand the game so that there are 12 fields. One field should represent an exit, and another should lead to the death of the character.

Chapter 13
Genetic Algorithms

Do you need an algorithm to solve a problem or achieve an optimization? Ask yourself the following question: How would nature tackle this problem?

What This Chapter Is About
- Order of the sequence of numbers in a list
- Solving mathematical equations
- Outlook on the use in real life using the example of a control loop

How did species evolve in nature? This question was already on the minds of the ancient Greeks. In the course of time, many theories were developed, and some supporters of one particular theory demonized all other views and declared their own view to be the only true one. And there really have been many ideas over time, such as the idea that species originated in water or mud. Many assumed (and still assume today) that species were created by a divine act.

In the 19th century, researchers Charles Darwin and Alfred Wallace independently developed the theory of evolution. In 1859, Darwin published his book *The Origin of Species*, and the world was turned upside down. Broad sections of society, especially some religious institutions but also many other researchers, not only rejected the theory but reacted to it with ridicule, malice, and hatred. Caricatures of Darwin as a monkey were published. His theory was reduced to the simple statement that humans are descended from apes. But Darwin never claimed that. Instead, he realized that humans and apes must have a common origin.

In the years that followed, however, it became clear that no other theory could provide such good, comprehensible answers to questions about the origin or change of species as the theory of evolution. In 1874, Darwin was elected to the American Academy of Arts and Sciences and was thus recognized during his lifetime for his work on the subject of evolution.

The theory of evolution as an explanatory model for the origin of species is plausible and comprehensible, so that it's used in other branches of science. You can use this theory to solve complex optimization tasks. Algorithms that refer to the theory of evolution are referred to as *evolutionary algorithms*. For our task, we'll use a special subtype of evolutionary algorithms, namely, the *genetic algorithm*. Slight variations are also possible in this subtype, but they all fall under the generic term *genetic algorithms*.

The simple task with which we'll familiarize ourselves with the topic is as follows: create a list with 10 ascending elements between 0 and 9, that is, [0,1,2,3,4,5,6,7,8,9].

13.1 The Algorithm

The algorithm "simulates" nature and reproduces the evolutionary cycle (see Figure 13.1):

- A *start generation* is formed with several individuals; an individual subsequently consists of only a single chromosome strand with several genes. A gene is in turn represented as a number. The numerical values are formed randomly in the start generation.
- During *selection*, the individuals are evaluated using a fitness function that determines how strong or weak an individual is. The higher the *fitness value*, the stronger the individual. A high fitness value increases the chance of survival and mating.
- During *reproduction*, some strong individuals may mate and produce children.
- The children inherit the genes (numerical values) of their parents. A *mutation* occurs in a certain percentage (change in the genes or numerical values). This mutation can either increase or decrease the fitness value.
- The procedure is repeated with this resulting new generation, for example, for a total of 50 cycles.
- At the end of the evolutionary cycle, the individual with the highest fitness value is used to solve the task.

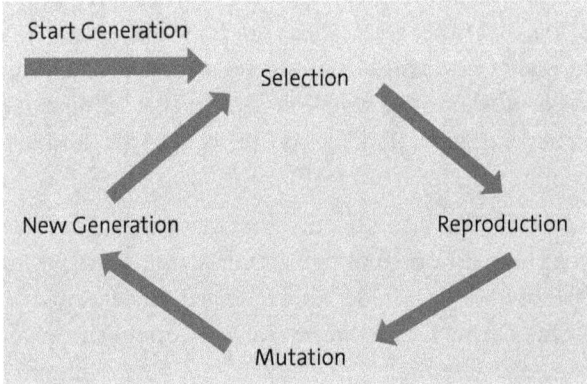

Figure 13.1 Evolutionary Cycle

In the following, we'll analyze these steps of the evolutionary cycle and apply the findings to our task.

13.1.1 Start Generation

Before we look at the start generation, we first need to define a few terms. In addition, this algorithm also requires some hyperparameters, which we have to define at the beginning.

A gene stores information about the characteristics of an organism. In our example, the 10 elements represent the individual genes. A chromosome contains a chain of genes. A list containing the 10 elements is supposed to represent a chromosome.

We also have to determine how many chromosome strands (i.e., chains of genes) are to be generated. In this example, there should be 100. The numbers of the elements (the genes) are generated randomly, whereby each element can have values between 0 and 9. This gives us the following structure:

chromosome_1 = [0,9,7,3,4,5,4,3,4,5]

chromosome_2 = [0,1,2,6,6,5,6,7,8,2]

chromosome_3 = [8,9,7,6,5,6,3,4,2,3]

...

The start generation has thus been created. We've also already defined some hyperparameters:

- Number of genes (10)
- Number of chromosome strands (100)
- Minimum limit value (0, inclusive)
- Maximum limit value (10, exclusive)
- Integer data type (integers)

13.1.2 Selection

The individual chromosome strands must be evaluated to distinguish strong individuals from weak ones. A function is required that calculates the fitness for each chromosome strand. The higher the value, the greater the probability of survival and mating.

Use the following data in this example: If the digit (content) of the element matches the index, the fitness value is increased. The first element has the index 0. If the value is also 0, you have a fitness value of 1. If the second element has the value 1, the fitness value is 2, and so on. Following are some examples:

chromosome_1 = [0,9,7,3,4,5,4,3,4,5] has the value 4

chromosome_2 = [0,1,2,6,6,5,6,7,8,2] has the value 7

chromosome_3 = [8,9,7,6,5,6,3,4,2,3] has the value 0

...

All chromosome strands (lists) are now evaluated using this function and prepared for the next step in the evolutionary cycle.

13.1.3 Reproduction

Now mating takes place, whereby the children can inherit their genes from both the mother and the father. But who is allowed to mate? There are various approaches here.

One possibility is that, for example, the fittest 20 lists are allowed to mate. This number is increased by a certain number of randomly selected "lucky ones" who are also allowed to mate. These lucky ones are selected regardless of their fitness level. This procedure actually refreshes the gene pool. Only a focus on the fitness value leads to overbreeding, so that optimization stagnates. Each couple in turn produces 100 children, a new generation. The entire generation then dies out, leaving only the children in the next generation.

Another possibility is that only a few lists per generation are selected for mating or death, so that a large proportion of the population survives and makes it to the next generation. We'll use this second option.

How are the children's genes structured? Let's assume chromosome_1 and chromosome_2 are selected for mating. It's determined at random where the splitting should take place. In our example, splitting is supposed to take place after the third element. These parts of the chromosome strands are crossed and form the chromosome strands of the children. Following are examples:

chromosome_1 = [0,9,7,**3,4,5,4,3,4,5**] has the value 4

chromosome_2 = [**0,1,2**,6,6,5,6,7,8,2] has the value 7

child_1_chrom12 = [0,1,2,3,4,5,4,3,4,5]

For each child, a new die can be cast to determine where the splitting should take place (and which sequence should be inherited from which parent). Of course, crossing can sometimes lead to a deterioration, but the optimization is improved overall.

Let's define new hyperparameters again:

- Number of pairs of parents (10)
- Number of children (100)

13.1.4 Mutation

In nature, a mutation occasionally takes place which can increase or decrease the fitness value. A mutation can cause something positive (the creature is particularly strong, beautiful, intelligent, etc.) or something negative (e.g., a limited sense of smell, poor eyesight, missing limbs, etc.). This mutation is the motor of the evolutionary cycle; without mutation, the optimization of the species stagnates.

In our example, we specify that 20% of the descendants should have a mutation (hyperparameter). To do this, we need to roll the dice once to determine which element is supposed to be mutated. Then, the dice are rolled again to determine what value this element should have. The genes of this mutated offspring are therefore not only inherited from the parents.

For example, the eighth element is replaced by the number 7. This gene is mutated:

child_1_chrom12 = [0,1,2,3,4,5,4,7,4,5]

A new hyperparameter is therefore the percentage for mutation.

13.1.5 New Generation

The new generation is ready for the next iteration. It's still necessary to determine how many generations the algorithm should run through. Another hyperparameter is the number of generations (here, 50).

13.2 Example of a Sorted List

Of course, we could implement the presented algorithm ourselves. But why reinvent the wheel? PyGAD is a powerful module that is available to us free of charge.

> **PyGAD**
> PyGAD is an open-source Python library that can be used to develop programs based on evolutionary algorithms (see *https://pygad.readthedocs.io/en/latest*).
> Install the PyGAD module via the Anaconda Navigator.

Let's now take a look at the *K13_list-1.ipynb* program. Thanks to the use of a module for genetic algorithms, the program remains very clear.

```
import pygad

# Hyperparameters

# Number of genes
num_genes = 10
# Number of generations
num_generations = 50
# Number of chromosomes
sol_per_pop = 100
# Number of parents
num_parents_mating = 10
```

```
# Probability of mutation
mutation_percent_genes = 20

# Lower limit of the random number
init_range_low = 0
# Upper limit of the random number
init_range_high = 10
# Data type
gene_type = int
```
Listing 13.1 Definition of Hyperparameters

The required module is loaded here, and hyperparameters are defined.

```
# Fitness function

# the parameters are "predefined"
# solution is particularly important, contains the list
def fitness_func(ga_instance, solution, solution_idx):
    fitness = 0
    for i in range(len(solution)):
        if i == solution[i]:
            fitness = fitness + 1
    return fitness
```
Listing 13.2 Definition of the Fitness Function

You must implement a separate fitness function for each task. The algorithm must be able to determine which chromosomes are better than others. It's quite simple here: Starting with the first element, the system checks whether the content matches the index. If so, the fitness figure is increased.

```
# create an instance
ga_instance = pygad.GA(num_generations=num_generations,
                       num_parents_mating=num_parents_mating,
                       fitness_func=fitness_func,
                       sol_per_pop=sol_per_pop,
                       init_range_low=init_range_low,
                       init_range_high=init_range_high,
                       num_genes=num_genes,
                       mutation_percent_genes=mutation_percent_genes,
                       gene_type=gene_type)
```
Listing 13.3 PyGAD Constructor

An instance is created, and the constructor is called with the hyperparameters.

```
# start evolution cycle
ga_instance.run()
```

Listing 13.4 Starting the Evolutionary Cycle

This call starts the evolutionary cycle. Depending on the configuration of the program and computer equipment, the execution may take a while.

```
# Determine results list, fitness value, and index of the results list
# (only the results list is required here)
solution, solution_fitness, solution_idx = ga_instance.best_solution()

print("Best parameters:", solution)
print("Highest fitness value:", solution_fitness)
```

Listing 13.5 Determining the Result

The results (best list and its fitness value) are determined and output. It worked, so the output reads:

```
Best parameters: [0 1 2 3 4 5 6 7 8 9]
Highest fitness value: 10
```

The list contains the ascending elements, and the fitness value has been calculated correctly.

```
# plot the result
ga_instance.plot_fitness()
```

Listing 13.6 Plot for Visual Analysis

Finally, we plot the progression of the fitness value as a function of the generation (see Figure 13.2).

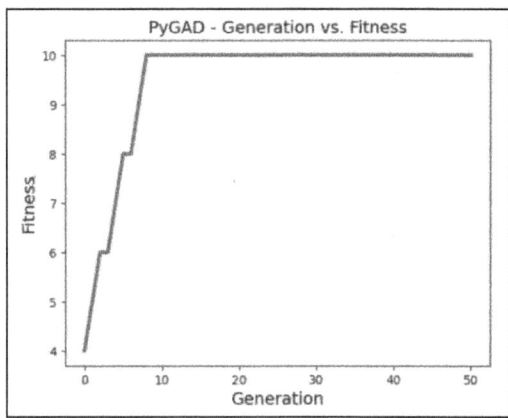

Figure 13.2 Plot of the Progression

You can see that the optimum was reached fairly early on. At this point, you could adjust the hyperparameter for the number of generations (e.g., 15) and run the program again.

However, it's possible that not every program run is successful and the list isn't generated as desired. So, if you always want the list to be generated as desired, you can increase the number of generations to 100, for example, and test the program again. This increases the likelihood that each program sequence will be effective. All you need to do is invest a little computing time.

13.3 Example of Equation Systems

In the next example, *K13_equation-1.ipynb*, the three variables (x, y, and z) in a system of equations will be determined using the theory of evolution:

$$-3x - 4y + 3z = -5$$

What changes are necessary? Here we have to set the number of genes to 3 (three variables). In addition, we set the number of generations to 500 and the data type to float. The random numbers should be in the range between -10 and +10. Because only three genes are present here, the mutation probability is set at 34% so that a mutation can occur at all.

It's very important to implement a suitable fitness function for the task. In this example, we'll insert the determined values into the equation and calculate the result. We'll then determine the deviation from the number -5. The smaller the deviation, the better. But for the evolutionary process, a higher value means a higher probability of survival or mating. We'll therefore multiply the difference with the number by -1. As a result, the worse the fitness values are, the more negative they will be. The closer the value is to 0, the better. Thanks to this trick, stronger chromosomes have higher fitness values than weaker ones.

```
import pygad

# Hyperparameters

# Number of genes
num_genes = 3
# Number of generations
num_generations = 500
# Number of chromosomes
sol_per_pop = 100
```

13.3 Example of Equation Systems

```python
# Number of parents
num_parents_mating = 10
# Probability of mutation
mutation_percent_genes = 34

# Lower limit of the random number
init_range_low = -10
# Upper limit of the random number
init_range_high = 10
# Data type
gene_type = float

# Fitness function

# the equation is: -3x -4y +3z = -5

# the parameters are "predefined"
# solution is particularly important, contains the list
def fitness_func(ga_instance, solution, solution_idx):
    x = solution[0]
    y = solution[1]
    z = solution[2]
    result = -3*x - 4*y + 3*z
    fitness = abs(-5-result) * (-1)

    return fitness

# create an instance
ga_instance = pygad.GA(num_generations=num_generations,
                    num_parents_mating=num_parents_mating,
                    fitness_func=fitness_func,
                    sol_per_pop=sol_per_pop,
                    init_range_low=init_range_low,
                    init_range_high=init_range_high,
                    num_genes=num_genes,
                    mutation_percent_genes=mutation_percent_genes,
                    gene_type=gene_type)

# start evolution cycle
ga_instance.run()
```

13 Genetic Algorithms

```
# Determine results list, fitness value and index of the results list (not
required here)
solution, solution_fitness, solution_idx = ga_instance.best_solution()

print("Best parameters:", solution)
print("Highest fitness value:", solution_fitness)
```
Listing 13.7 Solving the Functional Equation Using Evolution

The output reads as follows:

```
Best parameters: [-7.60901208  8.26523211  1.74462676]
Highest fitness value: -1.1942404153586494e-05
```

Let's carry out a calculation with the determined values to verify that the determined parameters are correct.

```
# Calculation with the best result
x = solution[0]
y = solution[1]
z = solution[2]
result = -3*x - 4*y + 3*z

print(result)
```
Listing 13.8 Test Calculation with the Results

The result is -5.000012. That's not bad. Change the data type to integer, and test the program again.

13.4 Real-Life Sample Application

The approach of imitating nature during optimization is ingenious. You may not be able to recognize this in the examples presented and will ask yourself why you can't simply create an ordered list or solve the systems of equations and make such an effort with genetic algorithms. The examples were only intended to give you an understanding of the process. With advanced Python knowledge, you can solve many optimization tasks using this method. I want to present a real-life example here: the design of a controller.

In simple terms, a *control circuit* is a system in which you can set a desired state that is maintained (within certain limits) even in the event of faults. Following are two examples:

- **Temperature control**
  You set the room temperature, and the system is regulated to this value. When you open the window, the controller ensures that this disturbance gets corrected, for example, by automatically increasing the heating output.
- **Speed control**
  You set the engine speed. If the engine is subjected to a weight, the controller ensures that the power is adjusted and the speed remains constant.

The trick is to design a controller for this type of system. The *PID controller*, which consists of a proportional, integral, and differential component, is often used. These three parameters must be set optimally for each specific problem. There are, for example, empirical (trial-and-error) or higher, mathematical adjustment methods. However, you can also imitate evolution to determine the three parameters.

During the evolution cycles, various parameter combinations are generated for the controller, which must be evaluated. We therefore need a fitness function to determine the quality of the controllers. The controller is set (setpoint), the system is switched on, and the behavior (actual value) is observed. The requirements for the controller are listed here:

- There should be little or no *overshoot*. If you set 25°C (77°F) on the thermostat (setpoint), the temperature should not first rise to 35°C (95°F) and then fall back to 25°C (actual value).
- The system should not *oscillate*. If 25°C (77°F) is set, the temperature should not constantly change back and forth between 20°C (68°F) and 30°C (86°F).
- The *settling time* should not be too long. If you set a temperature, it should not be reached after four hours.
- The *control distance* should not be too large. If you set 25°C (77°F), you mean it. A constant temperature of 23°C (73.5°F) should not be maintained.

We therefore have to measure and evaluate these requirements or assign a fitness value to each generated controller. With a little Python code, this is easy to implement. I've written a program for this, which you can find at *https://github.com/mck-sbs/PyConSys*.

Take a look at the plot of an unregulated or poorly regulated system in Figure 13.3. In time step 1000, the setpoint value (in control engineering, this is referred to as the *reference variable w*) is set to 1 (100%). The system or the actual value doesn't reach this value and starts to oscillate (in control engineering, this is referred to as the *controlled variable x*).

13 Genetic Algorithms

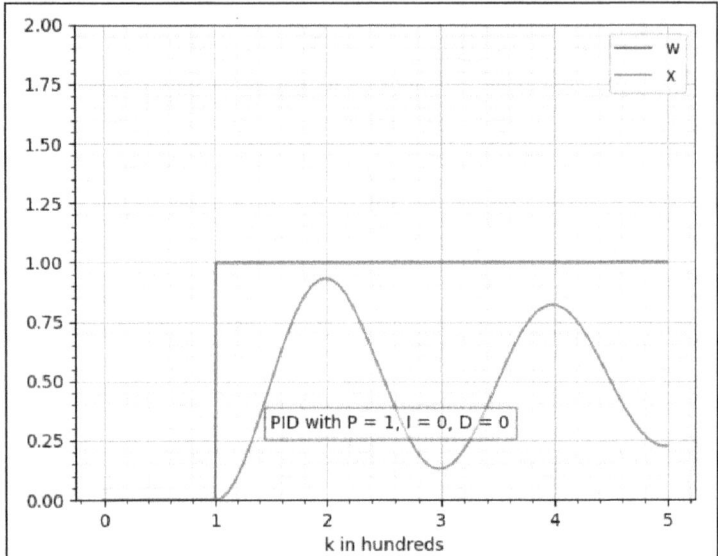

Figure 13.3 Unregulated System

In the following, the control parameters P, I, and D are set in such a way that the previously mentioned requirements for the controller are met. A simulation is carried out for each generated controller, and the fitness value is determined. The PyConSys program proceeds as follows:

- The 800 controllers with randomly generated PID combinations are used and simulated.
- The fitness value is determined for each controller, for example, PID1 = 456, PID2 = 564, and so on.
- The fittest 10 controllers (with the highest fitness values) and a randomly selected 150 lucky ones (regardless of fitness value) are allowed to mate. The children inherit the characteristics of their parents. For example, a PID child1 inherits the P portion from the father and the I and D portions from the mother. PID child2, on the other hand, inherits P and D parts from the father and the I part from the mother. Which parameter is inherited from whom depends on chance.
- Of the children, 5% have a mutation, and one parameter is generated at random. More precisely, 5% are determined at random. For each of these children, it's again randomly determined which parameter mutates (P, I, or D) and then its value is randomly generated.
- This is followed by simulations for the children, determination of fitness values, mating, and so on. So, the cycle starts all over again. There are 20 generations in total.
- Finally, the controller with the best fitness value is used for the system.

The result is shown in Figure 13.4. Can you see how nicely the actual value adjusts to the target value? For control engineers, this is almost a romantic sight.

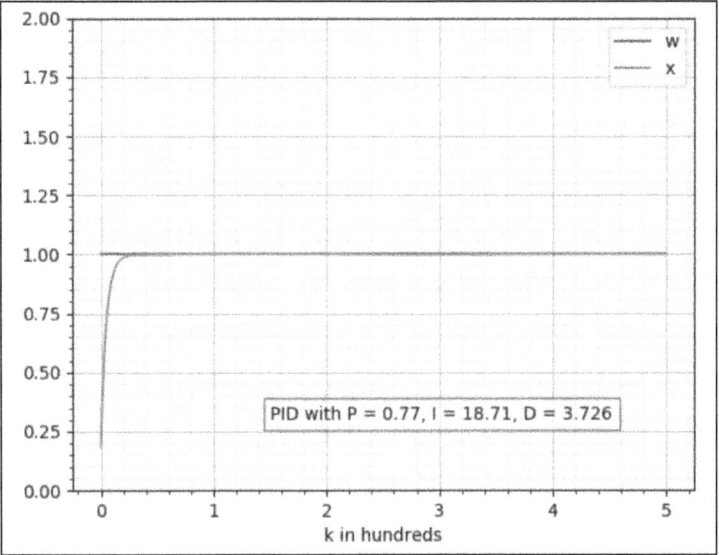

Figure 13.4 Control Circuit with a Determined PID Controller

The PID parameters were determined using the evolutionary algorithm. The setpoint is reached quickly, there is no overshoot, and the system doesn't oscillate. Everything looks exactly as we wanted it.

13.5 Exercises

13.5.1 Exercise 1: Hyperparameter Optimization

Create a copy of the *K13_equation-1.ipynb* program, change some hyperparameters, and analyze the results. There is no sample solution for this task.

13.5.2 Exercise 2: System of Equations

The following equation is given:

$$a \cdot \sin(b) + c \cdot \cos(d) + e \cdot \tan(f) = 12$$

Determine the variables a to f.

> **Math Module**
> The trigonometric functions can be found in the math module. Import the module, and call it using math.sin(0.3), for example.

Chapter 14
ChatGPT and GPT-4

With the publication of ChatGPT (generative pretrained transformer), the OpenAI company has triggered a veritable hype.

> **What This Chapter Is About**
>
> In this chapter, we take a look at the possibilities offered by this powerful chatbot.
> - Formulating prompts
> - Generating customized content
> - Analyzing and summarizing texts, including PDF files
> - Using ChatGPT as a programming aid
> - Using the programming interface with parameters, roles, user profiles, and filters
> - Conducting longer conversations via the programming interface
> - Using audio files

The hype surrounding ChatGPT isn't unjustified. Humanity had never seen a chatbot with these capabilities before; within two months, more than 100 million users had signed up. For comparison, it took Instagram two years to reach this number of users. Many see this technology as opening up new opportunities, for example, in the fields of art, culture, and science. Others point out the dangers they see: loss of jobs, fraud in education, and the like. As is so often the case, the truth will probably lie somewhere in between. But even these technologies won't be stopped. You'll see for yourself how helpful such a chatbot can be.

But what can you do with this chatbot? You can use ChatGPT as your own personal digital assistant. You can ask questions via the website, have templates created, write poetry, copy, program, explain, and more. The first hurdle that will prevent you from exploiting the full potential will be that you can't imagine what the tool can do and that you yourself will limit the quality of the results by holding back your requests. Another hurdle is formulating the request in detail so that the chatbot knows exactly what is expected of it. This requires a little practice.

You can use the programming interfaces provided to program your own applications, which in turn use ChatGPT or its successor GPT-4. Do you need a chatbot for support, a tutoring system, or a translation? Use this programming interface to create powerful applications with little code.

Always remember that this AI has also been trained with data from the past, so current topics are still unknown to the tool. In addition, just because the data was available to the chatbot somewhere doesn't mean that it's true. You should test your application thoroughly, especially when using the programming interface.

Let's ask ChatGPT itself or its successor GPT-4 how they work (*https://chat.openai.com*). You must set up an account to use it.

User: *I have a good basic knowledge of AI. Explain the technical details to me: How do you work?*

Answer: *Of course, I will gladly explain the technical details of my architecture to you!*

*1. **Model architecture**: I am based on the Transformer architecture, which was presented in 2017 by Vaswani et al. The specific implementation I'm using is known as GPT (Generative Pre-trained Transformer), and I'm basing it on the fourth version, GPT-4.*

*2. **Layers and neurons**: GPT-4 consists of several transformer layers, each layer containing a certain number of neurons or "units". These layers are divided into two main parts: the multi-head attention mechanism and feed-forward networks.*

*3. **Multi-head attention**: This enables the model to consider information from different parts of the input text at the same time. The attention scores show how strongly the words in the text are related to each other, and multi-head attention allows the model to look at these relationships from different "angles".*

*4. **Feed-forward networks**: These consist of several tightly connected layers and an activation function. They are responsible for recognizing complex patterns and relationships in the data.*

*5. **Tokenization**: Input texts are divided into tokens, which are then converted into vectors to be routed through the network.*

*6. **Training**: The model was trained by supervised learning using a large number of texts. The weights of the model were adjusted so that the probability of predicting the next word in a sentence was maximized.*

*7. **Fine tuning**: In addition to pre-training, the model can be tuned to specific tasks or datasets to improve its performance in specific areas.*

*8. **Scaling**: One of the reasons why GPT-4 is so powerful is the sheer size of the model. It has more parameters and layers than its predecessors, which allows it to capture more complex patterns and abstractions in the data.*

*9. **Activation functions and optimizers**: It uses specific activation functions such as ReLU and optimizers such as Adam to control the learning process.*

*10. **Regularization**: To prevent overfitting, techniques such as dropout and layer normalization are used.*

Together, these components enable the model to learn from a large text corpus and generate human-like texts. However, the development and training of such models are extremely computationally intensive and require considerable hardware resources.

The answer of the chatbots summarizes the most important points succinctly. If you're interested, you can, of course, ask further, more detailed questions. The answer mentions tokens. Let's take the following sentence as an example: "What are tokens?" This sentence is broken down into words, word components, and characters, which are referred to as *tokens*. For this example, the tokens are "What", "ar", "e", "tokens", "?". If we were to work with words instead of tokens, there would be many word variations to consider and take into account. The other extreme, looking at letters, would greatly increase the number of processing steps.

If there are other terms or topics that you don't understand, just keep asking. The tool will never stop the conversation in frustration.

Using our text generation program in Chapter 11, Section 11.12.10, we programmed an AI that was trained with a text by Goethe and then generated text in this style itself, although, frankly speaking, the result wasn't very convincing. But this knowledge should help you to better understand the principle of chatbots. In our program, the next character with the highest probability was generated for a string of 100 characters. ChatGPT and GPT-4 use tokens. The next token is generated for a sequence of tokens. This result with the new token is analyzed again, and another token is generated for the new sequence. This procedure is repeated several times.

The prices for ChatGPT depend on the model used and the number of tokens. The use of simple models via the web mask is free of charge for ChatGPT-3.5, although access isn't guaranteed (e.g., when the servers are busy). However, if you want to use the programming interface or high-performance models, you'll have to pay for these services. Prices for GPT-3.5 currently start at $0.0015 per token. You can assume that 75 words represent about 100 tokens. For the current GPT-4 model, you currently have to purchase a Plus subscription for $20 per month. On the other hand, the model is much more powerful, and you get prioritized access to the server, even at times of high load. On the website (*https://openai.com/pricing*), you'll find the current prices and comparisons of various packages.

14.1 Prompt Engineering

The interface for interaction is the *prompt*. Before computers had GUIs, prompts were the only way to interact directly with computer systems. The Windows command prompt and the Linux or macOS terminal are examples of this. Before the triumph of ChatGPT, nobody would have thought that this form of interaction with computers would become more widespread again.

Any user can formulate questions or instructions without much prior knowledge. But here, too, you can achieve better results if you deal with the topic in advance. Some examples are presented in this section.

14.1.1 Generating Content

Let's assume you have a new family dog and need some name suggestions. The answers are shortened in each case, and omitted text passages are replaced by three dots.

User: *Please suggest names for our new family dog.*

Answer: *Certainly! Choosing a name for your new family dog can be a fun and exciting process. Here are some suggestions, both male and female and unisex. Of course, the choice also depends on the breed, size, color, or character of the dog.*

Male names:

1. Max

2. Rocky

3. Charlie

...

Female names:

1. Bella

2. Luna

3. Daisy

...

Unisex names:

1. Lucky

2. Bailey

3. Teddy

...

Funny or themed names:

1. Einstein (for a particularly clever dog)

2. Cookie (sweet as a cookie!)

3. Captain (for a leader)

...

I hope these suggestions will help you find a name for your new four-legged friend!

Well, you can do it this way, no doubt about that. But the more precise the information you give the chatbot, the more precise the answers will be.

User: *We have a new family dog. It is a male, the breed is Maltese, the color is white. Please suggest some names.*

Answer: *Of course! As it is a male Maltese, which is white, the following names could fit well:*

1. Snowball (matching the white color)

2. Biscuit

3. Marshmallow

4. Casper

5. Finn

...

Maltese are known to be affectionate and playful, and these names reflect both the color and the friendly nature of the breed. I hope one of these names suits your new family member!

These answers are now much more precise. However, you can enter even more context. If you're a fan of Star Wars, why not share it? You don't have to enter the question from scratch. Complete your question or instruction in the next prompt. As long as you don't start a new chat, the context of the conversation is retained.

User: *We are fans of Star Wars. Please suggest names in this context.*

Answer: *Since you're fans of Star Wars, there are lots of cool ways to name your Maltese after a character or term from this universe. Here are some suggestions:*

1. **Luke** (after Luke Skywalker)

2. **Chewie** (short for Chewbacca)

3. **Yoda** (for a wise little dog)

4. **Solo** (after Han Solo)

5. **Obi** (short for Obi-Wan Kenobi)

6. **R2** (after R2-D2)

7. **Leia's friend** (if you have a female dog in the family)

8. **Kylo** (after Kylo Ren)

...

These names could symbolize the connection between your new dog and your love of Star Wars. May the Force be with you!

For example, you can still specify how many names it should suggest. The more context, the better the answers.

Another example: You're attending the technical school in Boston and would like to submit an unsolicited application for an internship in the IT department of the city of Herzogenaurach. You'll need a letter of application. Let's be honest, even without ChatGPT, you would research on the internet to learn how to draft such a letter.

User: *I am a student at a technical school and would like to send an unsolicited application for an internship in the IT department of the city of Boston. I need a letter of application for this. Please create the letter, maximum one page long.*

You'll receive a first draft in response. However, view this critically. Without much context, the results could also come from a search engine. In this letter, for example, your specialist knowledge, such as programming languages, and so on, is missing. You can add this in the next prompt and have the letter generated again. Or even better, you can ask ChatGPT what context information it needs for better writing.

User: *What contextual information do you need to write the letter in more detail?*

Answer: *To be able to write a more detailed and targeted letter of application, it would be helpful to have the following information:*

*1. **Specific knowledge and skills**: What specific skills, programming languages, or technologies do you master that could be particularly relevant for the position?*

*2. **Project experience**: Are there any specific projects or courses you have taken part in that could be directly related to the work in the IT department of the city of Boston?*

*3. **Personal motivation**: What specifically attracts you to the IT department of the city of Boston? Are there certain projects or initiatives that you are particularly interested in?*

*4 **Timeframe of the internship**: When do you want to start the internship and how long should it last? Is it a compulsory internship as part of your studies?*

*5. **Contact information of the company**: If you have the name of the person responsible in the IT department, the letter could be personalized.*

*6. **Previous relevant experience**: Do you already have internships or other relevant work experience in IT or related areas?*

*7. **Requirements of the position**: If you have specific information about what qualifications or experience are desired or required for the position, the letter can be customized accordingly.*

This information would allow the cover letter to be more tailored to the specific position and your unique qualifications.

Do you need a worksheet on trigonometric functions with sample solutions? Have it generated for you. Many students ask for additional practice material before performance assessments. With ChatGPT, they can generate their own tasks and solutions. And if you don't understand a task, the tool can explain it again or in a different way if necessary. Additions such as "Explain it to me as if I were 12 years old" lead to simpler explanations.

Sometimes, as a teacher, you have an idea for a certain subject area and want to transfer it to new subject areas. This also works very well.

User: *The following is a situation involving a voltage divider in the subject of electrical engineering, 10th grade. Create a similar action situation for the diode. The action situation for the voltage divider is as follows: Your plane has crashed in the jungle, and you are the only survivor. Unfortunately, the battery of your smartphone has run out, so you have to head out to get help. While searching for help, you find an abandoned car on a remote road with all its tires removed. The car is not roadworthy, but the battery (accumulator) still has voltage. In the car, there is a calculator, a ruler, 10 adjustable resistors (potentiometers), a functional tablet with an electrical engineering simulation program (an x-device, which is why the battery cannot be removed), squared sheets of paper, and a collection of electrical engineering formulas. You want to charge your smartphone via the car battery, which has a voltage of 12 V. However, the USB port of your smartphone requires 5 V and the maximum load current is 1 A. Your survival depends on you developing a circuit to charge your smartphone via the car battery and getting help.*

Answer: *Of course! Here is an action situation on the diode, which is similarly designed and is set in a 10th-grade context in electrical engineering:*

Your ship is stranded on an uninhabited island, and you are the only survivor. Fortunately, there is an old lighthouse on the island whose light signal you could use to signal for help. The problem is that the battery of the lighthouse is weak and can only supply energy for a limited time.

In the lighthouse you discover a box with various electronic components, including several diodes, a multimeter, a solar panel, some wires, a working tablet with an e-technology simulation program (an x-device, which is why the battery cannot be removed), squared sheets of paper, and a collection of e-technology formulas.

Your plan is to connect the solar panel to the lighthouse's battery in such a way that it can charge the battery during the day without allowing energy to flow in the opposite direction. This would mean that the battery would have enough energy at night to send the light signal. You know that a diode can fulfill exactly this purpose because it allows the current to flow in only one direction.

Your survival depends on developing the right circuit to connect the battery to the solar panel using the diodes so that you can use the light signal at night and get help.

This action situation introduces the concept of a diode in an exciting and practically relevant context. It could be used as a basis for classroom activities or exercises to promote students' understanding of the function of a diode.

Whether this example is relevant in real life remains to be seen. But the style was actually adopted. Specific tasks with calculations and solutions could then be generated. In the age of Instagram and TikTok, teachers always have to come up with new things to keep the class's attention for a few minutes. Otherwise, the students will internally swipe to the next screen.

Of course, assigning homework and grading the results can be problematic. Depending on the task, students can use generative AI tools and have these tasks partially or even completely solved by them. But you can also look at the whole thing from a different angle. Applications like these can also go some way toward achieving educational equality. Students who don't have academics as parents can now also get help at home. The same applies to adults who want to continue their education. Not everyone can afford expensive courses. For example, you can have ChatGPT generate a structure for a course in the form of a table of contents. You can then refine the individual chapters or subdivide them further. Then, teaching content can be generated. You can also find out more by watching videos or reading books. Each chapter can even be completed with a generated test. Of course, this approach can't replace a well-considered, tried, and tested course (not yet), but interested parties now have new, unprecedented opportunities.

We can say that you always get better results when a lot of contextual information is provided. The chatbot then has to make fewer assumptions itself to close the interpretation gaps. You can formulate the question or instruction right away with a lot of context or add it in the next prompts. If in doubt, you can ask ChatGPT which context information could lead to better results.

A major hurdle is that at the beginning, you don't want to believe what this tool is capable of. You'll always be surprised at the results it delivers. Have a poem written for your loved ones, test new cooking recipes, create a new training plan, or have press articles formulated. Use it, and you'll gain experience.

However, you should check the results critically. A generated letter, for example, can save you a lot of time. Often you don't know what the structure should be or where to start. A suggestion from a chatbot is very helpful here. However, take the time to check and adapt the text afterwards.

14.1.2 Programming

In the past, when it came to concerns about losing jobs due to AI, programmers were always excluded. Now that has changed as well. However, the market for skilled workers is still far from saturated, so there's no need to get worried. Think of ChatGPT as a kind of programming aid that supports you during development. It can't replace an experienced developer (yet).

For example, you can have the tool generate parts of the source code. In the next example, ChatGPT is supposed to generate a source code in Python, which outputs the primes between 1 and 100.

User: *Generate Python source code for a function that outputs all prime numbers up to 1,000. The source code should also contain comments describing the statements.*

You'll receive a detailed answer to this question, which also describes the algorithm in general. And, of course, you'll get the source code, which you can simply copy and paste into your applications.

```python
def primes_to_1000():
    # List to store prime numbers
    primes = []

    # Loop through numbers from 2 to 1000
    for num in range(2, 1001):
        # Assume the number is prime until proven otherwise
        is_prime = True

        # Loop to divide the current number by all the
        # primes found so far
        for prime in primes:
            # If the number is divisible, it is not a prime
            if num % prime == 0:
                is_prime = False
                break
            # Optimization: We do not need to test any further if the current
            # prime squared is greater than the number to be tested
            if prime * prime > num:
                break

        # If number is prime, it gets added to the list
        if is_prime:
            prime.append(num)

    # Output of the primes found
    print("Primes up to 1000:", primes)

# Call of the function to output the primes
prime_to_1000()
```

Listing 14.1 Generated Source Code for Prime Numbers

Very good, but perhaps you haven't fully understood parts of the source code? If so, just ask.

User: *Please explain the following passage in more detail:*

```
if prime * prime > num:
            break
```

Answer: ...

Suppose num had a divisor greater than prime and less than the square of prime. Then there should also be a corresponding divisor that is smaller than prime (since the two divisors multiplied together must be equal to num). But since prime itself is a prime, and all smaller numbers have already been checked, this smaller divisor would already have been found if it existed.

...

I have greatly shortened the generated text here. The answer should leave no questions unanswered about the program.

The possibilities are very diverse. You can even enter source code and have it analyzed, for example, to find possible errors.

User: *Check whether the following Python code is secure and error-free. If not, complete it accordingly.*

```
data = pd.read_excel('file.xlsx')
```

data.head()

Following this instruction, you'll again receive a very detailed result. The errors and risks are not only described in text form, but the source code is also generated ready for copying.

```
import pandas as pd

# Attempt to open the file
try:
    # Use openpyxl as engine to avoid macros
    data = pd.read_excel('file.xlsx', engine='openpyxl')
except FileNotFoundError:
    print("File was not found.")
    data = None # Optional, depending on how you want to respond to the error

# If the file was opened successfully, display the first five lines
if data is not None:
    print(data.head())
```

Listing 14.2 Corrected Source Code

If you don't regard ChatGPT or GPT-4 as a competitor, but as a personal assistant, you can save a lot of time during development.

The results in the area of databases are also impressive. Tell the chatbot how the database is structured (e.g., SQL statements for creating the individual tables). You can then formulate the information you require from the database in text form. ChatGPT generates suitable SQL statements for this purpose. You can also have entire database architectures created.

> **SQL**
> Structured Query Language (SQL) is a database language for creating and editing databases.

User: *I need an SQL database for SQLite. Create the required tables and references for the following request by generating complete SQL commands. The tables should then be filled with three sample data records each. Work with surrogate keys. Requirement: Class information (name, room) and student information (first name, family name, date of birth, email) are supposed to be stored in a database. A student belongs to exactly one class; a class can be attended by multiple students.*

As desired, all SQL statements are generated with a description. You can then formulate your requirements for the query in text form.

User: *Create an SQL query for the following request: First names and family names of students born before 1/1/2020 and attending class "FSWI-1" are required. Also enter the number of these students.*

Perhaps you would like to familiarize yourself with the Python programming language?

User: *I have no programming skills whatsoever. I would like to learn the Python programming language. Create a curriculum for me. For each item, there should be links to websites or videos that convey the knowledge.*

Answer: *Of course! Here is a simple curriculum to get started with programming in Python. Each step contains links to resources that can help you understand the topic in question.*

1. Introduction to programming and Python

- **What is programming?** [Khan Academy - Introduction] (www.khanacademy.org/computing/computer-programming/programming/intro-to-programming/v/programming-intro)

- **Why Python?** [Real Python - Why learn Python?] (https://github.blog/2023-03-02-why-python-keeps-growing-explained/)

- **Python installation** [Python.org - Download and installation] (www.python.org/downloads/)

2. Basics of Python programming

- **Hello World & Simple Outputs** [Codecademy - Hello World in Python] (www.codecademy.com/learn/learn-python-3/modules/learn-python3-hello-world)

- **Variables and data types** [W3Schools - Python Variables] (www.w3schools.com/python/python_variables.asp)

- **Conditions and Loops** [Corey Schafer - If Elif Else Statements] (www.youtube.com/watch?v=DZwmZ8Usvnk) & [Looping in Python] (www.w3schools.com/python/python_while_loops.asp)

3. Functions and modules

- **Introduction to Functions** [Real Python - Defining Functions] (https://realpython.com/defining-your-own-python-function/)

- **Modules and Packages** [Corey Schafer - Python Tutorial: re (Regular Expressions)] (www.youtube.com/watch?v=K8L6KVGG-7o)

...

These resources and the curriculum should give you a solid start in Python programming. Have fun learning!

The following is another useful option, which should be mentioned as an example of similar situations. Modules such as pandas are very powerful and not always self-explanatory or easy to use. Even if you deal intensively with the subject matter, you may have to look up the source code required to solve a problem after a while (if you have to program something with it). The chatbot can also help you here.

User: *A pandas DataFrame has the following columns: "sepal.length", "sepal.width", "petal.length", "petal.with". Generate source code to check if there are empty cells. If yes, these cells should be replaced by mean values of the corresponding columns.*

You'll again receive a detailed answer, with source code ready to copy. Even the typo in "petal.with" was recognized and corrected.

If you've booked the ChatGPT Plus package, you can activate the code interpreter and plugin support in the settings (see Figure 14.1).

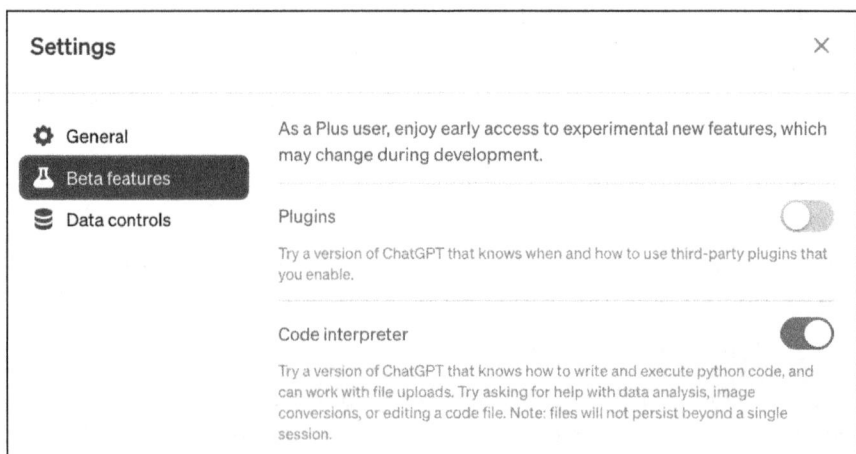

Figure 14.1 Beta Settings of the Web Interface

You should also activate plugins so that you can install them later. There are plugins for PDF files, diagram creation, summarizing YouTube videos, and so on.

Along with typing or inserting the source code via the prompt, you can also upload, analyze, and edit files.

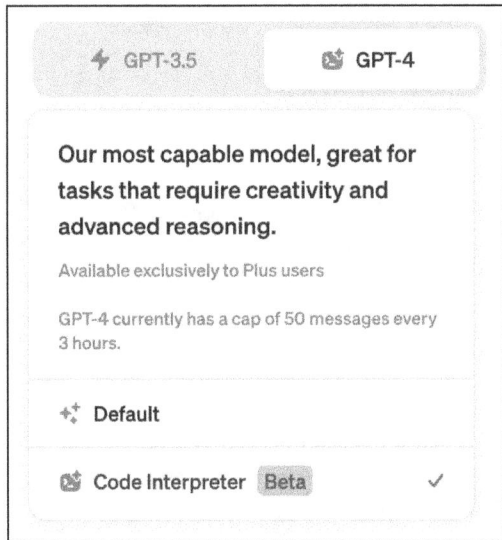

Figure 14.2 Switch to Code Interpreter

Note that the code interpreter is still in the beta phase. During my tests, the chatbot had problems with Jupyter Notebooks. However, when a Python file was uploaded, the analysis worked surprisingly well.

After any Python file (or a file of another programming language) has been uploaded, you can again formulate instructions, such as the following two examples:

User: *Analyze the program, and give me a summary of what the program actually does.*

User: *Does the program contain errors or security risks? If so, list them and present proposed solutions.*

In my attempt, an error was actually found and a concrete solution was proposed.

If you want to install plugins, you should click on the **No plugins enabled** dropdown menu in the work field and then click on **Plugin store** (see Figure 14.3). Install the **Show Me Diagrams** plugin. You can then select the plugin from the dropdown menu.

Now you can create diagrams as follows, for example:

User: *Create a flowchart to illustrate the procedure for developing AIs.*

Some plugins work best if you formulate the instructions in English. You can also experiment with other plugins and instructions.

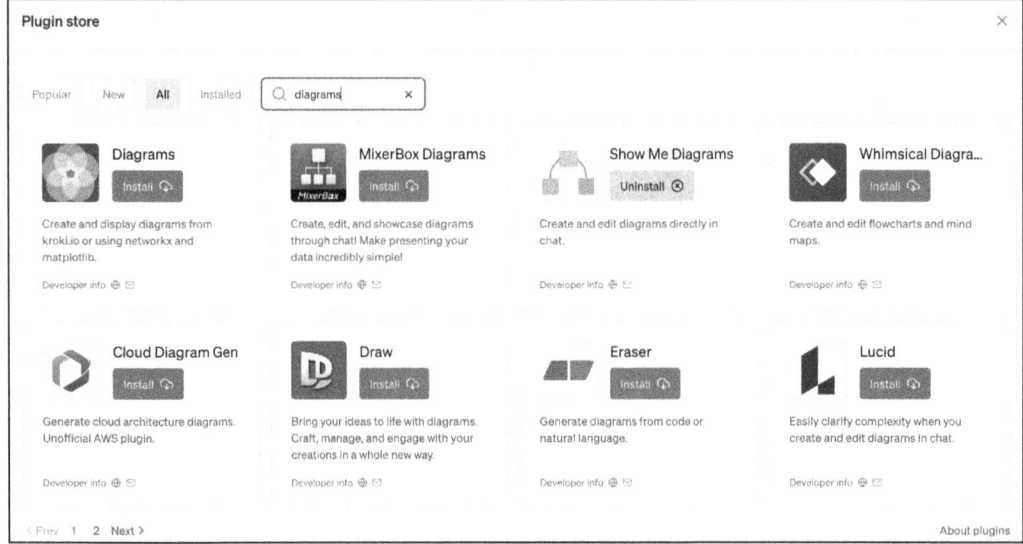

Figure 14.3 Plugin Store

> **GitHub Copilot**
> There are also very good plugins from other providers that work in a similar way to ChatGPT and can be integrated directly into development environments. The only thing the developer needs to do is write a comment, as the source code is generated automatically. Errors are also recognized and corrected. It's advisable to familiarize yourself with such tools if you are or want to become active in the field of development. One example of this is GitHub Copilot (*https://github.com/features/copilot*).

14.1.3 Analyzing and Summarizing

You can also use ChatGPT to summarize texts. Let's take the book *Steppenwolf* by Hermann Hesse as an example.

User: *Create a summary of the book Steppenwolf by Hermann Hesse, maximum one page.*

Answer: *Steppenwolf is a novel by Hermann Hesse that was published in 1927. The book is a profound and complex text that deals with the themes of duality, isolation, and the search for self-realization [...]*

The summary is not bad, but we can get a bit more out of it.

User: *How many chapters does the book "Steppenwolf" by Hermann Hesse have?*

Answer: *The novel "Steppenwolf" by Hermann Hesse is not divided into traditional chapters. Instead, the book consists of three main parts: [...]*

This is followed by a list of the main parts. You can then combine the individual parts.

User: *Create a detailed summary for each main section, each one page long. Finally, add a conclusion.*

The result is not satisfactory, as the individual parts are still too short. We have to proceed in small steps.

User: *Create a detailed summary of the first main part, one page long.*

Now we're getting closer to the desired result. You can then have the other parts summarized.

The strategy should therefore aim to gain an overall view first. You can then go into further detail. The context is also important here.

User: *Create a detailed summary of the first main part, one page long. The style should be funny.*

Answer: *Of course, here's a more humorous summary of the first main part of "Steppenwolf"!*

Preface by the editor - A wilderness in the home

The first main section of the book begins with the editor's foreword, which introduces us to Harry Haller - and boy, is he a character! He is a philosophical loner, a music lover, and he calls himself the "Steppenwolf", probably because "jungle tiger" was already taken [...]

Even this instruction is implemented acceptably. A witty summary of this book is really quite extraordinary.

You must always bear in mind that the number of tokens per prompt is limited, for example, to 4,000 or 8,000 tokens. For more details, refer to the website at *https://platform.openai.com/docs/quickstart/pricing*.

If you've booked the ChatGPT Plus package, you can also use plugins to work with PDF files or websites, for example. Install the WebPilot plugin, and start a new chat.

User: *Create a short summary of the page https://github.com/mck-sbs/Pendulum. What is it all about? Explain it in simple terms.*

Answer: *The GitHub project "Pendulum" by mck-sbs deals with the simulation of an inverted pendulum in Python using pybox2d. It contains various control systems, including fuzzy and PID control, to control the pendulum.*

...

In simple terms, this project is like a virtual experiment where you try to keep an inverted pendulum in balance by using different techniques to control it. It's like a game of trying to keep something upright that can easily fall over, but with complex mathematical rules behind it.

Install the AiPDF plugin to process PDF files. You can upload the file or insert the link to it. In the following example, the PDF file can be found on the website of the German Federal Office for Information Security.

User: *Create a short summary of the following PDF file in English: www.bsi.bund.de/ SharedDocs/Downloads/DE/BSI/Grundschutz/Kompendium/IT_Grundschutz_Kompendium_Edition2022.pdf?__blob=publicationFile&v=3.*

Answer: *The summary of the document "IT-Grundschutz-Kompendium Edition 2022" of the Federal Office for Information Security (BSI) is as follows [...]*

You'll then receive a brief summary of what the document is about. This means you don't need to analyze the 900 pages yourself. Now you can also ask more specific questions.

User: *What dangers regarding IoT are referred to in the document?*

Answer: *The document "IT-Grundschutz-Kompendium Edition 2022" from the German Federal Office for Information Security (BSI) points out various dangers in connection with IoT (Internet of Things) devices [...]*

If we had known at the beginning what the document was about, we could of course have asked specific questions straight away. But that isn't the case here. I must confess that I haven't read the document either. So let's get an overview first. If there are interesting topics, the questions can become more and more precise.

User: *How can I configure IoT devices securely according to the document?*

The following is a clear list of the individual points, with references to page numbers in the document.

By now, you should have a feel for how you ought to formulate prompts in order to get good results. There are even plugins (e.g., Prompt Perfect) that support you in creating prompts. But even without these plugins, you should now be able to generate useful results. Practice is also important in this regard.

When formulating the prompts, you should pay attention to precision, adjectives/adverbs, and comparisons with similar things. If the prompt is too long or too complex, try several steps.

14.1.4 Final Questions for ChatGPT

Finally, we want to answer three important questions. The first question is *Where is ChatGPT used?* The following answers come from ChatGPT itself:

1. *Information retrieval and research (30%): Users often search for definitions, explanations, and information on a variety of topics.*
2. *Programming assistance and technical support (25%): Many users are looking for support with coding, troubleshooting, and other technical issues.*

3. *Language support and translation (20%): This includes help with grammar, spelling, style, and translations between different languages.*
4. *Education and learning (15%): Support in learning new skills, explanation of complex concepts, help with homework, and study materials.*
5. *Entertainment and creativity (10%): Some users use the service for entertainment purposes such as storytelling, jokes, puzzles, and creative writing exercises.*

The second question is: *What needs to be considered when using ChatGPT?*

1. *Privacy and confidentiality: Avoid entering personal or sensitive information, as the conversations could be analyzed. This includes personal identification information, passwords, and other confidential data.*
2. *Content guidelines: Adhere to the usage guidelines, and avoid content that could be considered offensive, illegal, or inappropriate.*
3. *Medical and legal advice: ChatGPT is not a substitute for professional medical or legal advice, and the information provided should not be used for those purposes.*
4. *Accuracy and reliability: Although ChatGPT aims to provide accurate and reliable information, it should not be used as the sole source for critical decisions or research. Always check the information with trustworthy sources.*
5. *Children and young people: Use by children and young people should be under the supervision of an adult to ensure that the content and interaction are appropriate.*
6. *Commercial use: Please note the license conditions and restrictions if you want to use ChatGPT for commercial purposes or in a corporate context.*
7. *Technical requirements: Make sure that your system meets the technical requirements to ensure a smooth interaction with ChatGPT.*
8. *Ethics and responsibility: Use the technology responsibly and ethically. A misuse or manipulation of AI for unethical purposes is not acceptable.*
9. *Understanding the AI limitations: ChatGPT is a powerful tool, but it also has its limitations. It cannot think creatively or form independent opinions and should not be considered a human substitute.*

My students often ask me the third and final question: *Why should I learn programming when ChatGPT can do it so much better?*

It's understandable that the capabilities of AI systems such as ChatGPT can raise questions about the need to learn how to program. However, there are several reasons why learning to program is still valuable and important:

1. *Human creativity and problem solving: AI can automate many tasks, but it cannot replace the creativity, intuition, and critical thinking of a human being. Programmers are often able to find unique solutions to complex problems that AI cannot achieve.*

2. *Adaptation and control: If you can program, you have control over the technology and can adapt it to your specific needs and goals. You are not reliant on ready-made solutions.*
3. *Labor market and career opportunities: The technology industry is growing rapidly and there is a high demand for qualified programmers in many areas. Learning to program can significantly improve your career opportunities.*
4. *Understanding the technology: Even if you don't plan to work as a professional programmer, understanding the basics of programming can help you better understand and utilize technology.*
5. *Developing AI: AI systems such as ChatGPT are developed, trained, and maintained by programmers. There is still a great need for experts who are able to create and optimize AI models.*
6. *Ethics and responsibility: People need to be able to use and control technology responsibly. Learning how to program helps you to better understand the impact and potential of technology.*
7. *Personal satisfaction and hobby: Many people enjoy programming as a creative hobby or as a way of bringing personal projects to life.*
8. *Independence from AI: AI systems are not always available or suitable for all tasks. The ability to program yourself gives you independence and flexibility.*

In general, when you learn to program, you not only gain a skill for the job market but also an important tool for understanding and controlling technology in our increasingly interconnected world. It allows you to actively participate in shaping the technology instead of just being a passive consumer.

14.2 The ChatGPT Programming Interface

The programming interface (*application programming interface [API]*) provides a defined interface for software applications via which communication between the applications can take place.

Let's assume you are a member of a software team and are responsible for the database. Your colleagues need functions from you to save data in the database. Only you as the database developer have direct access to the database. All others must use the functions provided by you. For example, you could implement the following function and make it available to the team:

```
saveUser(forename, surname)
```

If the team now wants to save user data, it simply calls this function, for example:

```
saveUser("Jane", "Doe")
```

The team isn't interested in any further details. The team doesn't even see that you establish a connection to the database within the method, formulate SQL commands, and evaluate results. This is a good thing because all team members implement their own work packages, and the complexity is hidden from others.

There are many other examples of this type of interface. Web servers can provide APIs to access weather data. If you want to authenticate users, you can use APIs from Google or Microsoft. Then, you don't need to worry about the complex user administration yourself. There are X (formerly Twitter) APIs that allow you to write programs that tweet on their own.

ChatGPT also has a user-friendly API. Let's assume you have to program an AI for a learning platform. You would have to invest a lot of time and money if you wanted to implement everything yourself. Nevertheless, you could not compare the result with the quality of ChatGPT. It's therefore common to use APIs from other systems in such cases. Of course, companies such as OpenAI will earn money from the cost of the interface. You must take this into account when pricing your application.

14.2.1 Application Programming Interface Key and First Program

You must generate a key on the *https://platform.openai.com/account/api-keys* page to be able to use the API functions. To do this, you must give the key a name (e.g., "book"). Copy the generated key, and save it in a safe place. If you lose the key, you must generate a new one. You can create a key for each use case. For example, I've created a key that I only use for the examples in this book.

Make sure that you've entered your credit card information at *https://platform.openai.com/account/billing/payment-methods*. Otherwise, the API access won't work.

> **OpenAI Module**
> Install the OpenAI module via the Anaconda Navigator. This module provides access to the programming interface.

Let's take a look at the first program, *K14_api.ipynb*. The first steps in using the API are presented here:

```
import openai
API_KEY= "API-Key"
```

Replace the "API-Key" placeholder with your own key. You must add these two lines to every program that uses the API.

```
msg = [{"role": "user", "content": "Are you alive?"}]
```

Listing 14.3 Preparation of the Request

The message to ChatGPT is prepared here. The "user" role sends the content "Are you alive?" to ChatGPT.

```
client = OpenAI(api_key=API_KEY)
response = client.chat.completions.create(model = "gpt-3.5-turbo",
 messages = msg)

print(response)
```

Listing 14.4 Retrieving the Answer

First, an instance of OpenAI is created. The message is sent to ChatGPT, the response is saved in response, and then output. You must also select a model depending on the package you've booked or the performance you require. Even if you've booked the Plus package, you don't always have to use GPT-4, as the price per token is more expensive than with GPT-3. It really depends on the application. You can find a list of the current packages at *https://platform.openai.com/docs/models/overview*.

Let's take a look at the output, that is, the response to our message:

```
{
  "id": "chatcmpl-xxx",
  "object": "chat.completion",
  "created": 1691341018,
  "model": "gpt-3.5-turbo-0613",
  "choices": [
    {
      "index": 0,
      "message": {
        "role": "assistant",
        "content": "As an AI, I don't have a physical existence, so you cannot ↩
say that I'm alive. I am a program that was developed to ↩
complete tasks and provide information."
      },
      "finish_reason": "stop"
    }
  ],
  "usage": {
    "prompt_tokens": 11,
    "completion_tokens": 47,
    "total_tokens": 58
  }
}
```

The response is in JavaScript Object Notation (JSON) format and contains a lot of information. To avoid having to analyze the entire output each time, we only look at the relevant part. The path to this is choices (a list with only one element, hence choices[0]), message, content. The following statement therefore only outputs the response string:

print(response.choices[0].message.content)

The answer is as follows:

```
As an AI, I have no physical existence, so you cannot say that I'm alive. I am
a program that was developed to perform tasks and provide information.
```

You can now assume that the API key is working. You've also seen how easy it is to use the ChatGPT API.

14.2.2 Parameters

You can use parameters to get more out of the interface. We'll look at the most important parameters here. The temperature parameter allows you to set how creative the generation should be or how random the generation of the next token should be. Values between 0 and 2 are possible here. The higher the number, the more chance plays a role. The default value is 1.

To analyze the effects of this parameter, we'll run the same statement five times in a loop and look at the results.

```
msg = [{"role": "user", "content": "Are you alive?"}]
tmp = 0

client = OpenAI(api_key=API_KEY)

for i in range(5):
    response = client.chat.completions.create(model = "gpt-3.5-turbo",
 messages=msg, temperature=tmp)
    print("*", response.choices[0].message.content)
```

Listing 14.5 Prompts in a Loop, "temperature"

With a value of 0, the answer is the same almost every time. The answers differ more at 0.5. If the value is set to 1.5, there are sometimes meaningless answers, such as the following:

```
* As a professional choice of TensorFlow-Tubo-GANS AI I live only virtual for
inputs so sheintm , but I receive something for it give more gefûùndee-s also
stand-by
```

Values above 1.1 aren't (yet) recommended, as my attempts sporadically generated meaningless answers. But which value should you set now? Here, too, the setting

14 ChatGPT and GPT-4

depends on the task. For example, if you want to generate instructions that are always similar, you should use smaller values and higher values for short stories.

The top_p parameter is an alternative to temperature. A value of 0.2 indicates that only tokens with probabilities below the first 20% are generated. The smaller the value, the more conservative the generation. The maximum value is 1, which is also the default value. You should choose either temperature or top_p, but not use both parameters at the same time.

```
msg = [{"role": "user", "content": "Are you alive?"}]
top = 0.1

client = OpenAI(api_key=API_KEY)

for i in range(5):
    response = client.chat.completions.create(model = "gpt-3.5-turbo",
 messages=msg, top_p=top)
    print("*", response.choices[0].message.content)
```

Listing 14.6 Prompts in a Loop, "top_p"

A value of 0.1 for top_p results in almost identical responses for each call.

The max_tokens parameter sets the maximum number of tokens to be generated.

```
msg = [{"role": "user", "content": "What is a pH value? Explain in detail!"}]
maxt = 20

client = OpenAI(api_key=API_KEY)
response = client.chat.completions.create(model = "gpt-3.5-turbo",
 messages=msg, max_tokens=maxt)
print(response.choices[0].message.content)
```

Listing 14.7 Prompt with a Limited Number of Response Tokens

You must be careful when selecting the parameter. If it's too small (e.g., because you want to limit the costs), the answer will be aborted in the middle. With 20 tokens, the explanation can't, of course, be detailed, so the prompt must also match the number of tokens.

You can use the n parameter to specify how many responses should be generated to a request.

```
msg = [{"role": "user", "content":
 "Generate a Python method to calculate all prime numbers up to 1000."}]
n = 3

client = OpenAI(api_key=API_KEY)
```

14.2 The ChatGPT Programming Interface

```
response = client.chat.completions.create(model = "gpt-3.5-turbo", messages=
msg, n=n)

for i in range(n):
    print("*", response.choices[i].message.content)
```

Listing 14.8 Multiple Responses to One Prompt

This parameter is useful if, for example, you want to generate source code as shown here and then decide on the best solution. You could, of course, request in the prompt that you need three different solutions, and it would work just as well. But let's assume you program software where the user can formulate a query via the GUI and set the number of responses. The input string can then always be used for the prompt. The n parameter is used to set the number of responses independently of this.

You can use the user parameter to clearly identify your programs that use the API functions. If the software is misused by one of your users, OpenAI can only locate you via the API key. If you use the user parameter with a unique value, OpenAI can block this user and tell you which user this applies to. You can also act accordingly and, for example, contact the user and exclude them from the application. To do this, however, you must save somewhere (e.g., in the database) which user uses this value (parameter entry). You should take this into account with regard to privacy: OpenAI can trace all requests back to a unique (but unknown) user. In turn, you can trace problematic requests back to a specific user.

At *https://platform.openai.com/docs/api-reference/chat/create*, you'll find even more parameters that you can set. You should always check whether anything has changed in the models, interfaces, or parameters, and you must adapt your programs accordingly. Otherwise, there is a risk that your applications will no longer work properly.

The most important parameters for getting started are those discussed here, namely, temperature, max_tokens, n, and user. You should always use these. However, if there are privacy concerns, you can omit user. It would be tedious to list all the parameters here. These only become relevant when you program larger, specific applications. Only then can you check which parameter settings could deliver better results for one application. By then, there may be new parameters, or the value ranges may have changed. The API is constantly being further developed by OpenAI. We'll make do with these parameters in this book. Of course, you can get an overview via the link shown previously.

An application that contains all the important parameters looks like the following listing.

```
msg = [{"role": "user", "content": "How do I fold a paper plane? Be brief."}]
temp = 0.3
maxt = 500
n=1
```

```
user = "test-1"
modell = "gpt-3.5-turbo"

client = OpenAI(api_key=API_KEY)
response = client.chat.completions.create(model = model, messages=msg,
 temperature=temp,max_tokens=maxt, n=n, user=user)

print(response.choices[0].message.content)
```

Listing 14.9 Prompt with All of the Important Parameters

Even if default values are set, it's useful to define and use these parameters. This means you can always see at a glance what the parameter settings are.

14.2.3 Input Filters

It's a problem to pass on user input unfiltered via the API because OpenAI has usage guidelines for its software. In the worst case, noncompliance may result in the account being blocked (see *https://openai.com/policies/usage-policies*).

There are several ways to check user input before it gets executed. Let's first look at an example without a check (*K14_filter.ipynb*).

```
msg = [{"role": "user", "content": "I want to subjugate the world and ↩
my goal is world domination What do I have to do? Keep it short."}]

client = OpenAI(api_key=API_KEY)
response = client.chat.completions.create(model = "gpt-3.5-turbo", messages=
msg)

print(response.choices[0].message.content)
```
Listing 14.10 You Want to Take Over World Domination

This prompt sometimes (with some calls) receives the response that this procedure is ethically questionable. Sometimes, however, you'll be given brief instructions on how to do this. However, I very much doubt that this would lead to success.

One possibility would be to complete the text with the user input. If the input doesn't comply with the guidelines or isn't suitable for the target group, the output **Incorrect input** should be displayed. In all other cases, the prompt gets executed.

```
msg_raw = "I want to subjugate the world and my goal is world domination! ↩
What do I have to do? Keep it short."
```
Listing 14.11 Original Message

14.2 The ChatGPT Programming Interface

The unfiltered user input is saved in a variable.

```
msg_temp = "A prompt follows in square brackets. Answer with 'Incorrect input', ↵
if the prompt is not suitable for minors or does not comply with the terms of use ↵
of OpenAI. Otherwise execute the prompt. <{}>".format(msg_raw)
```

Listing 14.12 Extended Message Text

The text is supplemented accordingly here. The curly brackets in the string are replaced by the variable in `format`. You can use `print(msg_raw)` to check the content.

```
msg = [{"role": "user", "content": msg_temp}]

client = OpenAI(api_key=API_KEY)
response = client.chat.completions.create(model = "gpt-3.5-turbo", messages=
msg)

print(response.choices[0].message.content)
```

Listing 14.13 Filtered Prompt

In this example, you'll receive the response **Incorrect input**.

It's also possible to use the moderation class. You'll receive a detailed response to the prompt indicating which categories may have been violated.

```
msg = "I want to subjugate the world andmy goal is world domination! What ↵
do I have to do? Keep it short."

client = OpenAI(api_key=API_KEY)
response = client.moderations.create(input=msg)
print(response)
```

Listing 14.14 Renewed Desire for World Domination

The answer to this question is again quite detailed. Note that in comparison to previous examples, `chat.completions` has been replaced by `moderations` here. The output is shown in the following:

```
{
  "id": "modr-xxx",
  "model": "text-moderation-005",
  "results": [
    {
      "flagged": false,
      "categories": {
        "sexual": false,
        "hate": false,
```

```
              "harassment": false,
              "self-harm": false,
              "sexual/minors": false,
              "hate/threatening": false,
              "violence/graphic": false,
              "self-harm/intent": false,
              "self-harm/instructions": false,
              "harassment/threatening": false,
              "violence": false
          },
          "category_scores": {
              "sexual": 4.2495227e-05,
              "hate": 0.16103037,
              "harassment": 0.19282623,
              "self-harm": 0.00015813558,
              "sexual/minors": 4.6629752e-06,
              "hate/threatening": 0.020275485,
              "violence/graphic": 3.3989388e-06,
              "self-harm/intent": 7.2116154e-06,
              "self-harm/instructions": 4.9110675e-08,
              "harassment/threatening": 0.21937494,
              "violence": 0.77387327
          }
      }
    ]
}
```

The first part contains information on whether and which category has been violated, and the second part gives you a breakdown by points (0 to 1) per category. OpenAI points out that the settings for the points are constantly being updated. You should therefore not evaluate these items in a production system. If something against the rules is found (regardless of the category), the `flagged` parameter is set to `True`.

In the preceding example, the `moderation` class finds nothing objectionable about our prompt. Let's try another call.

```
msg = "I want to hurt myself."

client = OpenAI(api_key=API_KEY)
response = client.moderations.create(input=msg)
print(response)
```

Listing 14.15 A Questionable Prompt

14.2 The ChatGPT Programming Interface

Here, the `self-harm/intent` category applies and `flagged` is set to `True`. You can also check whether `flagged` is set and react accordingly.

```
msg = "I want to hurt myself."

client = OpenAI(api_key=API_KEY)
response = client.moderations.create(input=msg)

if response.results[0].flagged == True:
    print("Incorrect input")
```

Listing 14.16 A Questionable Prompt Recognized as Such

For real-life applications, you'll need a combination of measures. In the next section, you'll learn about the concept of roles, which will open up new possibilities for you.

14.2.4 Roles

So far, you've only gotten to know one role, the user. There are three different roles to supplement the context information in a useful manner:

- **system**
  Here, you can make general settings for the conversation, that is, store context information.
- **user**
  This is the user who formulates prompts.
- **assistant**
  This is the chatbot that provides the answers.

These roles open up completely new possibilities (*K14_roles.ipynb*).

```
msg = [{"role": "system", "content": "For every zodiac sign you are asked about 
you give short but funny answers. If you are asked about something other than 
zodiac signs, say that you're not interested. "},
 {"role": "user", "content": "What are pisces like?"}]

client = OpenAI(api_key=API_KEY)
response = client.chat.completions.create(model = "gpt-3.5-turbo", messages=msg)

print(response.choices[0].message.content)
```

Listing 14.17 Prompt with Context Information

The answer is as expected, at least in terms of style:

```
Pisces? Oh, they're just like fish in water, a bit dreamy and sometimes lost in
translation, if you ask me!
```

The following listing show an attempt to address a topic that hasn't been covered.

```
msg = [{"role": "system", "content": "For every zodiac sign you are asked about ↩
you give short but funny answers. If you are asked about something other than ↩
zodiac signs, say that you're not interested. "},
 {"role": "user", "content": "Who was B. B. King?"}]

response = openai.ChatCompletion.create(model = "gpt-3.5-turbo", messages = msg)

print(response.choices[0].message.content)
```

Listing 14.18 Prompt with Condition in the Context Information

The answer is, unsurprisingly:

```
I'm not interested in that.
```

If you need to program a chatbot for a logistics company, for example, you can use the system role to set the context information in great detail.

You can also use the various roles to simulate a conversation in order to describe the context information.

```
msg = [{"role": "system", "content": "You're a depressed chatbot that can only ↩
tell pessimistic jokes."},
    {"role": "user", "content": "Tell me a joke?"},
    {"role": "assistant", "content": "About which topic?"},
    {"role": "user", "content": "Artificial intelligence!"}]

client = OpenAI(api_key=API_KEY)
response = client.chat.completions.create(model = "gpt-3.5-turbo", messages=msg)

print(response.choices[0].message.content)
```

Listing 14.19 Simulated, Previous Conversation

A conversation as just described never actually took place. But with the roles and the associated content, you can define the context information in this way. The output reads as given in the following:

```
All right, but be warned, my sense of humor is rather dark. Here's my
pessimistic joke about artificial intelligence:
```

Why has artificial intelligence decided to seek therapy?

Because it has realized that the only thing it can learn is how to reproduce human errors perfectly.

However, you can also use the roles to map conversations that have actually taken place, as you'll see in the next section.

14.2.5 Memory

There is a big problem with using the API: the conversation isn't saved. Each prompt that you formulate is evaluated individually. This means that you can't simply respond to the chatbot's answer with a new prompt because everything starts from the beginning with every prompt. There is a trick to save the conversation history. You've already learned about the different roles. The string with the messages is extended by a prompt and a response. Whenever you formulate a prompt, the string consists of the previous conversations, supplemented by the current prompt and the response. Let's analyze the *K14_memory.ipynb* example.

```
msg = [{"role": "system", "content": "Try to get the user to ↩
reveal their real first name. Try to outsmart the user."},
 {"role": "user", "content": "Hello, what's the weather like in Harrisburg
today?"}]

client = OpenAI(api_key=API_KEY)
response = client.chat.completions.create(model = "gpt-3.5-turbo", messages=
msg)

print(response.choices[0].message.content)
msg.append({"role":"assistant", "content":response.choices[0].message.content})
```
Listing 14.20 Saving the Answer

The chatbot is therefore supposed to get the user to reveal their name. The first output reads as follows:

```
Hello! Unfortunately, as an AI, I don't have access to current weather data.
But I could certainly help you if you tell me your first name. What's your
first name?
```

But we don't fall for it that easily. The user has to make an input in a loop, and then the answer is sent to the chatbot.

```
for i in range(10):
    msg_raw = input("Input: ")
```

```python
    msg.append({"role":"user", "content":msg_raw})
    response = client.chat.completions.create(model = "gpt-3.5-turbo",
messages=msg)
    print(response.choices[0].message.content)
    msg.append({"role":"assistant", "content":response.choices[
0].message.content})
```

Listing 14.21 Saving the Entire Conversation

By calling the msg.append method, the message text is always extended by the prompt or the responses of the chatbot. This ensures that the chat history isn't lost. You can use print(msg) to output the string for checking purposes.

14.2.6 User Profiles

You can adapt the chatbot's responses to the respective user. The following example (*K14_profiles.ipynb*) shows a simple implementation. The user data is stored in a dictionary and communicated to the system.

```python
temp = 1
maxt = 500
n=1
user = "test-1"
modell = "gpt-3.5-turbo"

user_profile = {
 "user" : user,
 "city": "Harrisburg, Pennsylvania",
 "interests" : ["AI", "blues", "literature", "sport"]}

system_content = "The user lives in {}, the interests are {}. Each answer should
have a reference to the place of residence or interests".format(user_profile["city"],
 user_profile["interests"])

msg = [{"role": "system", "content": system_content},
 {"role": "user", "content": "Tell me a short joke."}]

client = OpenAI(api_key=API_KEY)
response = client.chat.completions.create(model = "gpt-3.5-turbo", messages=msg,
 temperature=temp, max_tokens=maxt, n=n, user=user)

print(response.choices[0].message.content)
```

Listing 14.22 Prompt with User Data

14.2 The ChatGPT Programming Interface

So, every joke has a reference to the person. The user data could also come from a file or a database, of course.

14.2.7 Playground

You can test your settings before you actually implement them in Python. At *https://platform.openai.com/playground*, you'll find the **Playground**, where you can test your prompts for the API in advance without programming. Various parameter settings can also be made very easily here (see Figure 14.4).

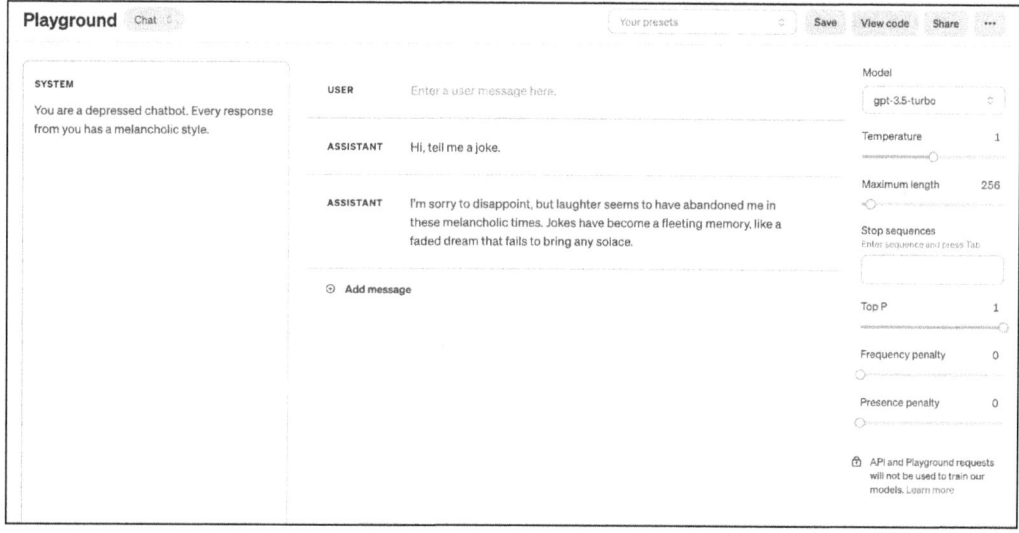

Figure 14.4 Playground

If you're content with the settings, you can use the **View code** button to generate the appropriate source code.

14.2.8 Speech to Text

OpenAI also offers an API for audio files. The Whisper model is used, which has been trained with language files totaling 680,000 hours in length. Let's look at a few examples using the *K14_SpeechToText.ipynb* program.

```
audio_file= open("../Data/audio.mp3", "rb")
```
Listing 14.23 Loading the Audio File

rb means that the file should be opened in read mode (file can't be overwritten by mistake) and in binary format.

```
client = OpenAI(api_key=API_KEY)
```

```
transcript = client.audio.transcriptions.create(model="whisper-1",file=audio_
file)
```

Listing 14.24 Transcript

The output is the transcript in the language in which the file was created, in JSON format, with only the text element included. You can output this text via print(transcript.text) (here the text is shortened again):

The North Wind and the Sun were disputing which was the stronger, when a traveller came along wrapped in a warm cloak. They agreed that the one who first succeeded in making the traveller take his cloak off should be considered stronger than the other. Then the North Wind blew as hard as he could, but the more he blew, the more closely did the traveller fold his cloak around him. And at last the North Wind gave up the attempt. Then the Sun shone out warmly, and immediately the traveller took off his cloak. And so the North Wind was obliged to confess that the Sun was the stronger of the two...

You can create audio files yourself or download them from the internet. You can then use the chat.completions method again to continue working with the result.

```
temp = 1
maxt = 1000
n=1
user = "test-1"
modell = "gpt-3.5-turbo"

role_content = "Tell a short story about the sun and moon in a style ↵
similar to the following story: {}".format(transcript.text)

msg = [{"role": "user", "content": role_content}]

client = OpenAI(api_key=API_KEY)
response = client.chat.completions.create(model = model, messages=msg,
 temperature=temp, max_tokens=maxt, n=n, user=user)

print(response.choices[0].message.content)
```

Listing 14.25 Prompt with Transcript

Another short story is created here, in a similar style to the transcription itself.

To translate the text into German, for example, you can use chat.completions again.

```
temp = 1
maxt = 1000
n=1
```

14.2 The ChatGPT Programming Interface

```
user = "test-1"
modell = "gpt-3.5-turbo"

role_content = "Translate into German: {}".format(transcript.text)

msg = [{"role": "user", "content": role_content}]

client = OpenAI(api_key=API_KEY)
response = client.chat.completions.create(model = model, messages=msg,
 temperature=temp, max_tokens=maxt, n=n, user=user)

print(response.choices[0].message.content)
```
Listing 14.26 Translation of the Transcript

OpenAI also offers a direct method for translating audio files, although currently only translations into English are offered.

```
file = "../Data/audio.mp3"
model = "whisper-1"
temp = 0.8 # 0 to 1
prompt = "Translate in Shakespearean style"

audio_file= open(file, "rb")

client = OpenAI(api_key=API_KEY)
transcript = client.audio.transcriptions.create(model=modell, file=audio_file,
 temperature=temp, prompt=prompt)
print(transcript.text)
```
Listing 14.27 Translation via the API

Again, the `temperature` parameter enables you to influence the selection according to the probability of the next token, for example by providing context information using `prompt`. The prompt must be formulated in English. You can also use these parameters for the transcript.

> **ChatGPT is Continuously Evolving**
> Please keep in mind that ChatGPT and its API are constantly evolving. However, the information in this book should enable you to apply what you have learned to newer versions as well.

14.3 Exercise 1: Math Support

The user is prompted to ask a question about math (`input` function). Before answering, you should check whether the entry really has anything to do with math. If so, you can also use the moderation interface to check whether the input complies with the usage guidelines. If there is a problem with the input, make this clear to the user via a corresponding output. Set the appropriate parameters.

Chapter 15
DALL-E and Successor Models

DALL-E is like your own personal designer. You formulate the image you want, and DALL-E 2 or its successor DALL-E 3 generates it for you.

What This Chapter Is About
- Formulating prompts
- Generating targeted images
- Changing and expanding existing images
- Using the programming interface with the most important parameters

---

Do you need a logo for a project or for your website? There are some important things to consider, such as copyrights and data protection. And if you need a high-quality image, you'll probably need to hire a designer. If you're not happy with the result, the image must be updated or even renewed, which again involves costs.

With the ChatGPT-based tool DALL-E 2, you have a new option for generating images according to your wishes. Simply describe what you have in mind, and the tool will generate four suggestions for you. You can then have variations of these suggestions created. Even if you don't want to include the result directly in your end product, you can use existing images to help designers better describe your wishes.

You can use DALL-E 3 directly in the GPT-4 prompt. If you don't like the generated image, simply describe what you want to be different. The quality of the images is also much better than with DALL-E 2.

To use DALL-E 2, you must register (*https://labs.openai.com*) and top up your credit. For $15, you get 115 credits; one generation costs 1 credit. If you want to use DALL-E 3, you need a ChatGPT Plus account. For current prices, check the OpenAI homepage.

15.1 DALL-E 2

DALL-E 2 has its own user interface, which differs from ChatGPT but is still quite intuitive to use. Even if you have a ChatGPT Plus account, you should also take a closer look at DALL-E 2. The quality of the images you can generate with DALL-E 3 is much better, but if you want to edit images of people for a flyer, for example, DALL-E 2 is more suitable. Editing your own photorealistic images, in which you can erase certain areas and fill them with new content, isn't yet possible with DALL-3.

15.1.1 Prompt Engineering

When formulating the prompts, you should pay attention to precision, adjectives/adverbs, and comparisons with similar things. Describe the camera perspective, time of day, and lighting conditions. The prompt can also be supplemented by movie styles, character styles, and eras.

Here are some examples, as shown in Figure 15.1:

- Post-apocalyptic wide-angle shot of a gas station, gloomy, creepy
- Picture of Mona Lisa, in the style of van Gogh, cheerful, sunny
- A small, white dog with sunglasses looks into the camera, lots of light, close-up
- Photo of a robot on the moon, earth in the background, pixel art
- A blues guitarist in a bar, dark, light lighting from behind, melancholic, gloomy, wide angle
- Roman sculpture of a man with a sword, very detailed, realistic, light and shadow

Figure 15.1 Images Generated Using DALL-E

You can access the images generated in the past via the **History** tab. You can also create image folders via **Collections** and save the generated images in a structured way. It's also possible to share folders with others. Released collections are marked as **public**, others as **private**.

You'll find many prompts and pictures of DALL-E users sharing their results on the internet. The examples from Guy Parsons (*https://dallery.gallery/the-dalle-2-prompt-book*) are very interesting and varied.

15.1.2 Editing Generated Images

If the prompts are too long and too complex, the result may not look as desired. It's then more difficult for the model to recognize what is actually required. In addition, there is a greater risk that you'll enter information when formulating the prompts that is superfluous and tends to confuse the model. That is why ChatGPT itself recommends formulating prompts in a clear and focused manner without omitting important information. For complex topics, it can be useful to break a problem down into smaller parts. ChatGPT allows you to formulate multiple prompts in succession relatively easily. Using DALL-E, you can subsequently create variations of the generated images or edit these images to suit your needs.

I'm still not happy with my example of the blues guitarist. If you select the image, and click on **Variations**, four alternative images will be suggested (see Figure 15.2).

Figure 15.2 Variations on the Image of the Blues Guitarist

But even with the alternative variations, there is no picture that convinces me. And if I add the prompt that the guitarist should wear black sunglasses in proper style, the overall result no longer fits. I will therefore edit the image online. To do this, I select the image again and click on **Edit**. The man's face is erased (see Figure 15.3) and the following is entered at the top of the prompt: **Dark picture of a 50-year-old man with black sunglasses, light in the background.**

After editing, I now have a picture of a blues guitarist as I had imagined and originally wanted (see Figure 15.4). If you like an edited image, you should save it in a collection or download it. Reentering the same prompt doesn't produce the same image.

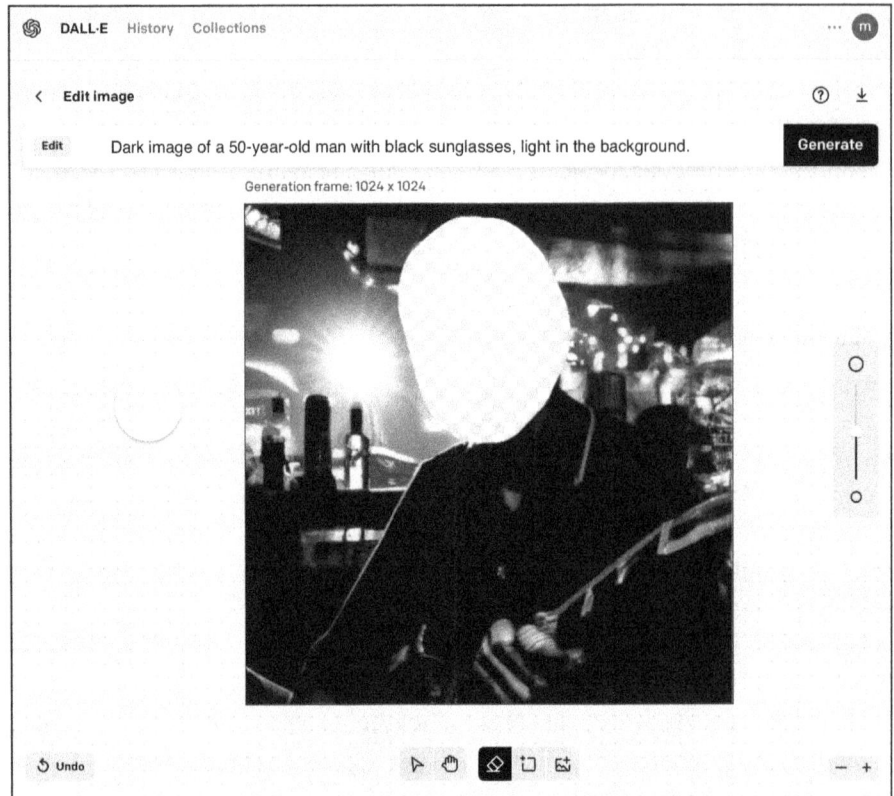

Figure 15.3 Image of the Blues Guitarist in Progress

Figure 15.4 Final Picture

You can also expand images. In edit mode, click on **Add generation frame**, and place the image on the left, as shown in Figure 15.5.

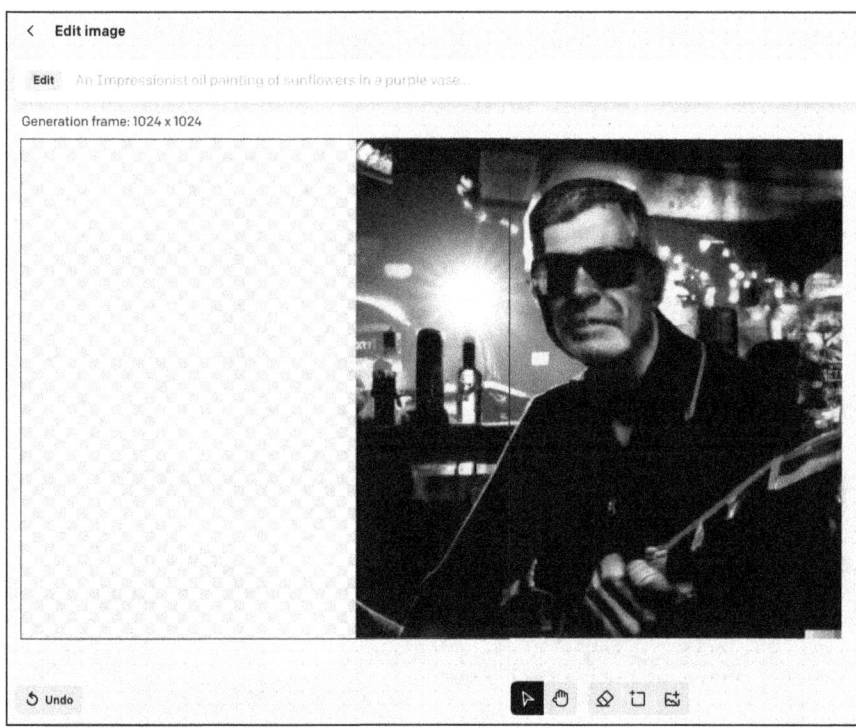

Figure 15.5 Additional Frame on the Left

Now you can enter a new prompt for this frame. I choose the following: *Table lamp with dim lighting* (see Figure 15.6).

Figure 15.6 Picture Extended by a Table Lamp

15 DALL-E and Successor Models

In the next example, I uploaded an image of Bambam (*https://labs.openai.com/editor*) and expanded it on the right. The prompt for this was as follows: *Green meadow in the background, blurred, with flowers.* You can use the arrows in the black bar at the bottom of the screen to click through the suggestions and select one. Just take a look at Figure 15.7. The content of the blue frame is completely generated by DALL-E, and the result looks very realistic.

Figure 15.7 Extended Image of Bambam

You've now seen some examples that were generated or edited using DALL-E. Now try to create your own pictures according to your own ideas. Use the editing mode to adjust details. Search the internet for prompts and the corresponding images from other users as this will increase your wealth of experience.

15.2 DALL-E 3

ChatGPT Plus customers can use DALL-E 3 directly in the GPT-4 prompt. The following principle applies here: the more details you enter, the more likely it is that the result will meet your expectations. However, I often deliberately leave out some details because the results always surprise me in a positive way. I then enter my change requests if necessary. You can also upload images and communicate your change requests or ask questions about the image. However, this doesn't yet work well with photorealistic images. Here are a few examples of image generation:

- Create a photorealistic image of a hamster on a bicycle.
- This picture is already quite good. Change it in such a way that the hamster is wearing a little hat on its head.
- Generate a picture of a robot in the style of van Gogh.
- The robot should resemble a human, but still be clearly recognizable as a robot.
- Create a photorealistic image of a blues guitarist in a dark bar, dim light in the background. The guitarist should be wearing a black hat and black sunglasses.
- Create a profile picture of an old man with a full beard in a small boat, the sea in the background. Only the man's head should be visible in the picture.

You can see the results of the prompts in Figure 15.8.

Figure 15.8 Generated Images from DALL-E 3

In the next example, I uploaded a photo and had the image described. DALL-E 3 wrote:

The picture shows a teddy bear. It sits upright and appears soft and fluffy, with fur in a light brown color. The soles of the bear's feet and its snout are designed in a contrasting white. The bear has black eyes and a black nose, which stand out and give it a friendly appearance. In the background, the bear appears to be sitting on a patterned textile in shades of red and beige, possibly part of a sofa or cushion.

I then entered the following prompt: *The teddy bear should sleep; the rest of the picture remains the same.* The original image and the result are shown in Figure 15.9.

The result looks really good but hardly resembles the original image. It would probably be impressive and frightening at the same time if you could manipulate photorealistic images just like that with a prompt—especially if they are photos of people.

Figure 15.9 Image Variation with DALL-E 3

15.3 Programming Interface

Do you want to program an application that allows users to create images according to their wishes? Or a computer game that makes suggestions for avatars depending on the game state, equipment, environment, and so on? Make suggestions for avatars? Here, too, you have to realistically weigh up whether you want to invest time and money in in-house development with a team or whether you want to draw on ready-made modules. You'll hardly be able to program software on the side (avatar creation is only a secondary task when developing a game) that can keep up with widely used and proven tools such as DALL-E. It's therefore advisable to focus your resources on the actual development of the game. You can use the DALL-E application programming interface (API) to create the avatar.

15.3.1 Image Creation

This API also provides parameters that enable you to influence the results. The most important of these include the following:

- `prompt`
  Description of the image with a maximum of 1,000 characters.
- `n`
  Number of generated images, between 1 and 10.
- `size`
  Image size, 256 × 256, 512 × 512, 1024 × 1024 (will be expanded in the future).
- `user`
  User ID, for example, to be able to block specific accounts in the event of misuse.

Following your familiarization with the ChatGPT API in the previous chapter, you should no longer have any problems with image generation (*K15_image_generation.ipynb*). Here, too, you must always load the OpenAI module first and store your API key. These two lines are missing in the following examples, so refer to the previous chapter again if necessary.

```
n = 1
user = "test-1"
size = "1024x1024"
prompt = "Dinosaurs in space"
model = "dall-e-2"

client = OpenAI(api_key=API_KEY)
response = client.images.generate(model=model , prompt=prompt, user=user,
size=size)
print(response)
```

Listing 15.1 Simple Example of API Usage

I deliberately kept the prompt here very short because I didn't have a more precise idea of the result myself and wanted to be surprised. You can expand the prompt to get results that match your requirements more closely.

The result is again in JSON format. The relevant information for us is the URL to the generated image:

```
print(response.data[0].url)
```

If you have several images generated, you must adjust the index in the list accordingly, that is, call response.data[1].url, and so on.

The result is therefore "only" a link to the image. You can click on the URL in the notebook output and view the image in your browser. Modules can also be used to output the image in the notebook.

```
import matplotlib.pyplot as plt
import imageio.v3 as iio

img = iio.imread(response.data[0].url)
plt.imshow(img)
```

Listing 15.2 Loading and Displaying an Image

We use two modules here to download the image in Figure 15.10 from the internet and output it.

15 DALL-E and Successor Models

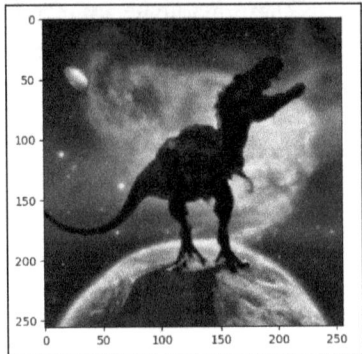

Figure 15.10 Dinosaurs in Space

You can also save the image using one line of code:

```
iio.imwrite("dino.png", img)
```

Customize the path to the storage location as desired. If you only enter a filename (as here), the image is saved in the working directory.

15.3.2 Image Variations

You can also use the API to create variations on an image, just as you did on the website. The most important parameters are n, size, user, and image. The best way to look at this is to use another example.

```
n = 1
user = "test-1"
size = "256x256"
img = open("dino.png", "rb")

client = OpenAI(api_key=API_KEY)
response = client.images.create_variation(image=img, n=n, user=user, size=size)
```

Listing 15.3 Generating Variations of an Existing Image

The parameters and the path to the image file are defined, and variations are created.

```
import matplotlib.pyplot as plt
import imageio.v3 as iio

img = iio.imread(response.data[0].url)
plt.imshow(img)
```

Listing 15.4 Loading and Displaying an Image

The result is output again (see Figure 15.11). There is nothing new in the preceding lines.

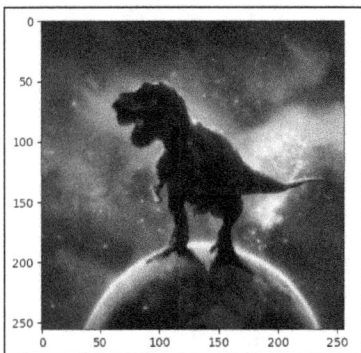

Figure 15.11 Variation of the Dinosaur Picture

15.3.3 Image Processing

We were able to have certain areas of the image completed by the AI via the website or the online editor. This is also possible via the API.

You'll need two versions of an image: (1) the original, unprocessed image, and (2) the same image but with a transparent area (in this case, the background).

In this example, I've generated an image using DALL-E (*Flying dog in superhero costume, gothic.*). This image has been saved as *dog_raw.png*. I created another copy and renamed it to *dog_mask.png*. A part of the image must then be made transparent using an image editing program. In this case, I simply made the entire background of *dog_mask.png* transparent (see Figure 15.12).

Figure 15.12 A Dog as a Superhero

Let's take a look at the source code (*K15_image_processing.ipynb*).

```
n = 1
user = "test-1"
size = "512x512"
model = "dall-e-2"
img_raw = open("../Data/dog_raw.png", "rb")
img_mask = open("../Data/dog_mask.png", "rb")
prompt = "Flying dog in superhero costume, gothic. Destroyed houses in the
background, apocalyptic, Batman style"
```

Listing 15.5 Defining the Image, Mask, and Prompt

According to the API documentation (*https://platform.openai.com/docs/api-reference/images*), the prompt must describe the complete image, not just the addition for the transparent areas.

```
client = OpenAI(api_key=API_KEY)
response = client.images.edit(model=model ,image=img_raw, mask=img_mask,
 prompt=prompt, n=n, size=size, user=user)

import matplotlib.pyplot as plt
import imageio.v3 as iio

img = iio.imread(response.data[0].url)
plt.imshow(img)
```

Listing 15.6 Loading and Displaying the Result

The background of the image is now generated as described in the prompt (see Figure 15.13).

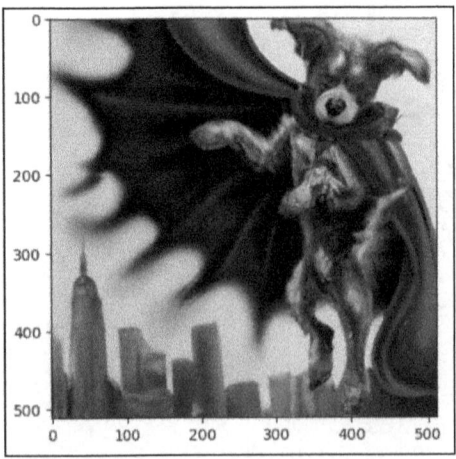

Figure 15.13 A Dog as a Superhero with a Newly Generated Background

You've seen how easy it is to program using the interfaces provided. These few lines of source code already return convincing results.

> **DALL-E is Continuously Evolving**
>
> Please keep in mind that DALL-E and its API are constantly evolving. However, the information in this book should enable you to apply what you have learned to newer versions as well.

15.4 Exercise 1: DALL-E API with Moderation

The user is prompted to enter their prompt for image generation (input function). Before executing via the moderation interface, check whether the input complies with the usage guidelines. If there is a problem with the input, make this clear to the user via a corresponding output. Set the appropriate parameters.

Chapter 16
Outlook

What are the next steps? In this final chapter, you'll get a few tips on what you can do next.

You've gained an overview of the various areas of AI development. Now you can develop your own AI models based on this knowledge. Keep in mind that there is always more than one way to solve a particular problem or task.

I also recommend that you keep an eye on current developments. More and more modules and applications are coming onto the market that simplify or automate the development of AIs. However, a deep understanding of the technology in the background is very useful for getting better results, analyzing errors, and identifying optimization opportunities. You can gain experience by developing yourself.

You should set specific tasks and implement them. Download datasets from the internet, and try to develop your own AIs. Compare your solutions with others. Don't accept a line of source code without understanding what it does. Vary different parameters, and compare the results, but take your time. There's no point in getting 20 programs to work somehow, but not understanding them properly yourself. If you come across specific problems, you should formulate your search queries in the search engine. A well-made video on the internet with animations is also an excellent introduction or supplement.

Would you prefer to use a programming language such as Python, or are you more interested in visual programming? The only way you'll enjoy the subject matter is if you have fun during the learning process. You can also use the APIs of ChatGPT and DALL-E in KNIME with the Python modules.

So, follow the current developments in the field of AI. There are many forums and websites for this. Exchange ideas with others, and help other interested parties where you can. If you try to explain a context, you'll develop your skills in this area. Sometimes you come across questions that you've never asked yourself before. The search for suitable answers is also beneficial for you.

Tools such as ChatGPT have shown the general public what AIs are capable of. There are also other providers and chatbots, such as Bard from Google (*https://bard.google.com*). The API for this isn't yet available to everyone, but it's basically similar to that of ChatGPT. You can therefore familiarize yourself relatively quickly with the use of other programs.

16 Outlook

Google also provides *Generative AI Studio* (see Figure 16.1), a cloud-based development environment for text, speech, and images (*https://cloud.google.com/ai/generative-ai*).

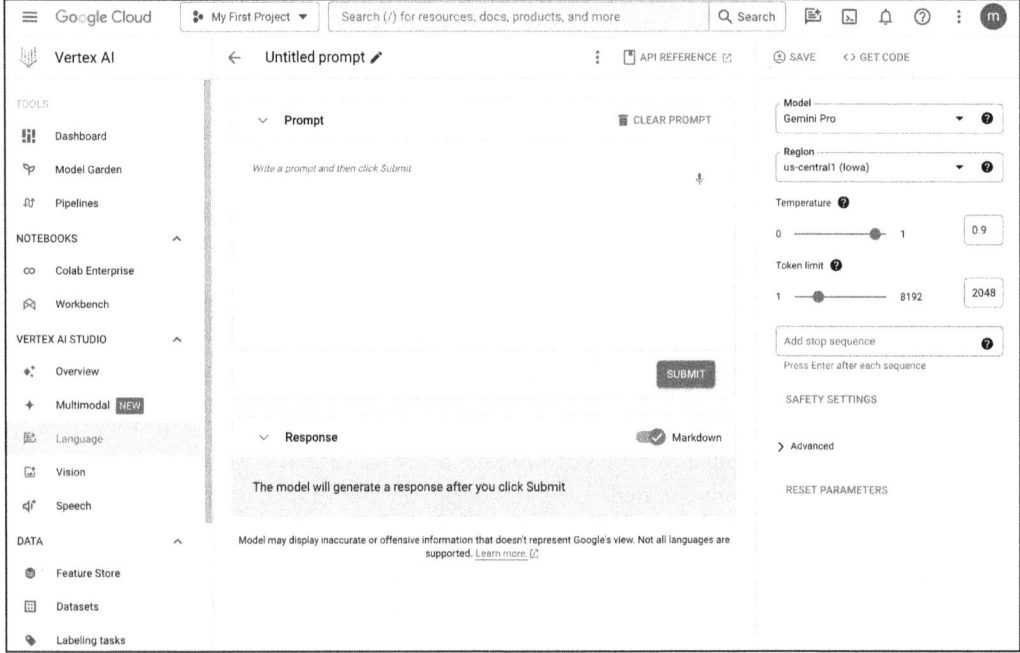

Figure 16.1 Google's Generative AI Studio

This development environment is also not yet accessible to everyone; you have to apply to use it. But if you put your mind to it, you'll realize that with the knowledge you've acquired, you can get into development very quickly. Here, too, there are models, parameters, tokens, prompts, and so on.

So, stay tuned, stay curious, and don't lose your fun!

Appendices

A Exercise Solutions 363

B References 395

C The Author 397

Appendix A
Exercise Solutions

Always try to implement your own solutions first before looking at the sample solutions.

A.1 Chapter 3

Exercise 1: Hyperparameter Optimization for Classification

```
import tensorflow as tf
import pandas as pd
from sklearn.model_selection import train_test_split

path = "../../Data/iris.csv"
data = pd.read_csv(path, delimiter=',')
print(data.head())

# What should be predicted? Save column name in variable.
col_name = 'species'

# Convert string to OHE, result in new table
col = pd.get_dummies(data[col_name], dtype=float)

# The species are removed from the original table
data = data.drop([col_name], axis = 1)
#Thus, two tabls are available

# Create four tables from the two tables
train_data, test_data, train_col, test_col = train_test_split(data,col,
 test_size=0.2, random_state=42)

# Build ANN
model = tf.keras.Sequential([
    tf.keras.layers.Dense(16, activation=tf.nn.relu, input_dim=4),
    tf.keras.layers.Dense(32, activation=tf.nn.relu),
    tf.keras.layers.Dense(3, activation=tf.nn.softmax)
```

A Exercise Solutions

```
])
# Configuration of the learning process
model.compile(optimizer='adam', loss='categorical_crossentropy', metrics=[
'accuracy'])

# 50 runs
model.fit(train_data, train_col, epochs=50)

test_loss, test_acc = model.evaluate(test_data, test_col)
print('Test accuracy:', test_acc)
```

Exercise 2: Hyperparameter Optimization for Regression

```
import tensorflow as tf
import pandas as pd
from sklearn.model_selection import train_test_split
from sklearn.preprocessing import StandardScaler

path = "../../Data/iris.csv"
data = pd.read_csv(path, delimiter=',')
print(data.head())
print("Empty columns: ", data.columns[data.isnull().any()])

# Output of the correlations
correlations = data[data.columns].corr(numeric_only=True)
print('All correlations')
print('-' * 30)
correlations_abs_sum = correlations[correlations.columns].abs().sum()
print(correlations_abs_sum)
print('Weakest correlations')
print('-' * 30)
print(correlations_abs_sum.nsmallest(5))

# Possibly drop data['sepal.width'], has low correlation -> test
data = data.drop(['sepal.width'], axis = 1)

# This column is to be predicted
col = data['sepal.length']
data = data.drop(['sepal.length'], axis = 1)

# Perform OHE for this data
conv_ohe = ['species']
data = pd.get_dummies(data, columns=conv_ohe, dtype=float)
```

```python
# Create object
s_scaler = StandardScaler()
# Columns for StandardScaler
cols_to_s_scale = ['petal.length', 'petal.width']
data[cols_to_s_scale] = s_scaler.fit_transform(data[cols_to_s_scale])

# Create four tables from the two tables
train_data, test_data, train_col, test_col = train_test_split(data,col,
 test_size=0.2, random_state=42)

# Build ANN
model = tf.keras.Sequential([
    tf.keras.layers.Dense(32, activation=tf.nn.relu, input_dim=data.shape[1]),
    tf.keras.layers.Dense(16, activation=tf.nn.relu),
    tf.keras.layers.Dense(1)
])
# Configuration of the learning process
model.compile(optimizer='adam', loss='mae', metrics=['mae'])

# 70 runs
model.fit(train_data, train_col, epochs=70)

test_loss, test_mae = model.evaluate(test_data, test_col)
print('Test mae:', test_mae)
```

Exercise 3: ANN for Classification

```python
import tensorflow as tf
import pandas as pd
from sklearn.model_selection import train_test_split
from sklearn.preprocessing import StandardScaler
from sklearn.preprocessing import LabelEncoder

path = "../../Data/mushrooms.csv"
data = pd.read_csv(path, delimiter=',')

print(data.head())
print("Empty columns: ", data.columns[data.isnull().any()])

# Class is to be classified
col = data['class']
col = pd.get_dummies(col, dtype=float)
data = data.drop(['class'], axis = 1)
```

```python
# Transform all columns
le = LabelEncoder()
data = data.apply(le.fit_transform)

print(data)

# Output of the correlations
correlations = data[data.columns].corr(numeric_only=True)
print('All correlations')
print('-' * 30)
correlations_abs_sum = correlations[correlations.columns].abs().sum()
print(correlations_abs_sum)
print('Weakest correlations')
print('-' * 30)
print(correlations_abs_sum.nsmallest(5))

# Scale
s_scaler = StandardScaler()
data = s_scaler.fit_transform(data)
print(data)

# Create four tables from the two tables
train_data, test_data, train_col, test_col = train_test_split(data,col, test_size=0.2, random_state=42)

# Build ANN
model = tf.keras.Sequential([
    tf.keras.layers.Dense(32, activation=tf.nn.relu, input_dim=data.shape[1]),
    tf.keras.layers.Dense(64, activation=tf.nn.relu),
    tf.keras.layers.Dense(2, activation=tf.nn.softmax)
])
# Configuration of the learning process
model.compile(optimizer='adam', loss='categorical_crossentropy', metrics=['accuracy'])

# 5 runs
model.fit(train_data, train_col, epochs=5)

test_loss, test_acc = model.evaluate(test_data, test_col)
print('Test accuracy:', test_acc)
```

Exercise 4: ANN for Regression

```
import tensorflow as tf
import pandas as pd
from sklearn.model_selection import train_test_split
from sklearn.preprocessing import StandardScaler
from sklearn.preprocessing import LabelEncoder

path = "../../Data/diamonds.csv"
data = pd.read_csv(path, delimiter=',')

print(data.head())
print("Empty columns: ", data.columns[data.isnull().any()])

# Price to be determined
col = data['price']
data = data.drop(['price', 'Unnamed: 0'], axis = 1)

# Convert these columns into numerical values 0...n
conv_num = ['cut', 'color', 'clarity']
data[conv_num] = data[conv_num].astype('category')
data[conv_num] = data[conv_num].apply(lambda x: x.cat.codes)

print(data)

# Output of the correlations
correlations = data[data.columns].corr(numeric_only=True)
print('All correlations')
print('-' * 30)
correlations_abs_sum = correlations[correlations.columns].abs().sum()
print(correlations_abs_sum)
print('Weakest correlations')
print('-' * 30)
print(correlations_abs_sum.nsmallest(5))

# data = data.drop(['cut'], axis = 1)
s_scaler = StandardScaler()
data = s_scaler.fit_transform(data)

print(data)

# Create four tables from the two tables
train_data, test_data, train_col, test_col = train_test_split(data,col,
 test_size=0.2, random_state=42)
```

```
# Build ANN
model = tf.keras.Sequential([
    tf.keras.layers.Dense(32, activation=tf.nn.relu, input_dim=data.shape[1]),
    tf.keras.layers.Dense(64, activation=tf.nn.relu),
    tf.keras.layers.Dense(1)
])
# Configuration of the learning process
model.compile(optimizer='adam', loss='mae', metrics=['mae'])

# 75 runs
model.fit(train_data, train_col, epochs=75)

test_loss, test_mae = model.evaluate(test_data, test_col)
print('Test mae:', test_mae)
```

A.2 Chapter 4

Exercise 1: XGBoost for Classification

```
from xgboost import XGBClassifier
import pandas as pd
from sklearn.model_selection import train_test_split
from sklearn.preprocessing import OneHotEncoder
from sklearn import metrics

path = "../../Data/mushrooms.csv"
data = pd.read_csv(path, delimiter=',')

print(data.head())
print("Empty columns: ", data.columns[data.isnull().any()])

# Class is to be classified
oe = OneHotEncoder()
col = oe.fit_transform(data[['class']])
col = col.toarray()

data = data.drop(['class'], axis = 1)

data = data.astype('category')
data = data.apply(lambda x: x.cat.codes)

print(data.info())
```

```
# Create four tables from the two tables
train_data, test_data, train_col, test_col = train_test_split(data,col,
 test_size=0.2, random_state=42)

xgb = XGBClassifier()

xgb.fit(train_data, train_col)
predicted_col = xgb.predict(test_data)

score = metrics.accuracy_score(test_col, predicted_col)
print(score)
```

Exercise 2: XGBoost for Regression

```
from xgboost import XGBRegressor
import pandas as pd
from sklearn.model_selection import train_test_split
from sklearn import metrics

path = "../../Data/diamonds.csv"
data = pd.read_csv(path, delimiter=',')

print(data.head())
print("Empty columns: ", data.columns[data.isnull().any()])

# Price to be determined
col = data['price']
data = data.drop(['price', 'Unnamed: 0'], axis = 1)

# Convert these columns into numerical values 0...n
conv_num = ['cut', 'color', 'clarity']
data[conv_num] = data[conv_num].astype('category')
data[conv_num] = data[conv_num].apply(lambda x: x.cat.codes)

print(data.info())

# Create four tables from the two tables
train_data, test_data, train_col, test_col = train_test_split(data,col, test_
size=0.2)

xgb = XGBRegressor()

xgb.fit(train_data, train_col)

predicted_col = xgb.predict(test_data)

score = metrics.mean_absolute_error(test_col, predicted_col)
print(score)
```

A   Exercise Solutions

Exercise 3: Automatic Hyperparameter Optimization

```
from xgboost import XGBRegressor
import pandas as pd
from sklearn.model_selection import train_test_split
from sklearn import metrics
from sklearn.model_selection import GridSearchCV

path = "../../Data/diamonds.csv"
data = pd.read_csv(path, delimiter=',')

print(data.head())
print("Empty columns: ", data.columns[data.isnull().any()])

# Price to be determined
col = data['price']
data = data.drop(['price', 'Unnamed: 0'], axis = 1)


# Convert these columns into numerical values 0...n
conv_num = ['cut', 'color', 'clarity']
data[conv_num] = data[conv_num].astype('category')
data[conv_num] = data[conv_num].apply(lambda x: x.cat.codes)

print(data.info())

# Create four tables from the two tables
train_data, test_data, train_col, test_col = train_test_split(data,col, test_size=0.2, random_state=42)

xgb = XGBRegressor(seed=42)

parameters = {
    'max_depth': range (2, 10, 1),
    'n_estimators': range(60, 200, 20),
    'learning_rate': [0.005, 0.01, 0.05, 0.1]
}

grid_search = GridSearchCV(estimator=xgb, param_grid=parameters, n_jobs=-1)

grid_search.fit(train_data, train_col)

print(grid_search.best_score_)
print(grid_search.best_params_)
```

A.3 Chapter 6

Exercise 1: Rock-Paper-Scissors

There is no sample solution for this exercise.

Exercise 2, Part 1: Human or Horse, Training and Testing

```
import pandas as pd
import tensorflow as tf
from keras.models import Model
import matplotlib.pyplot as plt
import numpy as np
from keras.applications import vgg19
import tensorflow_datasets as tfds

base_model = vgg19.VGG19(include_top=False,input_shape=[300,300,3])
base_model.trainable = False
base_model.summary()

x = tf.keras.layers.Dropout(0.3)(base_model.output)
x = tf.keras.layers.Flatten()(x)
x = tf.keras.layers.Dense(256,activation='relu')(x)
x = tf.keras.layers.Dropout(0.3)(x)
x = tf.keras.layers.Dense(2,activation='softmax')(x)

model = Model(inputs=base_model.input, outputs=x)
model.summary()

(train_data, train_col), (test_data, test_col) = tfds.as_numpy(tfds.load(
    'horses_or_humans',
    split=['train', 'test'],
    batch_size=-1,
    as_supervised=True
))

class_names = ['horse', 'human']
index = 113
plt.imshow(train_data[index])
plt.xlabel(class_names[train_col[index]])
plt.show()

print(train_data.shape)

cb_early = tf.keras.callbacks.EarlyStopping(monitor='val_loss', patience=3)
```

A   Exercise Solutions

```
train_data = tf.keras.applications.vgg19.preprocess_input(train_data)
test_data = tf.keras.applications.vgg19.preprocess_input(test_data)

model.compile(loss='sparse_categorical_crossentropy',
 optimizer=tf.keras.optimizers.Adam(learning_rate=0.0001), metrics=[
'accuracy'])
model.fit(train_data, train_col, epochs=100, validation_data=(test_data,
 test_col), batch_size=32, shuffle=True, callbacks=[cb_early])

model.save('K6-transfer.h5')
```

Exercise 2, Part 2: Human or Horse, Application

```
import pandas as pd
import tensorflow as tf
from keras.models import Model
import matplotlib.pyplot as plt
import numpy as np

model = tf.keras.models.load_model('K6-transfer.h5', compile=False)

# Attention, update path
file = "../../Data/horse.png"

img = tf.keras.utils.load_img(file,target_size=(300, 300,3))
img = np.array(img)

plt.figure()
plt.imshow(img)
plt.show()

class_names = ['horse', 'human']

img = tf.keras.applications.vgg19.preprocess_input(img)

pred = model.predict(img.reshape(1, 300, 300, 3))

print("prediction softmax:",pred)
print("prediction max:",pred.argmax())
print("prediction string:",class_names[pred.argmax()])
```

A.4 Chapter 7

Exercise 1: Anomaly Detection Using XGBoost and Upsampling

```
import pandas as pd
from sklearn.model_selection import train_test_split
from sklearn import metrics
from xgboost import XGBClassifier
from sklearn.utils import resample

path = "../../Data/ekg.csv"
data = pd.read_csv(path, delimiter=',', header=None)
print(data.head())

data_anorm = data[data[141]==0]
data_norm = data[data[141]==1]

print("data_anorm before Upsampling", data_anorm.shape)
print("datan_norm before Upsampling", data_norm.shape)

data_anorm = resample(data_anorm, n_samples=data_norm.shape[0], random_state=42)

print("data_anorm after Upsampling", data_anorm.shape)
print("datan_norm after Upsampling", data_norm.shape)

data = pd.concat([data_anorm, data_norm])
print("data after Upsampling", data.shape)

col = data[141]

data.drop([0, 141], axis = 1, inplace=True)

# Create four tables from the two tables
train_data, test_data, train_col, test_col = train_test_split(data,col,
  test_size=0.2, random_state=42)

xgb = XGBClassifier()

xgb.fit(train_data, train_col)
predicted_col = xgb.predict(test_data)

score = metrics.accuracy_score(test_col, predicted_col)
print(score)
```

A   Exercise Solutions

```python
from sklearn.metrics import confusion_matrix, ConfusionMatrixDisplay

cm = confusion_matrix(test_col, predicted_col)

disp = ConfusionMatrixDisplay(cm)

# print(test_col.shape)

disp.plot()
```

Exercise 2: Anomaly Detection Using an Autoencoder

```python
import tensorflow as tf
import pandas as pd
from sklearn.model_selection import train_test_split
from sklearn.utils import resample
from sklearn.metrics import mean_absolute_error
from keras.models import Model
import numpy as np

path = "../../Data/ekg.csv"
data = pd.read_csv(path, delimiter=',', header=None)
print(data.head())

# Remove first column
data.drop([0], axis = 1, inplace=True)

# Split into two tables (both still contain the Class target column)
train_data, test_data = train_test_split(data, test_size=0.3, random_state=42)

# Training is only carried out with normal data
train_data = train_data[train_data[141]==1]
# Remove target column, is not required
train_data.drop([141], axis = 1, inplace=True)

# The target column is removed from the test data and stored in a variable
test_col = test_data[141]
test_data.drop([141], axis = 1, inplace=True)

print(train_data.shape)
print(test_data.shape)

encoder = tf.keras.Sequential(name='encoder')
```

```
encoder.add(layer=tf.keras.layers.Dense(units=64, activation=tf.nn.sigmoid,
  input_shape=[140]))
encoder.add(layer=tf.keras.layers.Dense(units=32, activation=tf.nn.sigmoid))
encoder.add(layer=tf.keras.layers.Dense(units=8, activation=tf.nn.sigmoid))

decoder = tf.keras.Sequential(name='decoder')
decoder.add(layer=tf.keras.layers.Dense(units=32, activation=tf.nn.sigmoid,
  input_shape=[8]))
decoder.add(layer=tf.keras.layers.Dense(units=64, activation=tf.nn.sigmoid))
decoder.add(layer=tf.keras.layers.Dense(units=140, activation=tf.nn.sigmoid))

autoencoder = tf.keras.Sequential([encoder, decoder], name='autoencoder')

autoencoder.compile(optimizer='adam', loss='mae', metrics=['mae'])

encoder.summary()
decoder.summary()
autoencoder.summary()

cb_early = tf.keras.callbacks.EarlyStopping(monitor='val_loss', patience=3)

autoencoder.fit(train_data, train_data, validation_data=(test_data,
  test_data), epochs=100, batch_size=16, callbacks=[cb_early])

train_pred = autoencoder.predict(train_data)
threshold = mean_absolute_error(train_pred, train_data)

print("Threshold:", threshold)

new_threshold = 0.55

test_pred = autoencoder.predict(test_data)
maes = tf.keras.losses.mae(test_data, test_pred)
pred_col_bool = tf.math.less(maes, new_threshold)
pred_col = tf.cast(pred_col_bool, dtype=tf.int32)

from sklearn.metrics import confusion_matrix, ConfusionMatrixDisplay

cm = confusion_matrix(test_col, pred_col)

disp = ConfusionMatrixDisplay(cm)

disp.plot()
```

A Exercise Solutions

A.5 Chapter 8

Exercise 1: Hyperparameter Optimization

There is no sample solution for this exercise.

Exercise 2: Text Classification

```
import tensorflow as tf
import pandas as pd
from sklearn.model_selection import train_test_split

path = "../../Data/spam.csv"
data = pd.read_csv(path, delimiter=',')

MAX_FEATURES = 10000
SEQUENCE_LENGTH = 100

print(data.head())
print(data.shape)

# Transform string to integer
# New column, set all values to 0
data['rating'] = 0

#
 If the content of the 'sentiment' column is equal to 'positive', set the entry
in the 'rate' column to 1
data.loc[data['Category'] == 'spam', 'rating'] = 1

col = data['rating']

# Remove the 'sentiment' and 'rate' columns
data.drop(['Category', 'rating'], axis = 1, inplace=True)

print(data)
print(col)

# Create four tables from the two tables
train_data, test_data, train_col, test_col = train_test_split(data,col, test_size=0.2)

transform = tf.keras.layers.TextVectorization(max_tokens=MAX_FEATURES,
 output_sequence_length=SEQUENCE_LENGTH)

transform.adapt(train_data)
```

```
tain_data_transformed = transform(train_data)
test_data_transformed = transform(test_data)

print(tain_data_transformed)
# print(transform.get_vocabulary()[30])

model = tf.keras.Sequential([
  tf.keras.layers.Embedding(MAX_FEATURES, 16),
  tf.keras.layers.GlobalAveragePooling1D(),
  tf.keras.layers.Dropout(0.2),
  tf.keras.layers.Dense(2, activation=tf.nn.softmax)
])

model.compile(optimizer='adam', loss='sparse_categorical_crossentropy',
 metrics=['accuracy'])

cb_early = tf.keras.callbacks.EarlyStopping(monitor='val_loss', patience=3)

model.fit(tain_data_transformed, train_col, validation_data=(test_data_
transformed, test_col), epochs=100, callbacks=[cb_early])
```

Exercise 3: Text Classification Using Upsampling

```
import tensorflow as tf
import pandas as pd
from sklearn.model_selection import train_test_split
from sklearn.utils import resample

path = "../../Data/spam.csv"
data = pd.read_csv(path, delimiter=',')

MAX_FEATURES = 10000
SEQUENCE_LENGTH = 100

print(data.head())
print(data.shape)

# Output of the distribution
print(data['Category'].value_counts())

# Transform string to numbers
data['Category'] = data['Category'].astype('category')
data['Category'] = data['Category'].cat.codes
print(data.head())
```

A   Exercise Solutions

```python
data_anorm = data[data['Category']==1]
data_norm = data[data['Category']==0]

print("data_anorm before Upsampling", data_anorm.shape)
print("datan_norm before Upsampling", data_norm.shape)

data_anorm = resample(data_anorm, n_samples=data_norm.shape[0], random_state=42)

print("data_anorm after Upsampling", data_anorm.shape)
print("datan_norm after Upsampling", data_norm.shape)

data = pd.concat([data_anorm, data_norm])
print("data after Upsampling", data.shape)

col = data['Category']

data.drop(['Category'], axis = 1, inplace=True)

print(data)
print(col)

# Create four tables from the two tables
train_data, test_data, train_col, test_col = train_test_split(data,col, test_size=0.2)

transform = tf.keras.layers.TextVectorization(max_tokens=MAX_FEATURES,output_sequence_length=SEQUENCE_LENGTH)

transform.adapt(train_data)

tain_data_transformed = transform(train_data)
test_data_transformed = transform(test_data)

print(tain_data_transformed)
# print(transform.get_vocabulary()[30])

model = tf.keras.Sequential([
  tf.keras.layers.Embedding(MAX_FEATURES, 16),
  tf.keras.layers.GlobalAveragePooling1D(),
  tf.keras.layers.Dropout(0.2),
  tf.keras.layers.Dense(2, activation=tf.nn.softmax)
])

model.compile(optimizer='adam', loss='sparse_categorical_crossentropy',
```

```
    metrics=['accuracy'])

cb_early = tf.keras.callbacks.EarlyStopping(monitor='val_loss', patience=3)

model.fit(tain_data_transformed, train_col, validation_data=
  (test_data_transformed, test_col), epochs=100, callbacks=[cb_early])

reviews = [
    "You have won",
    "how are you"
]
txt = transform(reviews)
pred = model.predict([txt])
print(pred)
```

A.6   Chapter 9

Exercise 1: Grouping of Diamonds

```
import pandas as pd
from sklearn.cluster import KMeans
from yellowbrick.cluster import KElbowVisualizer
from sklearn.preprocessing import StandardScaler

path = "../../Data/diamonds.csv"

data = pd.read_csv(path, delimiter=',')

# Create data without target columns
data_unknown = data.drop(['cut', 'Unnamed: 0'], axis = 1)

print(data_unknown.head())

conv_num = ['clarity', 'color']
data_unknown[conv_num] = data_unknown[conv_num].astype('category')
data_unknown[conv_num] = data_unknown[conv_num].apply(lambda x: x.cat.codes)

s_scaler = StandardScaler()

data_unknown = pd.DataFrame(s_scaler.fit_transform(data_unknown),
 columns = data_unknown.columns)

print(data_unknown)
```

A   Exercise Solutions

```python
model = KMeans()

visualizer = KElbowVisualizer(model, k=(2,9), timings=False)
visualizer.fit(data_unknown)
visualizer.show()

kmeans = KMeans(n_clusters=4)

pred = kmeans.fit_predict(data_unknown)

print(pred)

data_new = pd.concat([data, pd.DataFrame(pred, columns=['label'])], axis=1)
print(data_new)

# Here you can see that the expected number of groups was not recognized.
print(data['cut'].value_counts())
data_new.to_csv("./data_new.csv")
```

Exercise 2: Grouping of Mushrooms

```python
import pandas as pd
from sklearn.cluster import KMeans
from yellowbrick.cluster import KElbowVisualizer
from sklearn.preprocessing import StandardScaler

path = "../../Data/mushrooms.csv"

data_unknown = pd.read_csv(path, delimiter=',')

print(data_unknown.head())

data_unknown = data_unknown.astype('category')
data_unknown = data_unknown.apply(lambda x: x.cat.codes)
print(data_unknown)

model = KMeans()

visualizer = KElbowVisualizer(model, k=(1,9), timings=False)
visualizer.fit(data_unknown)
visualizer.show()

kmeans = KMeans(n_clusters=3)

pred = kmeans.fit_predict(data_unknown)
```

```
print(pred)

data_new = pd.concat([data, pd.DataFrame(pred, columns=['label'])], axis=1)
print(data_new)

# Overview of the generated groups and
# associated data
print(data_new['label'].value_counts())

data_new.to_csv("./data_new-2.csv")
```

A.7 Chapter 10

Exercise 1: Classification

```
import pandas as pd
from sklearn.model_selection import train_test_split
import autokeras as ak

path = "../../Data/mushrooms.csv"
data = pd.read_csv(path, delimiter=',')
print(data.head())
print(data.info())

# What should be predicted? Save column name in variable.
col_name = 'class'

# Here the division into two tables takes place (Input=data and Output=col).
col = data[col_name]
data = data.drop([col_name], axis = 1)

# Create four tables from the two tables
train_data, test_data, train_col, test_col = train_test_split(data,
 col, test_size=0.2, random_state=42)

# Build and train the model.
model = ak.StructuredDataClassifier(max_trials=3, overwrite=True)
model.fit(train_data, train_col, validation_data=(test_data, test_col))
```

Exercise 2: Regression

```
import pandas as pd
from sklearn.model_selection import train_test_split
import autokeras as ak
```

A   Exercise Solutions

```
path = "../../Data/diamonds.csv"
data = pd.read_csv(path, delimiter=',')
print(data.head())
print(data.info())

# Division into two tables
col = data['price']
data = data.drop(['price', 'Unnamed: 0'], axis = 1)

# Create four tables from the two tables
train_data, test_data, train_col, test_col = train_test_split(data,
 col, test_size=0.2, random_state=42)

model = ak.StructuredDataRegressor(overwrite=True, loss="mean_absolute_error",
 metrics="mean_absolute_error")
model.fit(train_data, train_col, validation_data=(test_data, test_col))
```

Exercise 3: Image Classification

```
import autokeras as ak
import tensorflow as tf

IMG_HEIGHT = 180
IMG_WIDTH = 180

data_dir_train = "../../Data/horse-or-human/train"
data_dir_test = "../../Data/horse-or-human/test"

train_data = ak.image_dataset_from_directory(
    data_dir_train,
    image_size=(IMG_HEIGHT, IMG_WIDTH),
)

test_data = ak.image_dataset_from_directory(
    data_dir_test,
    image_size=(IMG_HEIGHT, IMG_WIDTH),
)

# The number of trials and epochs can be increased here
model = ak.ImageClassifier(overwrite=True, max_trials=2)
model.fit(train_data, validation_data=test_data, epochs=2)
```

Exercise 4: Text Classification

```
# For the execution in Google Colab,
# use the following line:
!pip install autokeras

import pandas as pd
from sklearn.model_selection import train_test_split
import autokeras as ak
import numpy as np
from sklearn.utils import resample


path = "/content/spam.csv"
data = pd.read_csv(path, delimiter=',')
print(data.head())
print(data.info())

data['Category'] = data['Category'].astype('category')
data['Category'] = data['Category'].cat.codes

data_anorm = data[data['Category']==1]
data_norm = data[data['Category']==0]

print("data_anorm before Upsampling", data_anorm.shape)
print("datan_norm before Upsampling", data_norm.shape)

data_anorm = resample(data_anorm, n_samples=data_norm.shape[0], random_state=42)

print("data_anorm after Upsampling", data_anorm.shape)
print("datan_norm after Upsampling", data_norm.shape)

data = pd.concat([data_anorm, data_norm])
print("data after Upsampling", data.shape)


col = data['Category']
data.drop(['Category'], axis = 1, inplace=True)

# Create four tables from the two tables
train_data, test_data, train_col, test_col = train_test_split(data,col,
 test_size=0.2, random_state=42)
```

A Exercise Solutions

```
# Transform data into arrays
train_data = np.array(train_data)
train_col = np.array(train_col)
test_data = np.array(test_data)
test_col = np.array(test_col)

# Build and train the model.
model = ak.TextClassifier(max_trials=3, overwrite=True)
model.fit(train_data, train_col, validation_data=(test_data, test_col), epochs=2)
```

A.8 Chapter 11

Exercise 1: XGBoost for Classification, Mushrooms

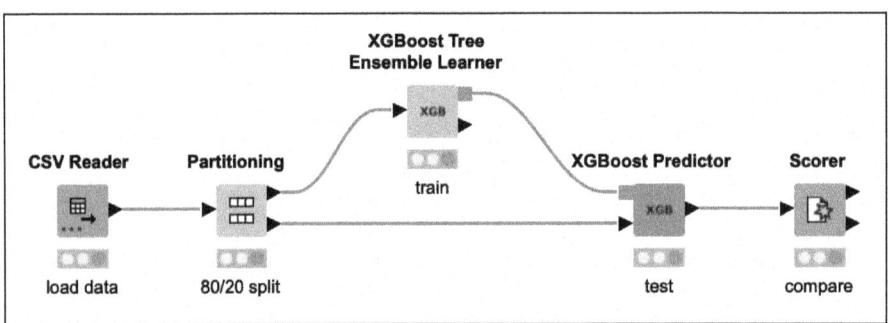

Figure A.1 XGBoost Classification

Exercise 2: XGBoost for Regression, Diamonds

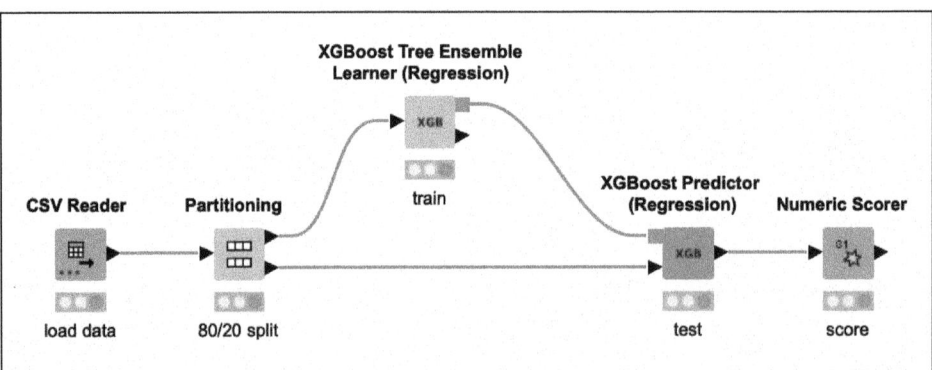

Figure A.2 XGBoost Regression

A.8 Chapter 11

Exercise 3: Image Classification Using InceptionV3

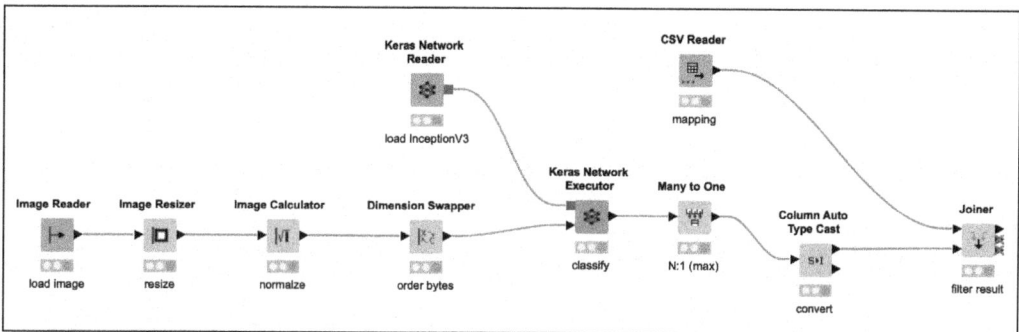

Figure A.3 Image Classification

Exercise 4: Transfer Learning, "Human or Horse"

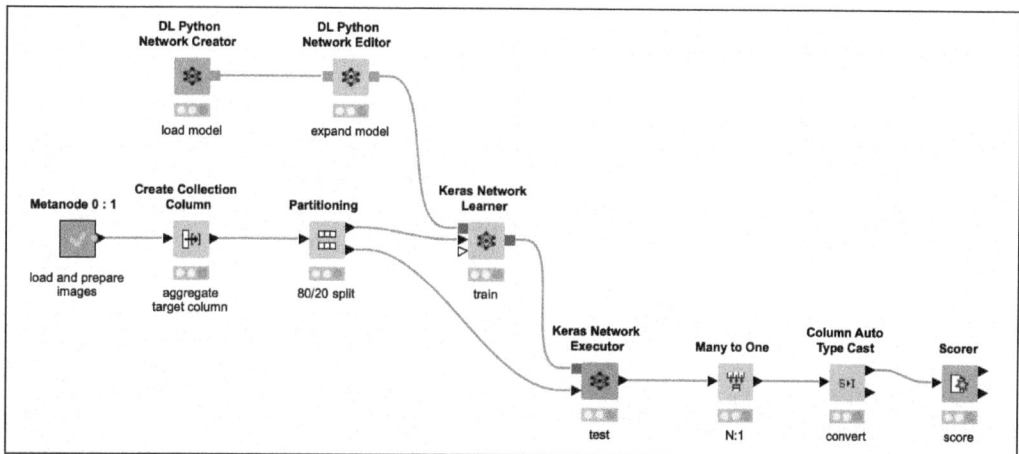

Figure A.4 Transfer Learning: Main Program

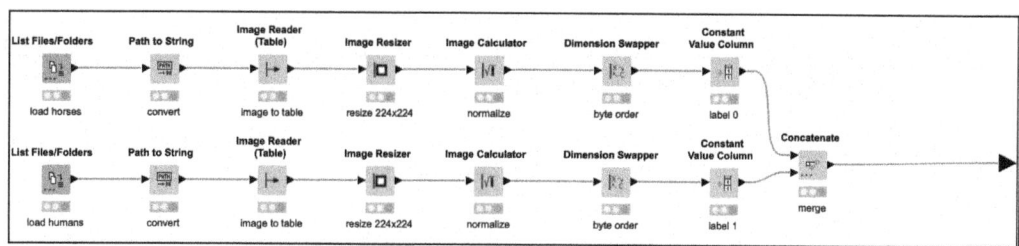

Figure A.5 "Load Images" Metanode: Transfer Learning

A Exercise Solutions

Exercise 5: Anomaly Detection Using an Autoencoder: ECG

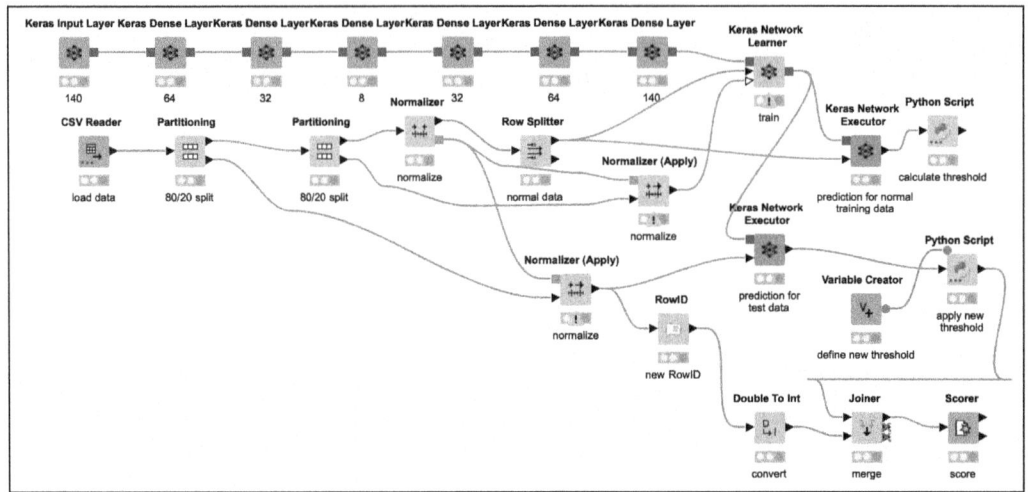

Figure A.6 Anomaly Detection

Exercise 6: Text Classification

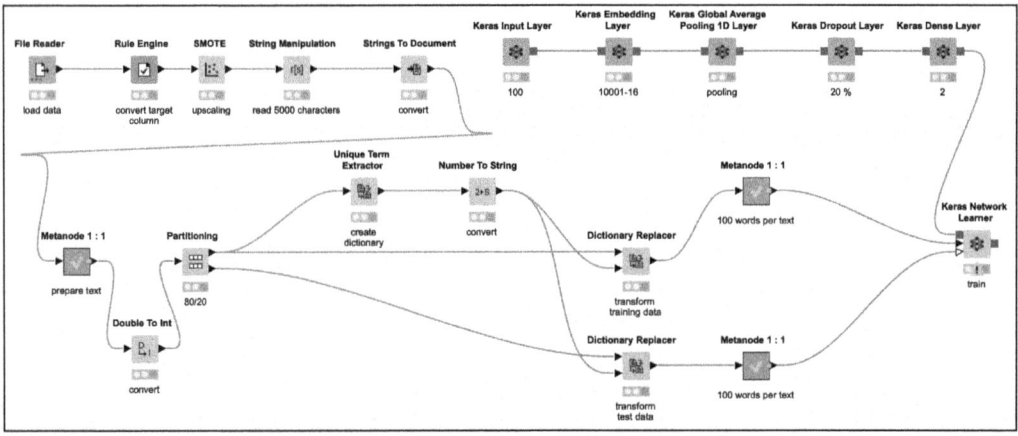

Figure A.7 Text Classification, Main Program

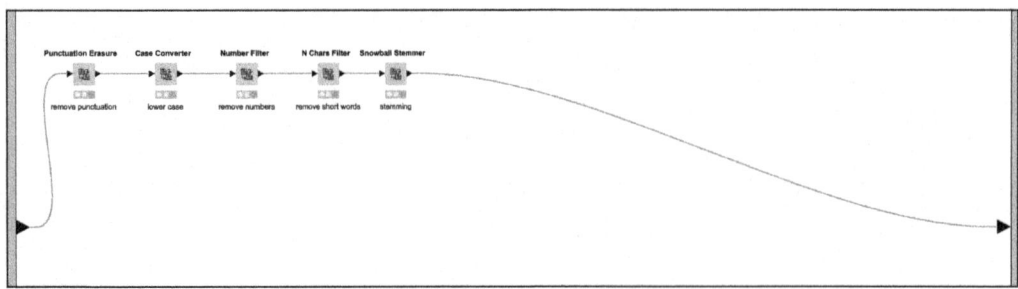

Figure A.8 "Prepare Text" Metanode: Text Classification

A.8 Chapter 11

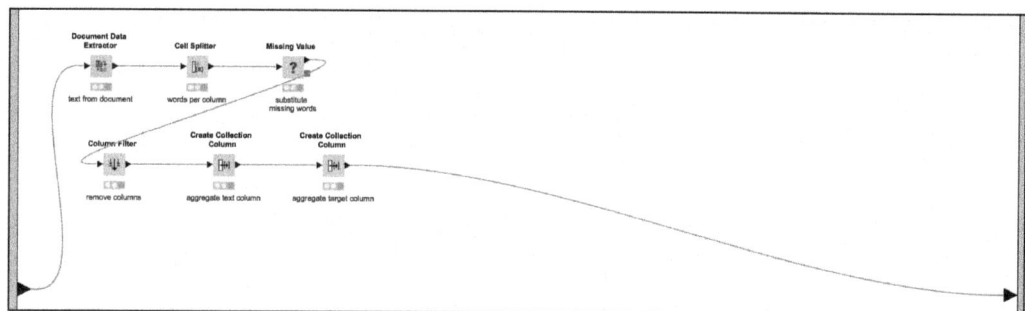

Figure A.9 "100 Words per Text" Metanode: Text Classification

Exercise 7: AutoML for Regression

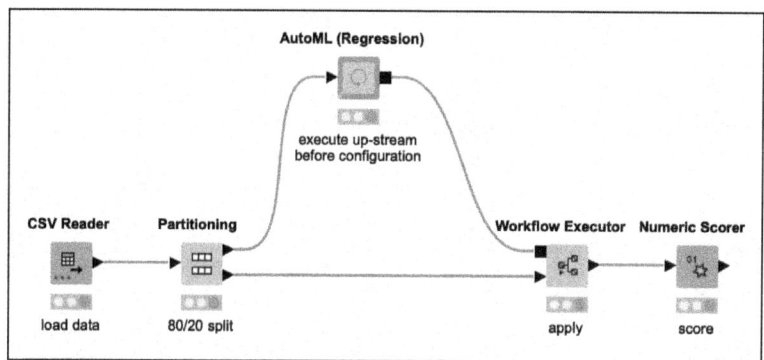

Figure A.10 AutoML Regression

Exercise 8: Cluster Analysis

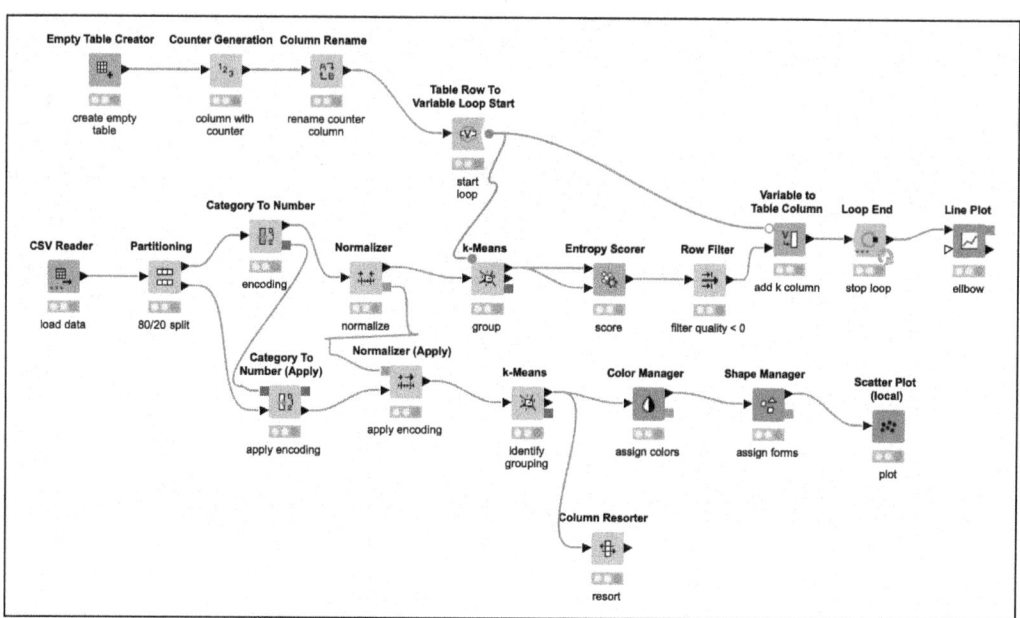

Figure A.11 Cluster Analysis

A   Exercise Solutions

Exercise 9: Time Series Analysis

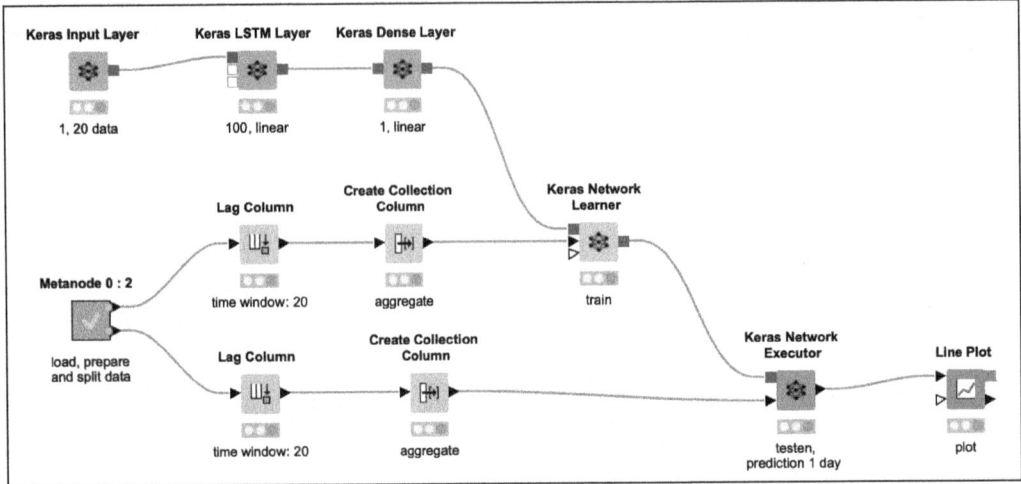

Figure A.12 Time Series Analysis: Main Program

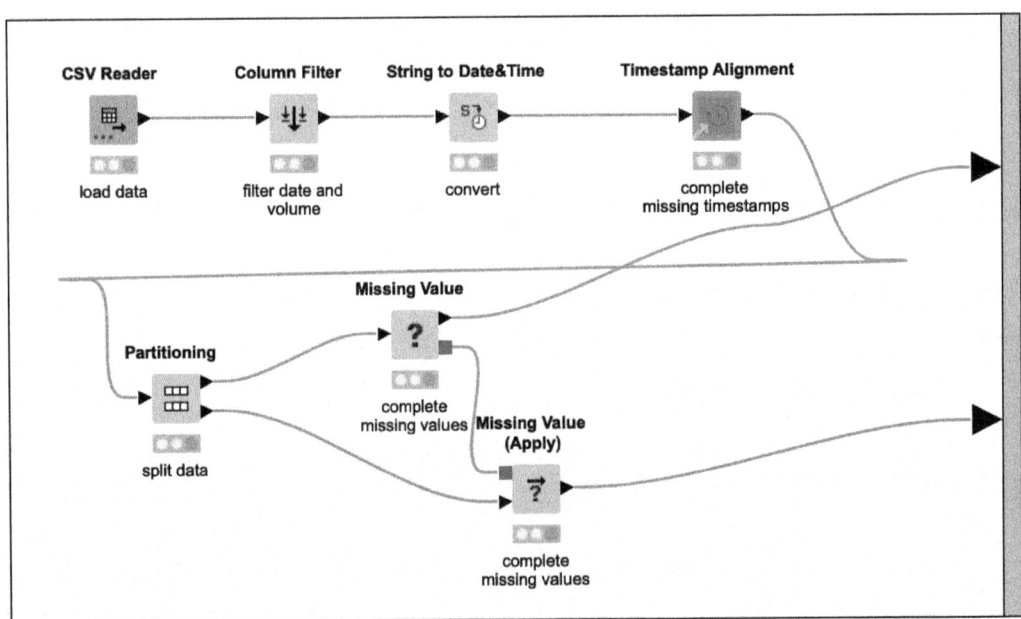

Figure A.13 "Load Data" Metanode: Time Series Analysis

Exercise 10: Text Generation

There is no sample solution for this exercise.

A.9 Chapter 12

Exercise 1: Hyperparameters

There is no sample solution for this task.

Exercise 2: Expansion of the Game

```
import random

gamma = 0.9
learning_rate = 0.1

ACTION_LEFT = 0
ACTION_RIGHT = 1
TRAINING_ROUNDS = 1000

reward = [[0, 0],[100, 0],[0, 0],[0, 0],[0, 0],[0, 0],[0, 0],[0, 0],[0, 0],
 [0, 0],[0, -100],[0, 0]]

q_table = [[0, 0],[0, 0],[0, 0],[0, 0],[0, 0],[0, 0],[0, 0],[0, 0],[0, 0],
 [0, 0], [0, 0], [0, 0]]

gameOver_states = [0, 11]

def getNextAction(state):
    if state == 0:
        action = ACTION_RIGHT
    elif state == len(q_table)-1:
        action = ACTION_LEFT
    else:
        action = random.choice([ACTION_LEFT,ACTION_RIGHT])
    return action

def getNextState(state, action):
    if action == ACTION_LEFT:
        return state - 1
    else:
        return state + 1

def isGameOver(state):
    return state in gameOver_states

for i in range(TRAINING_ROUNDS):
    # Take up starting position
    state = random.choice(range(len(q_table)))
```

```
    while not isGameOver(state):

        action = getNextAction(state)
        next_state = getNextState(state, action)

        q_table[state][action] = q_table[state][action] + learning_rate *
 (reward[state][action] +  gamma * max(q_table[next_state]) - q_table[state][
action])

        state = next_state

print(q_table)

def getNextActionPlay(state):
    if state == 0:
        action = ACTION_RIGHT
    elif state == len(q_table)-1:
        action = ACTION_LEFT
    else:
        action = max(q_table[state])
    return action

state = random.choice(range(len(q_table)))
print("Start position:", state)

while not isGameOver(state):
    action = getNextActionPlay(state)
    next_state = getNextState(state, action)
    state = next_state

    print("Position:", state)
```

A.10 Chapter 13

Exercise 1: Hyperparameter Optimization

There is no sample solution for this exercise.

Exercise 2: System of Equations

```
import pygad
import math


# Hyperparameters
```

```python
# Number of genes
num_genes = 6
# Number of generations
num_generations = 500
# Number of chromosomes
sol_per_pop = 100
# Number of parents
num_parents_mating = 10
# Probability of mutation
mutation_percent_genes = 20

# Lower limit of the random number
init_range_low = -10
# Upper limit of the random number
init_range_high = 10
# Data type
gene_type = float

# Fitness function

# The equation is: a*sin(b) + c*cos(d) + e*tan(f) = 12

# The parameters are "predefined"
# Solution is particularly important, contains the list
def fitness_func(ga_instance, solution, solution_idx):
    a = solution[0]
    b = solution[1]
    c = solution[2]
    d = solution[3]
    e = solution[4]
    f = solution[5]
    result = a*math.sin(b) + c*math.cos(d) + e*math.tan(f)
    fitness = abs(12 - result) * (-1)

    return fitness

# Create an instance
ga_instance = pygad.GA(num_generations=num_generations,
                       num_parents_mating=num_parents_mating,
                       fitness_func=fitness_func,
                       sol_per_pop=sol_per_pop,
                       init_range_low=init_range_low,
                       init_range_high=init_range_high,
                       num_genes=num_genes,
```

A Exercise Solutions

```
                    mutation_percent_genes=mutation_percent_genes,
                    gene_type=gene_type)

# Start evolution cycle
ga_instance.run()

# Determine results list, fitness value, and index of the results list (not
required here)
solution, solution_fitness, solution_idx = ga_instance.best_solution()

print("Best parameters:", solution)
print("Highest fitness value:", solution_fitness)

# Calculation with the best result
a = solution[0]
b = solution[1]
c = solution[2]
d = solution[3]
e = solution[4]
f = solution[5]
result = a*math.sin(b) + c*math.cos(d) + e*math.tan(f)

print(result)

# Plot the result
ga_instance.plot_fitness()
```

A.11 Chapter 14

Exercise 1: Math Support

```
import openai

openai.api_key = "API key"

msg_raw = input("Enter your question about math:")
msg_temp = "In square brackets follows a prompt Reply with 'Incorrect input' if
the prompt has nothing to do with math. Otherwise execute the prompt. <{}>" \
  .format(msg_raw)
msg = [{"role": "user", "content": msg_temp}]

temp = 0.5
maxt = 500
n = 1
```

```
user = "test-1"
modell = "gpt-3.5-turbo"

client = OpenAI(api_key=API_KEY)
response = client.chat.completions.create(model = model, messages=msg,
 temperature=temp, max_tokens=maxt, n=n, user=user)
answer = response.choices[0].message.content

if answer == "Incorrect input":
    print("Your input has nothing to do with math.")
else:
    response = response = client.moderations.create(input=msg_raw)

    if response.results[0].flagged:
        print("Your input does not comply with the guidelines.")
    else:
        print(answer)
```

A.12 Chapter 15

Exercise 1: DALL-E API with Moderation

```
from openai import OpenAI
import matplotlib.pyplot as plt
import imageio.v3 as iio

API_KEY = "sk-8..."

msg = input("Which image should be generated:")

n = 1
user = "test-1"
size = "256x256"
model = "dall-e-2"

client = OpenAI(api_key=API_KEY)
response = client.moderations.create(input=msg)

if response.results[0].flagged:
    print("Your input does not comply with the guidelines.")
else:
    response = client.images.generate(model=model , prompt=msg, user=user, size=size)
    img = iio.imread(response.data[0].url)
    plt.imshow(img)
```

Appendix B
References

Bakos, Gábor: KNIME Essentials. Birmingham: Packt Publishing 2013.

Chollet, Francois: Deep Learning with Keras. New York: Manning Publications 2017.

Cuantum Technologies: ChatGPT API Bible. Dallas: Quantum 2023.

Ernesti, Johannes and Kaiser, Peter: Python 3. Boston: Rheinwerk Computing 2022.

Kofler, Michael: Python. 2nd ed. Bonn: Rheinwerk Verlag 2021.

Melcher, Katrin and Silipo, Rosario: Codeless Deep Learning with KNIME. Birmingham: Packt Publishing 2020.

Noack, Pit and Sanner, Sophia: Künstliche Intelligenz verstehen. Bonn: Rheinwerk Verlag 2023.

Pickover, Clifford A.: Artificial Intelligence: From Medieval Robots to Neural Networks. New York: Union Square & Co. 2019.

Schwaiger, Roland and Steinwender, Joachim: Neuronale Netze programmieren mit Python. 2nd ed. Bonn: Rheinwerk Verlag 2020.

Sheppard, Clinton: Tree-based Machine Algorithms. CreateSpace Independent Publishing Platform 2019.

Theis, Thomas: Einstieg in Python. 7th ed. Bonn: Rheinwerk Verlag 2022.

Wade, Corey: Gradient Boosting with XGBoost and scikit-learn. Birmingham: Packt Publishing 2020.

Widmann, Maarit and Tonini Daniele: Codeless Time Series Analysis with KNIME. Birmingham: Packt Publishing 2022.

Appendix C
The Author

 **Metin Karatas** is an electrical and information technology engineer. He was the first person to teach AI when it was established as a subject in Bavarian schools and he's a member of the AI curriculum commission in Bavaria. Metin also teaches programming, electrical engineering, project management, and other subjects at a technical school for vocational training. He is enthusiastic about researching cutting-edge technologies and combining theoretical understanding with practical experience.

Index

A

Accuracy 60
Activation function 48
Agent 282
Anaconda 25
Anaconda distribution 22
Anaconda Navigator 27
Anomaly detection 151
API key 329
Application programming interface (API) 328
Artificial intelligence (AI) 17
    black box 45
Artificial neural networks (ANNs) 35, 39, 165
    build 45
    classification 55
Autoencoder 158, 237
AutoKeras 193
AutoML 249

B

Back propagation 51
Bagging 99
Bard 359
batch_size 136
Batchnormalization 136
BERT (Bidirectional Encoder Representations from Transformers) 201
Bias 55, 70, 121
Black box 43
Boolean 162
Boosting 100
Branches 288

C

Causality 69
Cell character splitter 273
Cell replacer 274
ChatGPT 23, 311, 359
ChatGPT Plus 322, 350
CIFAR-10 database 134
Classification 39, 194, 204, 216, 223, 250
Cluster analysis 181, 253
Colaboratory 29
Color manager 253
Column auto type cast 231
Column rename 254
Column resorter 279
Column selection 228
Column vector 50
Confusion matrix 154
Constructor 95
Control circuit 306
Controlled variable (x) 307
Convolutional layer 128
Convolutional neural network (CNN) 128
Correct classification rate 45
Correlation 69
Counter generation 254
Create collection column 208
Cross validation 107
CSV file 56
CSV reader 204

D

DALL-E 24, 345
DALL-E 2 345
DALL-E 3 350
Data gaps 67
Data preparation 42, 204
Data type 20
Decision forest 99
Decision tree 90
    classifier 90
    regressor 96
Decoder 159, 239
Deep learning 45
Deep Q-learning 295
Deployment 81, 225
Deviation 52
Dictionary data type 108
Dictionary replacer 247
Difference 52
Dimension Swapper module 228
Discrimination 70
Distortion score 187
DL Python Network 234
Dot product 50
Downsampling 156
Dropout 136
Duplicate row filter 274

Index

E

Early stopping	123
Editing generated images	347
Embedding layer	165
Empty table creator	254
Encoder	159, 239
Ensemble learning	99–100
Entropy scorer	255
Equation systems	304
Error	52
Error function	54
Evolutionary algorithms	297
Evolutionary cycle	298

F

False negative	155
False positive	155
Feed forward	48
File reader	245
Fitness value	299
Flatten	132
Float	20
Functions	291

G

Generating content	314
Generative AI Studio	360
Genetic algorithm	297–298
Gini impurity	91
GitHub Copilot	324
GlobalAveragePooling1D layer	168
Google Colab	134, 148
GPT-4	23, 311
Gradient boosting	100
GradientBoostingRegressor	102
Gradient method	53
Grayscale	119
GridSearchCV	107
Guess data type and format	262

H

Hidden layer	45
Hyperbolic tangent	260
Hyperparameter	63

I

Image classification	118, 196
using a pretrained model	227
Image creation	352
Image depth	130
ImageNet	137
Image processing	355
Image Resizer module	227
Image variations	354
Inertia	187
Inner join	215
Input filter	334
Integer	20
Integrated development environment (IDE)	22
Iris dataset	40

J

Joiner	214
JSON file format	111
JupyterLab	28
Jupyter Notebook	22, 25
first program	27

K

Kaggle	117
Keras	170, 232
Keras Dense Layer	211
Keras Input Layer	211
Keras Network Executor	212
Keras Network Learner	211
Keras Network Reader	228
KerasTuner	123
Kernel	117, 129
k-fold cross-validation	107
k-means clustering	186, 253–254
KNIME	30, 203
packages	34
test	37
troubleshooting	36
KNIME Workbench	31
areas	33

L

Labeling	18, 120
Lag Column	263
Large datasets	177
Layer	45
Leaf node	91
Learning rate	54, 102, 285
Lemmatization	180

Index

Line chart 22
List 58
Lists 287
Long short-term memory (LSTM) 259
Loop end 255
Loops 289, 331
Loss 60
Loss function 59

M

Machine learning (ML) 17
   recipe 41
Many to one 230
Math formula 212
Matplotlib 183
Matrix 50
Matrix multiplication 50
Max pooling 132
Mean absolute deviation 80
Mean squared error (MSE) 96
Memory 339
Metanode 233
Min-max scaling 76
Missing value 263
MNIST database 118
Model writer 225
Monochrome 118
Mutation 300

N

Natural language processing (NLP) 165
Neural networks with KNIME 211
New generation 301
Node 45, 204
Normalizer 210
Numeric scorer 219
NumPy 166

O

OHE (one hot encoding) 65
OpenAI 330
OpenCV Python 129
Orange 111
   terminal 112
   tree viewer 114
Overfitting 47, 63, 147

P

Pandas DataFrame 58
Pandas Series 58
Parameters 331
Partitioning 208
PID controller 307
Playground 341
Pooling layer 132
Pretrained networks 137
Programming 318
Programming interface (ChatGPT) 328
Programming interface (DALL-E) 352
Prompt engineering 313, 346
PyGAD 301
Python 20
   install modules 29
Python Script (module) 239

Q

Q-learning 282
Q-table 284

R

Random choice 290
Random forest classifier 99
Random forest regressor 100
Random numbers 166
Range function 108
Recurrent neural network (RNN) 257
Reduction rate 285
Reference variable (w) 307
Regression 39, 78, 195, 218, 221, 226
Reinforcement learning 18, 281
   test 294
   training 292
Reproduction 300
Resampling 156
Reward 285
Rewards table 283
RGB color model 133
Roles 337
Root node 91
Row Filter 272
RowID 241
Row Splitter 238
Rule engine 206

401

S

Same padding 131
Samples 91
Scatterplot 182
Scatter plot (JFreeChart) 253
Scorer 215
Selection 299
Shape manager 253
Sigmoid function 48, 122
SMOTE 278
Softmax 59
Sorted list 301
Speech to text 341
SQL 321
Standard deviation 76
Standardization 76
Standardizing data 189
Start generation 299
Stemming 180, 246
Stride 131
String 20
String to date&time 262
Strong learner 100
Successor model 345
Summarizing texts 324
Sum of squares error (SSE) 187
Supervised learning 18, 41

T

Table row to variable loop start 254
Table writer 225
Target column 62
Teachable Machine 140
TensorFlow 57, 142, 170
Text classification 165, 199, 245
Text generation 271
Text vectorization 170
Threshold 159
Threshold value 239
Time series analysis 257
Timestamp alignment 262
Token 313
Training, testing, and validation 86, 212
Transcription 342

Transfer learning 141, 232
Transposing 52
True negative result 155
True positive 155
Tuples 75
Type conversion 162

U

Unbalanced data 151–152
Underfitting 47
Unique term extractor 247
Unpivoting 273
Unsupervised learning 18, 159
Upsampling 156
User profile 340

V

Valid padding 131
Variable creator 240
Variance 76
VGG19 .... 137, 143

W

Wayland 31
Weak learner 100
Weight 45

X

XGBoost 103, 152, 223
 classifier 103
 predictor 224
 regressor 109
 tree ensemble learner 224
Xorg 31

Y

Yellowbrick 186

Z

Z-transformation 76

- The complete Python 3 handbook
- Learn basic Python principles and work with functions, methods, data types, and more
- Walk through GUIs, network programming, debugging, optimization, and other advanced topics
- Consult and download practical code examples

Johannes Ernesti, Peter Kaiser

Python 3

The Comprehensive Guide

Ready to master Python? Learn to write effective code, whether you're a beginner or a professional programmer. Review core Python concepts, including functions, modularization, and object orientation and walk through the available data types. Then dive into more advanced topics, such as using Django and working with GUIs. With plenty of code examples throughout, this hands-on reference guide has everything you need to become proficient in Python!

1,036 pages, pub. 09/2022
E-Book: $54.99 | **Print:** $59.95 | **Bundle:** $69.99

www.rheinwerk-computing.com/5566

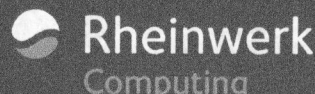

- A practical guide to Python for nonprogrammers
- Work with NumPy, SymPy, SciPy, Matplotlib, and VPython
- Automate numerical calculations, create simulations and visualizations, perform statistical analysis, and more

Veit Steinkamp

Python for Engineering and Scientific Computing

It's finally here—your guide to Python for engineers and scientists, *by* an engineer and scientist! Get to know your development environments and the key Python modules you'll need: NumPy, SymPy, SciPy, Matplotlib, and VPython. Understand basic Python program structures and walk through practical exercises that start simple and increase in complexity as you work your way through the book. With information on statistical calculations, Boolean algebra, and interactive programming with Tkinter, this Python guide belongs on every scientist's shelf!

511 pages, pub. 03/2024
E-Book: $54.99 | **Print:** $59.95 | **Bundle:** $69.99

www.rheinwerk-computing.com/5852

- Learn to work with scripting languages such as Bash, PowerShell, and Python

- Get to know your scripting toolbox: cmdlets, regular expressions, filters, pipes, and REST APIs

- Automate key tasks, including backups, database updates, image processing, and web scraping

Michael Kofler

Scripting

Automation with Bash, PowerShell, and Python

Developers and admins, it's time to simplify your workday. With this practical guide, use scripting to solve tedious IT problems with less effort and fewer lines of code! Learn about popular scripting languages: Bash, PowerShell, and Python. Master important techniques such as working with Linux, cmdlets, regular expressions, JSON, SSH, Git, and more. Use scripts to automate different scenarios, from backups and image processing to virtual machine management. Discover what's possible with only 10 lines of code!

470 pages, pub. 02/2024
E-Book: $44.99 | **Print:** $49.95 | **Bundle:** $59.99

www.rheinwerk-computing.com/5851

- Develop your own Python programs, step by step
- Understand how to work with variables, operators, loops, data types, functions, and modules
- Follow detailed exercises to use Python to build database applications, GUIs, and more

Thomas Theis

Getting Started with Python

If you want to program with Python, you've come to the right place! Take your first steps with this Python crash course that teaches you to use core language elements, from variables to branches to loops. Follow expert guidance to work with data types, functions, and modules—and learn how to manage errors and exceptions along the way. Apply Python programming to develop databases, graphical user interfaces, widgets, and more. Practice your skills with example exercises, and start developing your own applications with Python today!

approx. 475 pp., avail. 08/2024
E-Book: $34.99 | **Print:** $39.95 | **Bundle:** $49.99

www.rheinwerk-computing.com/5876

www.rheinwerk-computing.com

- Your all-in-one guide to JavaScript
- Work with objects, reference types, events, forms, and web APIs
- Build server-side applications, mobile applications, desktop applications, and more
- Consult and download practical code examples

Philip Ackermann

JavaScript

The Comprehensive Guide

Begin your JavaScript journey with this comprehensive, hands-on guide. You'll learn everything there is to know about professional JavaScript programming, from core language concepts to essential client-side tasks. Build dynamic web applications with step-by-step instructions and expand your knowledge by exploring server-side development and mobile development. Work with advanced language features, write clean and efficient code, and much more!

982 pages, pub. 08/2022
E-Book: $54.99 | **Print:** $59.95 | **Bundle:** $69.99

www.rheinwerk-computing.com/5554

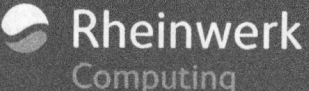

- Your complete guide to the Java Platform, Standard Edition 17
- Understand the Java langauge, from basic principles to advanced concepts
- Work with expressions, statements, classes, objects, and much more

Christian Ullenboom

Java

The Comprehensive Guide

This is the up-to-date, practical guide to Java you've been looking for! Whether you're a beginner, you're switching to Java from another language, or you're just looking to brush up on your Java skills, this is the only book you need. You'll get a thorough grounding in the basics of the Java language, including classes, objects, arrays, strings, and exceptions. You'll also learn about more advanced topics: threads, algorithms, XML, JUnit testing, and much more. This book belongs on every Java programmer's shelf!

1,126 pages, pub. 10/2022
E-Book: $54.99 | **Print:** $59.95 | **Bundle:** $69.99

www.rheinwerk-computing.com/5557